AF598033

METHODS IN MOLECULAR BIOLOGY™

Series Editor
John M. Walker
School of Life Sciences
University of Hertfordshire
Hatfield, Hertfordshire, AL10 9AB, UK

For further volumes:
http://www.springer.com/series/7651

Amino Acid Analysis

Methods and Protocols

Edited by

Michail A. Alterman

Tumor Vaccines and Biotechnology Branch, CBER, FDA, Bethesda, MD, USA

Peter Hunziker

Functional Genomics Center, University of Zűrich, Zűrich, Switzerland

Editors
Michail A. Alterman
Tumor Vaccines and Biotechnology Branch
CBER, FDA
Bethesda, MD, USA
Michail.Alterman@fda.hhs.gov

Peter Hunziker
Functional Genomics Center
University of Zürich
Zürich, Switzerland
peter.hunziker@access.uzh.ch

ISSN 1064-3745 e-ISSN 1940-6029
ISBN 978-1-61779-444-5 e-ISBN 978-1-61779-445-2
DOI 10.1007/978-1-61779-445-2
Springer New York Dordrecht Heidelberg London

Library of Congress Control Number: 2011941583

Printed on acid-free paper

Humana Press is part of Springer Science+Business Media (www.springer.com)

Preface

Amino Acid Analysis (AAA) has been an integral part of analytical biochemistry for almost 60 years. AAA was originally developed by Moore and Stein and was at the very heart of their work on the mechanism of enzyme catalysis for which they were awarded a Nobel Prize in Chemistry in 1972. In a relatively short time since the previous AAA book in this series has been published (10 years), the variety of AAA methods changed dramatically with more methods shifting to the use of mass spectrometry (MS) as a detection method. At the same time, a number of old techniques acquired a new make-up, like combination of AccQ-Tag with UPLC and MS, instead of HPLC and fluorescence. Another new aspect is miniaturization. One of the chapters in this book describes an AAA in a single cell. However, the most important aspect is that AAA in this day and age should be viewed in the context of Metabolomics as a part of Systems Biology.

Historically, analysis of amino acids (AA) includes derivatization (pre, on-, or post-column) coupled with chromatographic separation. A wide variety of separation techniques were applied to separation of AA over the years. This list includes ion-exchange LC, reverse-phase HPLC, gas chromatography (GC), and capillary electrophoresis (CE). Recent advances in mass spectrometry (MS) led to the application of electrospray ionization coupled with LC or CE for AA detection which may also allow the analysis of underivatized AA. Two most recent technological advances in AAA include the application of MALDI TOF MS and TOF/TOF MS/MS and microfluidics. AAA techniques evolve and follow the bioanalytical technological advances.

Sample preparation for AAA plays a critical role in the successful implementation of AAA. Correspondingly, in this book a reader can find chapters describing general as well as specific approaches to the sample preparation. A number of chapters describe different applications of AAA. Some chapters describe specific applications of AAA in clinical chemistry as well as in food analysis, microbiology, and other biomedical fields. Separate chapters are devoted to the application of AAA for protein quantitation and chiral AAA.

Our goal was to present a spectrum of all available methods for readers to choose the method that most suits particular laboratory and needs. And, at the same time we attempted to present more than one method for each application or detection/separation approach so that again the readers can find the one that most suits their needs and available laboratory conditions.

What is unique about this book, and in essence about AAA itself, is that it is related and is of interest to anyone involved in biomedical research or, in general, in life sciences. One can find here techniques essential in medicine, or in drug metabolism, or cell biology, even in archeology, in meat industry, in marine biology, in agriculture, and the list goes on. All of the described techniques are multifaceted and in many cases can serve as a blueprint for the analysis of other chemically related classes of metabolites.

Bethesda, MD, USA — *Michail A. Alterman*
Zürich, Switzerland — *Peter Hunziker*

Contents

Contributors

P. Agrafiotou • *Laboratory of Physical Chemistry, Department of Chemistry, Aristotle University of Thessaloniki, Thessaloniki, Greece*

Peter G. Allingham • *Cooperative Research Centre for Sheep Industry Innovation, CJ Hawkins Homestead, University of New England, Armidale, NSW, Australia*

Michail A. Alterman • *Tumor Vaccines and Biotechnology Branch, CBER, FDA, Bethesda, MD, USA*

Jenny M. Armenta • *Waters Corporation, Beverly, MA, USA*

Abdulla A.-B. Badawy • *The Cardiff School of Health Sciences, University of Wales Institute Cardiff (UWIC), Western Avenue, Cardiff, Wales, UK*

Kristof Brijs • *Laboratory of Food Chemistry and Biochemistry, Leuven Food Science and Nutrition Research Centre (LFoRCe), Katholieke Universiteit Leuven, Leuven, Belgium*

Philip Britz-McKibbin • *Department of Chemistry and Chemical Biology, McMaster University, Hamilton, ON, Canada*

David M. Bunk • *Analytical Chemistry Division, National Institute of Standards and Technology, Gaithersburg, MD, USA*

Inge Celus • *Laboratory of Food Chemistry and Biochemistry, Leuven Food Science and Nutrition Research Centre (LFoRCe), Katholieke Universiteit Leuven, Leuven, Belgium*

Michelle L. Colgrave • *CSIRO Division of Livestock Industries, Queensland Bioscience Precinct, St Lucia, QLD, Australia*

Diego F. Cortés • *Virginia Bioinformatics Institute, Virginia Polytechnic Institute and State University, Blacksburg, VA, USA*

Jan A. Delcour • *Laboratory of Food Chemistry and Biochemistry, Leuven Food Science and Nutrition Research Centre (LFoRCe), Katholieke Universiteit Leuven, Leuven, Belgium*

Madeleine Dell'mour • *Department of Chemistry, University of Natural Resources and Life Sciences – BOKU Vienna, Vienna, Austria*

Katja Dettmer • *Institute of Functional Genomics, University of Regensburg, Regensburg, Germany*

Stephan R. Fagerer • *Department of Chemistry and Applied Biosciences, Laboratory of Organic Chemistry, ETH Zürich, Zürich, Switzerland*

Durk Fekkes • *Department of Psychiatry, Erasmus MC, University Medical Center Rotterdam, Rotterdam, Netherlands*

Iole Maria Di Gangi • *Department of Pediatrics, University of Padova, Padova, Italy*

Gino Giannaccini • *Department of Psychiatry, Neurobiology, Pharmacology and Biotechnology, University of Pisa, Pisa, Italy*

Giuseppe Giordano • *Department of Pediatrics, University of Padova, Padova, Italy*

NATALIA V. GOGICHAEVA • *Analytical Proteomics Laboratory, University of Kansas, Lawrence, KS, USA*
LIPING GU • *Department of Biology and Microbiology, South Dakota State University, Brookings, SD, USA*
ANTONINA GUCCIARDI • *Department of Pediatrics, University of Padua, Padova, Italy*
FRANK-MATHIAS GUTZKI • *Institute of Clinical Pharmacology, Hannover Medical School, Hannover, Germany*
STEPHAN HANN • *Department of Chemistry, University of Natural Resources and Life Sciences – BOKU Vienna, Vienna, Austria*
SONJA HESS • *Proteome Exploration Laboratory, California Institute of Technology, BI 211, MC139-74, Pasadena, CA, USA*
AKIYOSHI HIRAYAMA • *Institute for Advanced Biosciences, Keio University, Kakuganji, Tsuruoka, Yamagata, Japan*
RALF HOFFMANN • *Institute of Bioanalytical Chemistry, Faculty of Chemistry and Mineralogy and Center for Biotechnology and Biomedicine, Universität Leipzig, Leipzig, Germany*
DIANA HOFMANN • *Central Division of Analytical Chemistry/BioSpec, Jülich, Germany*
PETR HUŠEK • *Laboratory of Analytical Biochemistry, Biology Centre, Academy of Sciences of the Czech Republic, České Budějovice, Czech Republic*
SHINSUKE INAGAKI • *Laboratory of Analytical and Bio-Analytical Chemistry, School of Pharmaceutical Sciences, and Global COE Program, University of Shizuoka, Shizuoka, Japan*
ZORANA JELIĆ-IVANOVIĆ • *Department of Medical Biochemistry, Faculty of Pharmacy, University of Belgrade, Belgrade, Serbia*
A. DANIEL JONES • *Department of Biochemistry and Molecular Biology, and Department of Chemistry, Michigan State University, East Lansing, MI, USA*
ALUN JONES • *The University of Queensland, Institute for Molecular Bioscience, Brisbane, QLD, Australia*
HANNELORE KASPAR • *Institute of Functional Genomics, University of Regensburg, Regensburg, Germany*
MEGUMI KATO • *Bio-Medical Standard Section, National Metrology Institute of Japan (NMIJ), National Institute of Advanced Industrial Science and Technology (AIST), Tsukuba, Ibaraki, Japan*
GUNDA KOELLENSPERGER • *Department of Chemistry, University of Natural Resources and Life Sciences – BOKU Vienna, Vienna, Austria*
ROMAN KÖRNER • *Max-Planck Institute of Biochemistry, Martinsried, Germany*
YURI P. KOZMIN • *Shemyakin-Ovchinnikov Institute of Bioorganic Chemistry RAS. Ul. Miklukho-Maklaya, GSP Moscow, Russia*
BERT LAGRAIN • *Laboratory of Food Chemistry and Biochemistry, Leuven Food Science and Nutrition Research Centre (LFoRCe), Katholieke Universiteit Leuven, Leuven, Belgium*
LIEVE LAMBERTS • *Laboratory of Food Chemistry and Biochemistry, Leuven Food Science and Nutrition Research Centre (LFoRCe), Katholieke Universiteit Leuven, Leuven, Belgium*
TOBIAS LANGROCK • *Institute of Bioanalytical Chemistry, Faculty of Chemistry and Mineralogy and Center for Biotechnology and Biomedicine, Universität Leipzig, Leipzig, Germany*

ROBERT L. LAST • *Departments of Biochemistry and Molecular Biology and Department of Plant Biology, Michigan State University, East Lansing, MI, USA*
YI-MING LIU • *Department of Chemistry and Biochemistry, Jackson State University, Jackson, MS, USA*
MARK S. LOWENTHAL • *Analytical Chemistry Division, National Institute of Standards and Technology, Gaithersburg, MD, USA*
ANTONIO LUCACCHINI • *Department of Psychiatry, Neurobiology, Pharmacology and Biotechnology, University of Pisa, Pisa, Italy*
EMANUELE MARCONI • *DISTAAM, Università degli Studi del Molise, Campobasso, Italy*
JORGE C. MASINI • *Instituto de Química, Universidade de São Paulo, São Paulo, Brazil*
JAMES S. O. MCCULLAGH • *Department of Chemistry, University of Oxford, Oxford, UK*
MARIA CRISTINA MESSIA • *DISTAAM, Università degli Studi del Molise, Campobasso, Italy*
OLGA A. MIRGORODSKAYA • *Research Institute of Influenza, Saint-Petersburg, Russia*
ANJA MITSCHKE • *Institute of Clinical Pharmacology, Hannover Medical School, Hannover, Germany*
MAURO NATURALE • *Department of Pediatrics, University of Padova, Padova, Italy*
PETER J. OEFNER • *Institute of Functional Genomics, University of Regensburg, Regensburg, Germany*
MARCO OLDIGES • *IBG-1: Biotechnology, Forschungszentrum Jülich GmbH, Jülich, Germany*
LIONELLA PALEGO • *Department of Psychiatry, Neurobiology, Pharmacology and Biotechnology, University of Pisa, Pisa, Italy*
A. PAPPA-LOUISI • *Laboratory of Physical Chemistry, Department of Chemistry, Aristotle University of Thessaloniki, Thessaloniki, Greece*
MARILDA RIGOBELLO-MASINI • *Instituto de Química, Universidade de São Paulo, São Paulo, Brazil*
PETER ROEPSTORFF • *Department of Biochemistry and Molecular Biology, University of Southern Denmark, Odense M, Denmark*
INE ROMBOUTS • *Laboratory of Food Chemistry and Biochemistry, Leuven Food Science and Nutrition Research Centre (LFoRCe), Katholieke Universiteit Leuven, Leuven, Belgium*
CAROLINA SALAZAR • *Department of Biological Sciences, University of North Texas, College of Arts and Sciences, Denton, TX, USA*
VLADIMIR SHULAEV • *Department of Biological Sciences, University of North Texas, College of Arts and Sciences, Denton, TX, USA*
PETR ŠIMEK • *Laboratory of Analytical Biochemistry, Biology Centre, Academy of Sciences of the Czech Republic, České Budějovice, Czech Republic*
S. SOTIROPOULOS • *Laboratory of Physical Chemistry, Department of Chemistry, Aristotle University of Thessaloniki, Thessaloniki, Greece*
TOMOYOSHI SOGA • *Institute for Advanced Biosciences, Keio University, Kakuganji, Tsuruoka, Yamagata, Japan*
SLAVICA SPASIĆ • *Department of Medical Biochemistry, Faculty of Pharmacy, University of Belgrade, Belgrade, Serbia*
NADINE STEIN • *IBG-1: Biotechnology, Jülich, Germany*

AXEL P. STEVENS • *Institute of Functional Genomics, University of Regensburg, Regensburg, Germany*
AKIKO TAKATSU • *Bio-Medical Standard Section, National Metrology Institute of Japan (NMIJ), National Institute of Advanced Industrial Science and Technology (AIST), Tsukuba, Ibaraki, Japan*
BJÖRN THIELE • *Institute of Bio- and Geosciences, Forschungszentrum Jülich GmbH, ICG-3, Jülich, Germany*
TOSHIMASA TOYO'OKA • *Laboratory of Analytical and Bio-Analytical Chemistry, School of Pharmaceutical Sciences, and Global COE Program, University of Shizuoka, Surugaku, Shizuoka, Japan*
JENNIFER A. TRIPP • *Department of Chemistry and Biochemistry, San Francisco State University, San Francisco, CA, USA*
DIMITRIOS TSIKAS • *Institute of Clinical Pharmacology, Hannover Medical School, Hannover, Germany*
KERRI TYRRELL • *CSIRO Division of Livestock Industries, Queensland Bioscience Precinct, St Lucia, QLD, Australia*
HELENA ZAHRADNÍČKOVÁ • *Laboratory of Analytical Biochemistry, Biology Centre, Academy of Sciences of the Czech Republic, České Budějovice, Czech Republic*
SHULIN ZHAO • *Department of Chemistry and Biochemistry, Jackson State University, Jackson, MS, USA*
GORDANA D. ŽUNIĆ • *Institute for Medical Research, Military Medical Academy, Belgrade, Serbia*

Chapter 1

Rapid LC–MS/MS Profiling of Protein Amino Acids and Metabolically Related Compounds for Large-Scale Assessment of Metabolic Phenotypes

Liping Gu, A. Daniel Jones, and Robert L. Last

Abstract

Amino acids extracted from a biological matrix can be resolved and measured using a 6-min per sample method through high-performance liquid chromatography with a short C18 column and rapid gradient using the ion-pairing reagent perfluoroheptanoic acid. LC-tandem mass spectrometry with multiple reaction monitoring (MRM) transitions selective for each compound allows simultaneous quantification of the 20 proteinogenic amino acids and 5 metabolically related compounds. Distinct MRM transitions were also established for selective detection of the isomers leucine/isoleucine and threonine/homoserine.

Key words: High-performance liquid chromatography-tandem mass spectrometry, Ion-pairing reagent, Multiple reaction monitoring transition, Underivatized amino acids, Internal standard, Amino acid quantification, Amino acid analysis

1. Introduction

Comprehensive analysis of amino acids in large sample sets presents challenges due to their high water solubility, range of ionic characteristics, and lack of a universal and selective chromophore. High-performance liquid chromatography (HPLC) coupled with pre- (1–3) or postcolumn (4) derivatization and UV/Vis or fluorescence detection has established dominance as a standard method for amino acid analysis. Gas chromatography (GC) with flame ionization or mass spectrometry (MS) detection can also be employed for separating volatile derivatives of amino acids (5, 6). However, the need for derivatization often requires complete evaporation of solvent before derivatization. The derivatization process may

Michail A. Alterman and Peter Hunziker (eds.), *Amino Acid Analysis: Methods and Protocols*, Methods in Molecular Biology, vol. 828, DOI 10.1007/978-1-61779-445-2_1, © Springer Science+Business Media, LLC 2012

also compromise data quality due to inconsistent yields of a single derivative for each amino acid and derivative instability. For both HPLC and GC methods, resolving the suite of amino acids have required 20–60 min analysis time per sample, reducing their utility for analysis of large numbers of samples required for mutant screening or other large-scale projects.

LC coupled with tandem mass spectrometry (LC–MS/MS) provides a specific and sensitive technique for amino acid analysis without derivatization (7–10). The method described here adapts a 96-well microtiter plate format for sample preparation using aqueous extraction (11) followed by filtration to eliminate cell debris. Twenty proteinogenic amino acids as well as γ-aminobutyric acid (GABA), hydroxyproline (Hyp), and the biosynthetic pathway intermediates anthranilate, homoserine (Hse), and *S*-methylmethionine (SMM) are separated by reverse-phase chromatography using the ion-pairing agent perfluoroheptanoic acid in the mobile phase in a total 6 min assay time (12).

Detection and quantification are achieved using electrospray ionization/tandem mass spectrometry in multiple reaction monitoring (MRM) mode to monitor transitions of the protonated molecules to their specific product ions. Two internal standards, Val-d_8 and Phe-d_8, are spiked in the standards and seed samples allowing extraction normalization and loading control, respectively. The calibration curves for each compound are created by plotting amino acid standard concentrations as *x*-axis and peak area ratio of amino acid/Phe-d_8 as *y*-axis using linear regression and the endogenous concentrations of the 25 compounds calculated according to the slope and intercept from the standard curves and the peak area ratio of AA/Phe-d_8 in the extract. The described method is suitable for quantifying underivatized amino acids and related metabolites in *Arabidopsis* seed tissue with good reproducibility, high accuracy, and low intra- and interday variation. This approach may also fulfill the requirements for the diagnosis of amino acid metabolism-related disorders in clinical fields, analysis of fermentation processes, and for food analysis once matrix effects are evaluated for each new sample matrix. Increased accuracy for specific analytes may be achieved by including heavy isotope-labeled standards for one or more target metabolites (12).

2. Materials

2.1. Chemicals

1. 20 amino acids, anthranilate, GABA, Hyp, Hse, SMM, and DTT (Sigma, St. Louis, MO).
2. L-Phe-α,β,β,2,3,4,5,6-d_8 as internal standard for quantification (Cambridge Isotope Laboratories, Andover, MA).

3. L-Val-2,3,4,4,4,5,5,5-d_8 as internal standard for assessing recovery and loading control (Cambridge Isotope Laboratories, Andover, MA).
4. Perfluoroheptanoic acid (Sigma, St. Louis, MO).
5. Acetonitrile (HPLC grade, Sigma, St. Louis, MO).
6. Methanol (HPLC grade, Sigma, St. Louis, MO).
7. Isopropanol (HPLC grade, Sigma, St. Louis, MO).

2.2. Amino Acid Standard Stocks

1. 20 amino acids, anthranilate, GABA, Hyp, Hse, SMM, and DTT were each dissolved in ultrapure water to make individual stock solution at 0.5–100 mM concentrations, depending upon their water solubility, and stored at −20°C.
2. Single use (50 μL) aliquots of 1 mM (173 μg/mL) L-Phe-α,β,β,2,3,4,5,6-d_8 in water stored at −80°C (Internal Standard Solution 1).
3. Single use (50 μL) aliquots of 1 mM (125 μg/mL) L-Val-2,3,4,4,4,5,5,5-d_8 in water stored at −80°C (Internal Standard Solution 2).
4. Master mixture consisted of 100 μM 25 AA + 100 μM each of the 20 amino acids, anthranilate, GABA, Hyp, Hse, SMM, with the addition of 100 μM DTT (see Note 1), was prepared from the stocks with water and stored at −20°C.

2.3. Sample Preparation

1. 3-mm stainless steel beads (CCR Products, West Hartford, CT).
2. TissueLyser 3 mm Bead Dispenser, 96-well (Qiagen, Valencia, CA).
3. S2200 Dual Head paint shaker (Hero Products Group, Delta, BC, Canada) or an equivalent device.
4. 1.1-mL MicroTubes strips, MicroCaps strips, and MicroRacks (Dot Scientific, Burton, MI).
5. 96-Well MultiScreen® Solvinert filter plate with 0.45 μm low-binding hydrophilic polytetrafluoroethylene (PTFE) membrane and 96-well v-bottom collection plate (Millipore, Billerica, MA).
6. 96-Well full skirt PCR plates (Denville Scientific, Metuchen, NJ).
7. Aluminum sealing films (Excel Scientific, Victorville, CA).
8. Silicone sealing mats for 96-well microplates (Axygen Scientific, Union City, CA).
9. 12 mm (outer diameter) × 32 mm (length) 2-mL wide-opening screw-top vials (Agilent Technologies, Santa Clara, CA).
10. Screw caps with PTFE and rubber septa for above vials (Agilent Technologies).
11. 200-μL conical-bottom inserts for above vials (SUN SRI, Rockwood, TN).

12. Polypropylene 2-ml Vial Rack, 50 holes.
13. Matrix 1,250-μL eight-channel electronic pipette with expandable spacing (Thermo Fisher Scientific, Waltham, MA).
14. 10 μL eight-channel electronic pipette (Thermo Fisher Scientific).
15. 100 μL eight-channel electronic pipette (Thermo Fisher Scientific).

2.4. HPLC–MS/MS Components

1. Symmetry C18 3.5 μm (particle size), 2.1 mm (inner diameter) × 100 mm (length) analytical column (Waters, Milford, MA).
2. Waters Symmetry C18 3.5 μm, 2.1 × 10 mm guard column, and guard column holder (Waters, Milford, MA).
3. One-piece PEEK Direct-Connect column coupler for Waters fittings (Grace Davison Discovery Science, Deerfield, IL).
4. LC-20AD HPLC pumps and SIL-5000 auto injector with temperature-controlled stack (Shimadzu Scientific Instruments, Columbia, MD).
5. Quattro micro API mass spectrometer with MassLynx V4.0 software (Waters).

3. Methods

3.1. Standard Working Solutions

1. Prepare working standards from the master mixture at concentrations of 0, 0.5, 1.0, 5.0, 10, 20, and 50 μM 25 AA, plus 10 μM Phe-d_8 (see Note 2) and 10 μM Val-d_8 (see Note 3) as internal standards in water on the day of HPLC–MS/MS analysis.
2. Transfer 100 μL standard solutions into screw-cap vials with inserts.
3. Place standard vials into Vial Rack in the sample storage stack (set at 10°C) of Shimadzu SIL-5000 auto injector.

3.2. Amino Acid Extraction

1. Weigh approximately 3 mg of *Arabidopsis* seed (see Note 4) samples after removing nonseed materials, transfer via a funnel into 1.1-mL MicroTubes strips or their equivalent (see Note 5), and store in a humidity-controlled container in the cold room until processing.
2. Dispense two 3-mm stainless steel balls into each well with TissueLyser Bead Dispenser.
3. Add 270 μL extraction buffer containing 30 μL of 100 μM Phe-d_8 (Internal Standard Solution 1), 30 μL of 100 μM DTT,

and 210 μL of H_2O into each well using eight-channel electronic pipette (see Notes 6 and 7).

4. Seal tightly with MicroCaps strips.
5. Homogenize tissues by vigorous shaking with two 3-mm stainless steel balls for 5 min on a S2200 paint shaker (see Note 8).
6. Briefly spin down samples (at 3,220 × *g* for 2 min at 4°C with a swinging bucket centrifuge) and incubate in a water bath at 90°C for 10 min (see Note 9).
7. While the samples are incubated in the water bath, prewet filter plate by adding 100 μL of distilled water into each well and centrifuge the filter plate with the collection plate at 2,000 × *g* for 5 min at 4°C. Place the prewetted filter plate on a new collection plate.
8. Clarify seed extracts by centrifuge at 3,220 × *g* for 10 min at 4°C with a swinging bucket centrifuge.
9. Purify the extracts through low-binding hydrophilic PTFE membrane by carefully transferring ~240 μL of supernatant from step 8 (see Note 10) into the prewetted filter plate (step 8) and centrifuge at 2,000 × *g* for 30 min at 4°C (see Note 11).
10. While the samples are in the centrifuge, add 10 μL of 100 μM Val-d_8 (Internal Standard Solution 2) into each receiving well of full skirt PCR microplate.
11. Transfer 90 μL of filtrates from step 9 into 96-well plate containing 10 μL 100 μM Val-d_8 from step 10.

    Optional, an identical back-up sample plate may be prepared (see Note 12*).*
12. Seal the plates with aluminum sealing film (store at −80°C if needed, see Note 13).
13. Mix them by vortexing and briefly spin down.
14. Replace the aluminum film with a silicone sealing mat.
15. Place the sample plate in the sample storage stack (set at 10°C) of Shimadzu SIL-5000 auto injector.

3.3. HPLC Separation

1. Prepare 1 mM (364 mg/L) perfluoroheptanoic acid in water (Mobile phase A) and acetonitrile (Mobile phase B).
2. Assemble the analytical column and guard column in the correct orientation and set column oven temperature at 30°C (see Note 14).
3. Purge the HPLC system to eliminate bubbles and remove previous solvents from the tubing.
4. Set HPLC flow rate at 0.3 mL/min.
5. Condition the column for 30 min with 98% 1 mM perfluoroheptanoic acid (A) and 2% acetonitrile (B).

6. Program the HPLC gradient as following: 0–0.09 min, 98% A, 2% B; 0.1–2.29 min, 80% A, 20% B; 2.3–4.09 min, 60% A, 40% B; 4.1–6 min, 98% A, 2% B.
7. Program the inlet method to clean the syringe after each injection (2 × 100 μL isopropanol and 1 × 100 μL water).
8. Generate a sample list including sample name, type, position, injection volume (set as 10 μL), HPLC methods (see steps 6 and 7 *above*), and MS detection method (see Subheading 3.4 step 7 *below*).
9. Run the sample list.

   Optional, inject 5 μM amino acid standard before running the sample list, and insert an injection of blank (water) between the amino acid standards and the biological samples (see Notes 15 and 16*)*.
10. Enable the shut-down program so that the HPLC and MS will stop after finishing the sample list.
11. Clean the guard and analytic columns after every 96 chromatographic runs (see Notes 17 and 18 for detailed information) and store them at 25°C with both ends sealed.

3.4. MS/MS Parameters

1. Select positive ion mode for electrospray ionization.
2. Set the capillary voltage, extractor voltage, and rf lens at 3.17 kV, 4 V, and 0.3, respectively.
3. Set the flow rates of cone gas and desolvation gas at 20 and 400 L/h, respectively.
4. Set the source temperature to 110°C and desolvation temperature to 350°C.
5. Set argon collision gas to yield a pressure of 2×10^{-3} mbar measured at the collision cell manifold.
6. Verify that the mass scale for both the first and second mass analyzers is valid by infusing, via syringe pump, a 0.1% aqueous sodium formate solution and performing both Q1 and Q3 scans over m/z 20–300; if masses are not within 0.1 m/z unit of the theoretical values, recalibrate with the sodium formate solution.
7. Enter the optimized MRM transition (see Note 19), source cone voltages, collision cell voltages, and adequate dwell time for each analyte. The method composed of two ESI+ functions (0–1.8 and 1.8–6.0 min) covering full run time to allow for adequate dwell time for each analyte (see Table 1).

3.5. Data Acquisition with MassLynx and Processing with QuanLynx

1. Collect integrated peak area, peak heights, and retention time of extracted ion MRM chromatograms for each compound with MassLynx software. Pay specific attention to the internal standard signals of Val-d_8 to ensure loading accuracy.

Table 1
MRM transitions, optimized source cone voltages, collision cell voltages, and analyte retention times for LC–MS/MS analysis of amino acids and related metabolites

Compound	Precursor ion > product ion (*m/z*)	Cone voltage (V)	Collision voltage (V)	Retention time (min)	Function no.
Ala	90 > 44	18	15	1.56	1
Arg	175 > 70	26	20	3.96	2
Asn	133 > 87	26	20	1.18	1
Asp	134 > 74	18	15	1.16	1
Cys	122 > 76	18	15	1.39	1
Gln	147 > 130	18	15	1.26	1
Glu	148 > 84	18	15	1.32	1
Gly	76 > 30	18	40	1.30	1
His	156 > 110	18	15	3.52	2
Ile/Leu	132 > 86	18	15	2.76 and 2.95	2
Ile	132 > 69	18	15	2.76	2
Leu	132 > 30	18	15	2.95	2
Lys	147 > 84	18	15	3.76	2
Met	150 > 104	18	15	2.17	2
Phe	166 > 120	18	15	3.17	2
Pro	116 > 70	26	15	1.54	1
Ser	106 > 60	18	20	1.20	1
Thr	120 > 57	26	25	1.37	1
Trp	205 > 188	18	15	3.65	2
Tyr	182 > 136	18	15	2.11	2
Val	118 > 72	18	15	2.08	2
Anthranilate	138 > 120	18	15	3.11	2
GABA	104 > 87	18	15	2.43	2
Hyp	132 > 86	18	15	1.16	1
Hse	120 > 44	26	20	1.35	1
SMM	164 > 102	18	15	3.38	2
Phe-d_8	174 > 128	18	15	3.15	2
Val-d_8	126 > 80	18	15	2.08	2

2. Create the calibration curves for each compound by plotting amino acid standard concentrations as *x*-axis and peak area ratio of amino acid/Phe-d_8 as *y*-axis using linear regression.
3. Calculate the endogenous concentrations of the 25 compounds according to the slope and intercept from the standard curves and the peak area ratio of AA/Phe-d_8 in the extract (see Note 20).

4. Notes

1. DTT serves as a reducing reagent for preventing and reversing cysteine disulfide bond formation.
2. Phe-d_8 is used as internal standard to normalize extraction. The final concentration of 10 μM Phe-d_8 is spiked in both the standards and seed samples. It can be replaced by a less costly compound that is not found in the biological samples being analyzed and has similar physical–chemical properties as amino acids, such as norleucine, nitrotyrosine, and 2-aminoethyl-L-cysteine.
3. Val-d_8 is used to monitor recovery and loading accuracy. It can be replaced by other compound with similar physical–chemical properties as amino acids (see Note 2).
4. This method has been most extensively evaluated using *Arabidopsis* seeds as biological material. It has also been employed for the analysis of amino acids extracted from *Arabidopsis* leaf tissues (13), bacterial materials (Gu, unpublished data), and from acid hydrolysis of grain proteins (14).
5. 1 mL 96-deep well microplate and lid (VWR) can be used for sample collection instead of 1.1-mL MicroTubes strips and MicroCaps strips.
6. The ratio of 100 μL of extraction buffer/mg seeds was used for easy handling, although 200 μL of extraction buffer/mg seeds was tested to be an appropriate ratio to minimize matrix effects (12). The initial extract buffer (30 μL of 1 mM Phe-d_8, 30 μL of 1 mM DTT, and 210 μL of H_2O) prepared with 100/9 μM Phe-d_8 and 100/9 μM DTT will give rise to the final concentration of 10 μM after 10 μL of 100 μM Val-d_8 addition into 90 μL purified extract (see Subheading 3.2 step 9).
7. If the sample volume is less than 60 μL due to a limited amount of starting materials, use screw-cap vials with 100 μL insert (glass, Mandrel, w/PP Bottom Spring, SUN SRI, Rockwood, TN).

8. Grinding tissues using a paint shaker or other disruption devices requires extreme care. Two 1-in. thick foam sheets filled in both inner sides of the clamps helps to secure MicroRacks; place the MicroRacks near the edge of the clamps to increase agitation force. In addition, flip the sample plate horizontally by 180° after shaking for 5 min and grind for another 5 min for uniform homogenization.
9. To avoid having the MicroCaps strips pop up during the 90°C incubation, flip the MicroRack lid and place a heavy metal on top of the lid. Submerge Microtubes on two-third of the way into the 90°C water bath so that bath water does not get into the sample wells.
10. Do not let pipette tips touch seed pellets when transferring the supernatants to avoid clogging pipette tips by seed residues.
11. The duration of filtration by centrifuge at 2,000 × *g* at 4°C may vary depending on the biological tissues, seed extracts take about 20–50 min and leaf extracts require 5–10 min.
12. This extra sample plate is extremely helpful when problems happen during LC–MS/MS run, or some important findings need to be confirmed by repeating LC–MS/MS or using different methods.
13. Amino acids extracted from plants with water/90°C incubation treatment are fairly stable when stored from several days at 10°C up to 2 months at −80°C (12).
14. The guard column can be replaced by 0.5 μm stainless steel high pressure inline solvent filter and 0.5 μm (porosity) 0.062"(inner diameter) × 0.062" (thickness) × 0.25" (outer diameter) stainless steel frits (IDEX Health and Science, Oak Harbor, WA) (13). The retention time will shift from this replacement.
15. An injection of amino acid standard (5 μM) before running the sample list is used to make sure the signals and retention times are normal. New analytical and/or guard column should be installed if the intensity of the chromatogram valley between Ile and Leu is more than 40% of the peak top.
16. One or more injections of blank (water) between the amino acid standards and the biological samples can help to minimize the carry-over of some amino acids, particularly at 50 μM concentration.
17. Clean the analytical column with 100% methanol for 20 min, 50% methanol/50% water 20 min, and 80% methanol/20% water 20 min at the flow rate at 0.3 mL/min after every 96 chromatographic runs.
18. After every 96 chromatographic runs, sequentially sonicate the guard column in 15 mL isopropanol (HPLC grade) for 10 min

followed by 15 mL methanol for 10 min. Flush the guard column in both orientations with 100% methanol for 20 min, 50% methanol/50% water 20 min, and 80% methanol/20% water 20 min at the flow rate at 0.3 mL/min.

19. The product ion *m/z* yielding the strongest signal was used in MRM transition for quantifying each amino acid. Second MRM transitions for Ile and Leu confirmation were also included, respectively. Since Thr and Hse were unable to be separated chromatographically, isomer-selective product ions at *m/z* 57 (Thr) and 44 (Hse) were used for Thr and Hse detection.

20. The most accurate defining of amino acid absolute concentrations in the extracts can be achieved by spiking stable isotopically labeled version of each target amino acid as internal standards (12).

Acknowledgments

We thank Lijun Chen of the Michigan State University Mass Spectrometry Core and Yan Lu in the Last Laboratory for assistance in various stages of the project. This project was supported by National Science Foundation grant MCB-0519740 (to R.L.L.).

References

1. Lammens J, Verzele M (1978) Rapid and easy HPLC analysis of amino acids. Chromatographia 11:376–378
2. Deyl Z, Hyanek J, Horakova M (1986) Profiling of amino acids in body fluids and tissues by means of liquid chromatography. J Chromatogr 379:177–250
3. Sarwar G, Botting HG (1993) Evaluation of liquid chromatographic analysis of nutritionally important amino acids in food and physiological samples. J Chromatogr 615:1–22
4. Moore S, Spackman DH, Stein WH (1958) Automatic recording apparatus for use in the chromatography of amino acids. Fed Proc 17:1107–1115
5. Hardy JP, Kerrin SL (1972) Rapid determination of twenty amino acids by gas chromatography. Anal Chem 44:1497–1499
6. Kaspar H et al (2009) Urinary amino acid analysis: a comparison of iTRAQ-LC-MS/MS, GC-MS, and amino acid analyzer. J Chromatogr B Analyt Technol Biomed Life Sci 877:1838–1846
7. Chaimbault P et al (1999) Determination of 20 underivatized proteinic amino acids by ion-pairing chromatography and pneumatically assisted electrospray mass spectrometry. J Chromatogr A 855:191–202
8. Petritis K et al (2001) Simultaneous analysis of underivatized chiral amino acids by liquid chromatography-ionspray tandem mass spectrometry using a teicoplanin chiral stationary phase. J Chromatogr A 913:331–340
9. Petritis KN et al (1999) Ion-pair reversed-phase liquid chromatography for determination of polar underivatized amino acids using perfluorinated carboxylic acids as ion pairing agent. J Chromatogr A 833:147–155
10. Qu J et al (2002) Validated quantitation of underivatized amino acids in human blood samples by volatile ion-pair reversed-phase liquid chromatography coupled to isotope dilution tandem mass spectrometry. Anal Chem 74:2034–2040
11. Streatfield SJ et al (1999) The phosphoenolpyruvate/phosphate translocator is

required for phenolic metabolism, palisade cell development, and plastid-dependent nuclear gene expression. Plant Cell 1999, 11:1609–1622

12. Gu L, Jones AD, Last RL (2007) LC–MS/MS assay for protein amino acids and metabolically related compounds for large-scale screening of metabolic phenotypes. Anal Chem 79: 8067–8075

13. Lu Y et al (2008) New connections across pathways and cellular processes: industrialized mutant screening reveals novel associations between diverse phenotypes in Arabidopsis. Plant Phys 146:1482–1500

14. Bals B, Balan V, Dale BE (2009) Integrating alkaline extraction of proteins with enzymatic hydrolysis of cellulose from wet distiller's grains and solubles. Bioresour Technol 100:5876–5883

Chapter 2

Combination of an AccQ·Tag-Ultra Performance Liquid Chromatographic Method with Tandem Mass Spectrometry for the Analysis of Amino Acids

Carolina Salazar, Jenny M. Armenta, Diego F. Cortés, and Vladimir Shulaev

Abstract

Amino acid analysis is a powerful tool in life sciences. Current analytical methods used for the detection and quantitation of low abundance amino acids in complex samples face intrinsic challenges such as insufficient sensitivity, selectivity, and throughput. This chapter describes a protocol that makes use of AccQ·Tag chemical derivatization combined with the exceptional chromatographic resolution of ultra performance liquid chromatography (UPLC), and the sensitivity and selectivity of tandem mass spectrometry (MS/MS). The method has been fully implemented and validated using different tandem quadrupole detectors, and thoroughly tested for a variety of samples such as *Plasmodium falciparum*, human red blood cells, and *Arabidopsis thaliana* extracts. Compared to currently available methods for amino acid analysis, the AccQ·Tag UPLC-MS/MS method presented here provides enhanced sensitivity and reproducibility, and offers excellent performance within a short analysis time and a broad dynamic range of analyte concentration. The focus of this chapter is the application of this improved protocol for the compositional amino acid analysis in *A. thaliana* leaf extracts using the Xevo TQ for mass spectrometric detection.

Key words: 6-Aminoquinolyl-*N*-hydroxysuccinimidyl carbamate, AccQ·Tag, Amino acid analysis, UPLC-ESI-MS/MS

1. Introduction

The importance of quantitative analysis of amino acids in life sciences and food industry cannot be emphasized enough. Stein et al. (1) initially introduced amino acid analysis in the late 1950s, and since then many analytical platforms for amino acids analysis have been developed. These approaches include the combination of different separation strategies, such as capillary electrophoresis, liquid chromatography, or gas chromatography, coupled with ultraviolet,

Michail A. Alterman and Peter Hunziker (eds.), *Amino Acid Analysis: Methods and Protocols*, Methods in Molecular Biology, vol. 828, DOI 10.1007/978-1-61779-445-2_2, © Springer Science+Business Media, LLC 2012

fluorescence, electrochemical, and mass spectrometry detection systems (2–8). In general, currently employed amino acid analyses can be divided into three major categories: (a) direct analysis of free amino acids, (b) separation of free amino acids by ion-exchange chromatography followed by postcolumn derivatization (e.g., with ninhydrin or *o*-phthalaldehyde), and (c) precolumn derivatization techniques (e.g., with phenyl isothiocyanate; 7-flouro-4-nitrobenzo-2-oxa-1,3-diazole, or 6-aminoquinolyl-*N*-hydroxysuccinimidyl carbamate, AQC). Direct analysis methods are attractive due to high throughput and simplicity. Nevertheless, sensitivity of analysis can be compromised as a result of matrix interferences. On the other hand, precolumn derivatization techniques may be influenced by buffer salts in the samples and may also result in multiple derivatives of a given amino acid, which have to be considered in the result interpretation. Despite that, precolumn derivatization techniques are very sensitive and, unlike their postcolumn counterparts, usually require very small sample size per analysis and result in better throughput. In general, precolumn derivatization, combined with reversed-phase separation, offers greater efficiency, ease of use, and higher speed of analysis than the conventional ion-exchange techniques (9).

Among the instrumental methods based on precolumn derivatization for the analysis of free amino acids, high-performance liquid chromatography (HPLC) and ultra performance liquid chromatography (UPLC) combined with AccQ·Tag technology continue to gain acceptance and recognition for their improved limits of detection, superior efficiency and sensitivity, and reduced analysis time (4, 6–12). AccQ·Tag technology uses AQC which reacts with primary and secondary amines to form highly stable fluorescent urea derivatives. During the reaction the excess reagent is rapidly hydrolyzed to yield 6-aminoquinoline (AMQ), *N*-hydroxysuccinimide, and carbon dioxide. AccQ·Tag-derivatization followed by reverse-phase liquid chromatographic (RPLC) separation of the less-polar derivatives is readily amenable to mass spectrometry.

A method for amino acid analysis using precolumn AccQ·Tag derivatization followed by UPLC separation coupled to electrospray ionization tandem mass spectrometry (ESI-MS/MS) with multiple reaction monitoring (MRM) detection has been previously described by our research group using the Waters Acquity TQD system (9). This UPLC-MS/MS MRM quantitation method was further improved and implemented in the Waters Xevo TQ mass spectrometer for increased sensitivity in amino acid detection and quantitation (13). The latest method, herein described, resulted in improved detection limits ranging from 10.2 amol to 23.1 fmol (on column), as well as enhanced linear dynamic range. The application of this method to the analysis of amino acids in *Arabidopsis thaliana* is illustrated.

2. Materials

2.1. AccQ·Tag Ultra Amino Acid Derivatization

1. Four-block digital readout dryblock heater (VWR 13259-056 or similar).
2. Refrigerated benchtop shaker-incubator (Lab-Line MaxQ 4000, Barnstead or similar).
3. Limited volume polypropylene vial: 250 μL.
4. Clear glass, screw-thread vials: 2 mL.
5. AccQ·Tag Ultra Reagent Powder (Waters Corporation): AQC derivatizing reagent, dry powder.
6. AccQ·Tag Ultra Borate Buffer (Waters Corporation): Borate derivatization buffer used to ensure optimum pH (8.8) for derivatization.
7. AccQ·Tag Ultra Reagent Diluent (Waters Corporation): Acetonitrile (LC-MS grade) for reagent reconstitution.
8. Derivatizing reagent solution: Tap AccQ·Tag Ultra Reagent Powder vial and add 1 mL of AccQ·Tag Ultra Reagent Diluent. Vortex. Heat on top of heating block at 55°C for up to 15 min and vortex occasionally to complete solubilization (see Note 1).
9. Methanol-water mixture, 50:50 v/v: In a clean glass container, mix 50 mL of LC-MS grade water and 50 mL of LC-MS grade methanol.
10. Stable-isotope-labeled reference compounds (internal standards): L-asparagine-$^{15}N_2$; L-serine,2,3,3-d_3; L-glutamine-2,3,3,4,4-d_5; glycine-d_5; D-L-alanine-2,3,3,3-d_4; proline-2,5,5-d_3; methionine-methyl-d_3; tryptophan-2′,4′,5′,6′,7′-d_5(indole-d_5); leucine-d_{10}; valine-d_8; L-histidine (ring 2-^{13}C); L-glutamic acid-2,4,4-d_3; ornithine-3,3,4,4,5,5-d_6; lysine-3,3,4,4,5,5,6,6-d_8; phenyl-d_5-alanine (Cambridge Isotope Laboratories, Andover, MA, USA and CDN isotopes, Pointe-Claire, Quebec, Canada).
11. Internal standard stock solutions (1 mg/mL): Prepare 1 mg/mL solutions of individual internal standards in LC-MS grade water into an externally threaded cryotube vial. Store solutions at −80°C.
12. Internal standard mixture (30 μg/mL): In a 25-mL volumetric flask, mix 750 μL each from the stock solutions of internal standards. Make up to final volume with 50% v/v methanol–water mixture. Leave one aliquot for current use and store remaining aliquots at −80°C in amber borosilicate glass vials with screw caps.
13. Internal standard mixture (8 μg/mL): Add 2,670 μL of the 30 μg/mL internal standard mixture into a 10-mL volumetric flask. Make up to final volume with 50% v/v methanol–water

mixture. Divide the solution into working aliquots. Leave one aliquot for current use and store remaining aliquots at −80°C in amber borosilicate glass vials with screw caps.

14. Extraction buffer with internal standards (4 μg/mL): Add 6.7 mL of the 30 μg/mL internal standard mixture into a 50-mL volumetric flask. Make up to a final volume with 50% v/v methanol–water mixture. Divide the solution into working aliquots and store at −80°C in amber borosilicate glass vials with screw caps (see Note 2).
15. Amino acid calibration stock solution (0.5 μmol/mL in 0.2N lithium citrate buffer). Available from Sigma-Aldrich, Inc; Product No. A9906 (see Note 3).
16. Working calibration mixture (0.25 μmol/mL amino acids, 4 μg/mL internal standards): Mix 500 μL of amino acid stock solution (0.5 μmol/mL) with 500 μL of the 8 μg/mL internal standard mixture in a 2-mL clear glass vial.
17. Calibration curve standards: Prepare calibration curve standards (1 mL each) with concentration range from 0.25 μmol/mL to 476.8 fmol/mL by serial dilutions of the 0.25 μmol/mL working calibration amino acid mixture spiked with isotopically labeled internal standards (4 μg/mL). Use internal standard mixture (4 μg/mL) to make up the final volume of each standard to 1 mL and to keep a constant concentration of internal standards at 4 μg/mL in all the concentration levels. Divide each solution into working aliquots. Leave one aliquot of each standard for current use and store remaining aliquots at −80°C.

2.2. Direct Infusion ESI-MS/MS

1. LC-MS grade water or ultrapure water prepared by purifying deionized water in a water treatment system (18.2 MΩ-cm).
2. L-Amino acids, kit of 1 g each: L-alanine, L-arginine, L-asparagine, L-aspartic acid, L-cysteine, L-cystine, L-glutamic acid, L-glutamine, glycine, L-histidine, *trans*-4-hydroxy-L-proline, L-isoleucine, L-leucine, L-lysine, L-methionine, L-phenylalanine, L-proline, L-serine. L-threonine, L-tryptophan, L-tyrosine, and L-valine (Sigma-Aldrich, Co., St. Louis, MO; Cat. No. LAA-21).
3. Amino acid stock solutions (1 mg/mL): Prepare 1 mg/mL solutions of individual L-amino acids in LC-MS grade water.
4. Derivatized amino acid solutions for infusions (10 μg/mL): Prepare 100 μg/mL solution of each L-amino acid by appropriate dilution of its corresponding stock solution. Derivatize each amino acid using the AccQ·Tag derivatization kit. Final

amino acid concentration after derivatization is 10 μg/mL (see Note 4 and 5).

5. Waters Xevo TQ mass spectrometer with electrospray ionization (ESI) probe (or similar tandem quadrupole mass spectrometer).

2.3. Metabolite Extraction from Arabidopsis Leaf Tissue

1. Dry ice.
2. LC-MS grade methanol.
3. LC-MS grade water or ultrapure water (18.2 MΩ-cm).
4. Mixer Mill MM 300 and Mixer Mill Adapter Set (QIAGEN).
5. Bench top refrigerated microcentrifuge (Beckman Coulter, Inc., Brea, CA or similar).
6. Ultrasonic cleaner (Branson, model 3510R-MT or similar).
7. Microcentrifuge tubes with screw caps and O-rings: 1.5 mL.
8. Stainless steel beads: 2.3 mm.
9. Limited volume polypropylene vial: 250 μL.
10. Polypropylene caps with Red PTFE/White Silicone/Red PTFE septa.
11. Methanol–water mixture, 50:50 v/v.
12. Extraction buffer with internal standards (4 μg/mL).

2.4. UPLC-MS/MS Analysis

1. Waters Acquity UPLC system equipped with a binary solvent manager, an autosampler, a column heater, and interfaced to a Waters Xevo TQ mass spectrometer by means of an electrospray ionization probe (or similar UPLC system integrated with tandem quadrupole mass spectrometer).
2. AccQ·Tag Ultra column, 2.1 × 100 mm, 1.7 μm particles (Waters Corp.).
3. Acetonitrile (LC/MS grade).
4. LC-MS grade water or ultrapure water (18.2 MΩ-cm).
5. AccQ·Tag Ultra concentrate solvent A (Waters Corp.).
6. AccQ·Tag Ultra solvent B (Waters Corp.).
7. Eluent A: 10:90 v/v AccQ·Tag Ultra concentrate solvent A/water.
8. Eluent B: 100% AccQ·Tag Ultra solvent B.
9. Strong wash solvent: 70% acetonitrile in LC/MS water.
10. Weak wash solvent: 10% acetonitrile in LC/MS water.
11. Collision gas: dry, high purity argon (99.997%).
12. API gas: dry, oil-free nitrogen (with a purity of at least 95%).

3. Methods

3.1. Direct Infusion ESI-MS/MS and UPLC-MS/MS Analysis

1. Using the IntelliStart software, or any other available instrument specific software, determine the MRM transitions for each amino acid adduct by direct infusion of the derivatized L-amino acid solutions. Set the infusion flow rate to 20 μL/min, the desolvation temperature to 350°C, and the desolvation gas flow rate to 600 L/h. Select the option to automatically create a MS tune file which will be populated with the optimized cone voltages and collision energies for all transitions of each amino acid. The MRM transitions, cone voltages, and collision energies for a selected group of amino acids are listed in Table 1 (see Notes 6 and 7).
2. Create an LC method with the following gradient: 0–0.54 min (99.9% A), 5.74 min (90.0% A), 7.74 min (78.8% A), 8.04–8.64 min (40.4% A), 8.73–10 min (99.9% A) (see Note 8).
3. Perform a UPLC-MS/MS analysis of a representative sample or a mixture of amino acids using the MS tune file created earlier to determine the retention times of each targeted amino acid (column flow rate = 0.7 mL/min, column temperature = 55°C, autosampler temperature = 25°C, injection volume = 1 μL, desolvation temperature = 600°C, desolvation gas flow rate = 1,000 L/h, collision gas (argon) flow rate = 0.15 mL/min, dwell time = 0.01 s).
4. After determination of the retention times, proceed to the adjustment of the time window per MRM function. Divide the 10 min run time into time segments such that the minimum number of MRM transitions is observed per time window (see Note 9). If an Acquity UPLC system coupled to a Xevo TQ mass spectrometer is used, the optimized MRM method described in Table 1 can be used for the analysis of amino acids.

3.2. Metabolite Extraction from Arabidopsis Leaf Tissue

1. Transfer 5 mg of freeze-dried plant leaf tissue to a 1.5-mL microcentrifuge tube with screw caps/O-rings. Keep plant leaf tissue on dry ice before and after weighing.
2. Add two 2.3 mm stainless steel beads to the microcentrifuge tube.
3. Add 125 μL of extraction buffer (4 μg/mL).
4. Grind samples two times in the mixer mill and rotate adapter set 180° between repetitions (30 cycles/s, 30 s per repetition).
5. Incubate sample on dry ice for 5 min.
6. Sonicate samples for 1 min in ultrasonic cleaner.

Table 1
UPLC-MS/MS conditions for the determination of AccQ·Tag-amino acid derivatives[a]

Compound number	Amino acid/stable isotope labeled amino acid	Parent ion[b] (m/z)	Daughter ion[b] (m/z)	Cone voltage (V)	Collision energy (ev)	R_t (min)	Time window (min)	Internal standard
1	Hydroxyproline	302.11	171.01	24	21	1.49	1.03–2.02	32
2	Histidine	326.21	171.01	18	17	1.60	1.13–2.16	3
3	L-Histidine (ring 2-^{13}C)	327.21	171.01	18	17	1.64	1.13–2.16	
4	Asparagine	303.13	171.01	24	21	1.84	1.60–2.36	5
5	L-Asparagine-^{15}-N_2	305.27	171.01	24	21	1.87	1.39–2.39	
6	3-Methyl-histidine	340.21	171.01	24	21	1.97	1.50–2.10	3
7	Taurine	296.11	171.01	18	15	2.13	1.66–2.66	11
8	1-Methyl-histidine	340.21	171.01	24	21	2.20	2.08–2.50	3
9	L-Serine-2,3,3-d_3	279.11	171.01	25	19	2.48	2.04–3.05	
10	Serine	276.11	171.01	25	19	2.51	2.04–3.05	9
11	L-Glutamine-2,3,3,4,4-d_5	322.15	171.01	22	24	2.64	2.16–3.16	
12	Glutamine	317.21	171.01	22	24	2.67	2.16–3.16	11
13	Carnosine	397.21	171.01	24	21	2.74	2.27–3.27	3
14	Arginine	345.21	171.01	27	31	2.77	2.29–3.29	40
15	Glycine-d_5	248.25	171.01	27	21	2.88	2.42–3.42	
16	Glycine	246.08	171.01	27	21	2.88	2.39–3.39	15
17	Homoserine	290.12	171.01	24	21	3.02	2.40–3.40	9

(continued)

Table 1 (continued)

Compound number	Amino acid/stable isotope labeled amino acid	Parent ion[b] (m/z)	Daughter ion[b] (m/z)	Cone voltage (V)	Collision energy (ev)	R_t (min)	Time window (min)	Internal standard
18	Ethanolamine	232.09	171.01	24	21	3.04	2.58–3.58	15
19	Aspartic acid	304.11	171.01	27	23	3.24	2.76–3.76	21
20	Sarcosine	260.17	171.01	25	21	3.68	3.30–3.90	15
21	L-Glutamic acid-2,4,4-d_3	321.11	171.01	27	21	3.82	3.33–4.33	
22	Glutamic acid	318.11	171.01	27	21	3.84	3.35–4.34	21
23	Citrulline	346.21	171.01	22	24	3.87	3.39–4.38	11
24	B-Alanine	260.17	171.01	25	21	4.08	3.88–4.45	15
25	Threonine	290.11	171.01	25	21	4.30	3.81–4.80	9
26	D-L-Alanine-2,3,3,3-d_4	264.1	171.01	25	21	4.72	4.22–5.22	
27	L-Alanine	260.17	171.01	25	21	4.74	4.44–5.26	26
28	γ-Amino-*n*-butyric acid	274.11	171.01	25	21	4.92	4.43–5.25	15
29	α-Amino adipic acid	332.33	171.01	25	21	5.13	4.63–5.64	21
30	Creatinine	284.11	171.01	24	21	5.34	5.00–5.84	32
31	β-Aminoisobutryic acid	274.11	171.01	25	21	5.38	5.24–5.89	15
32	Proline-2,5,5-d_3	289.32	171.01	25	21	5.38	4.88–5.88	
33	Proline	286.16	171.01	25	21	5.39	4.91–5.91	32
35	δ-Hydroxylysine	503.21	171.01	28	21	5.62, 5.73	5.12–6.12	47
36	α-Amino-*n*-butyric acid	274.11	171.01	25	21	5.99	5.75–6.75	26
38	Cystathionine	563.6	171.01	28	21	6.10, 6.24	5.60–6.60	40

40	Ornithine-3,3,4,4,5,5-d_6	479.22	171.01	16	18	6.11	5.62–6.62	
42	Ornitine	473.22	171.01	16	18	6.14	5.65–6.65	40
44	Cystine	581.64	171.01	28	21	6.45	5.96–6.96	40
47	Lysine-3,3,4,4,5,5,6,6-d_8	495.58	171.01	18	18	6.52	6.03–7.03	
49	Lysine	487.21	171.01	18	18	6.55	6.05–7.05	47
50	Tyrosine	352.21	171.01	24	21	6.61	6.12–7.12	61
51	Methionine-methyl-d_3	323.13	171.01	27	21	6.75	6.25–7.25	
52	Methionine	320.21	171.01	27	21	6.77	6.27–7.27	51
53	Valine-d_8	296.21	171.01	28	21	6.88	6.39–7.39	
54	Valine	288.21	171.01	28	21	6.91	6.43–7.43	53
56	Homocystine	609.36	171.01	28	21	7.54	7.04–8.02	40
58	Leucine	302.21	171.01	28	21	7.67	7.18–8.18	60
59	Isoleucine	302.21	171.01	28	21	7.75	7.26–8.26	60
60	Leucine-d_{10}	312.24	171.01	28	21	7.78	7.13–8.13	
61	Phenyl-d_5-alanine	341.21	171.01	28	21	7.84	7.34–8.34	
62	Phenylalanine	336.21	171.01	28	21	7.86	7.37–8.36	61
63	Tryptophan-2′,4′,5′,6′,7′-d_5(indole-d_5)	380.21	171.01	28	21	7.94	7.44–8.44	
64	Tryptophan	375.21	171.01	28	21	7.96	7.44–8.44	63

[a] Conditions are specific for the Waters Acquity UPLC – Xevo TQ MS system

[b] $(M+H)^+$-adduct

7. Centrifuge sample at 13,000 rpm/17,900×*g* and 4°C for 5 min.
8. Transfer the supernatant to a limited volume vial. Be careful not to transfer any debris.
9. Re-extract the sample with 125 μL extraction buffer.
10. Perform second round of extraction (steps 4–7) and combine the extracts. Store the extract at –80°C if it is not derivatized immediately.

3.3. AccQ·Tag Ultra Amino Acid Derivatization

1. Place 70 μL of AccQ·Tag Ultra Borate Buffer in a 250-μL reduced volume polypropylene vial using a micropipette.
2. Add 10 μL of amino acid standard or sample extract and mix.
3. Add 20 μL of the derivatizing reagent solution, vortex immediately after addition.
4. Let the solution stand for 1 min at room temperature.
5. Heat the vial in a heating block or an incubator for 10 min at 55°C.
6. Remove samples from heating block and place in instrument for analysis (see Note 10).

3.4. Sample Analysis and Data Analysis

1. Start the UPLC-ESI-MS/MS system and equilibrate to initial conditions of the method (flow rate = 0.7 mL/min, 99.9% of eluent A, 0.1% of eluent B, column temperature = 55°C, sample temperature = 25°C).
2. Create a sample analysis sequence with a blank as the first run followed by quality control standard (a solution containing derivatized internal standards) followed by experimental samples.
3. Insert appropriate number of blanks between the experimental samples or calibration standards to avoid carryover. The sequence should end with a quality control sample followed by a blank sample. Representative total ion chromatograms obtained by UPLC-ESI-MS/MS analysis of amino acids in *A. thaliana* leaf extract are shown in Fig. 1.
4. Perform the UPLC-ESI-MS/MS analysis with the chromatographic and MS parameters specified in Table 1.
5. For data analysis use instrument specific software. If the Acquity UPLC – Xevo TQ MS system is being used, conduct the data

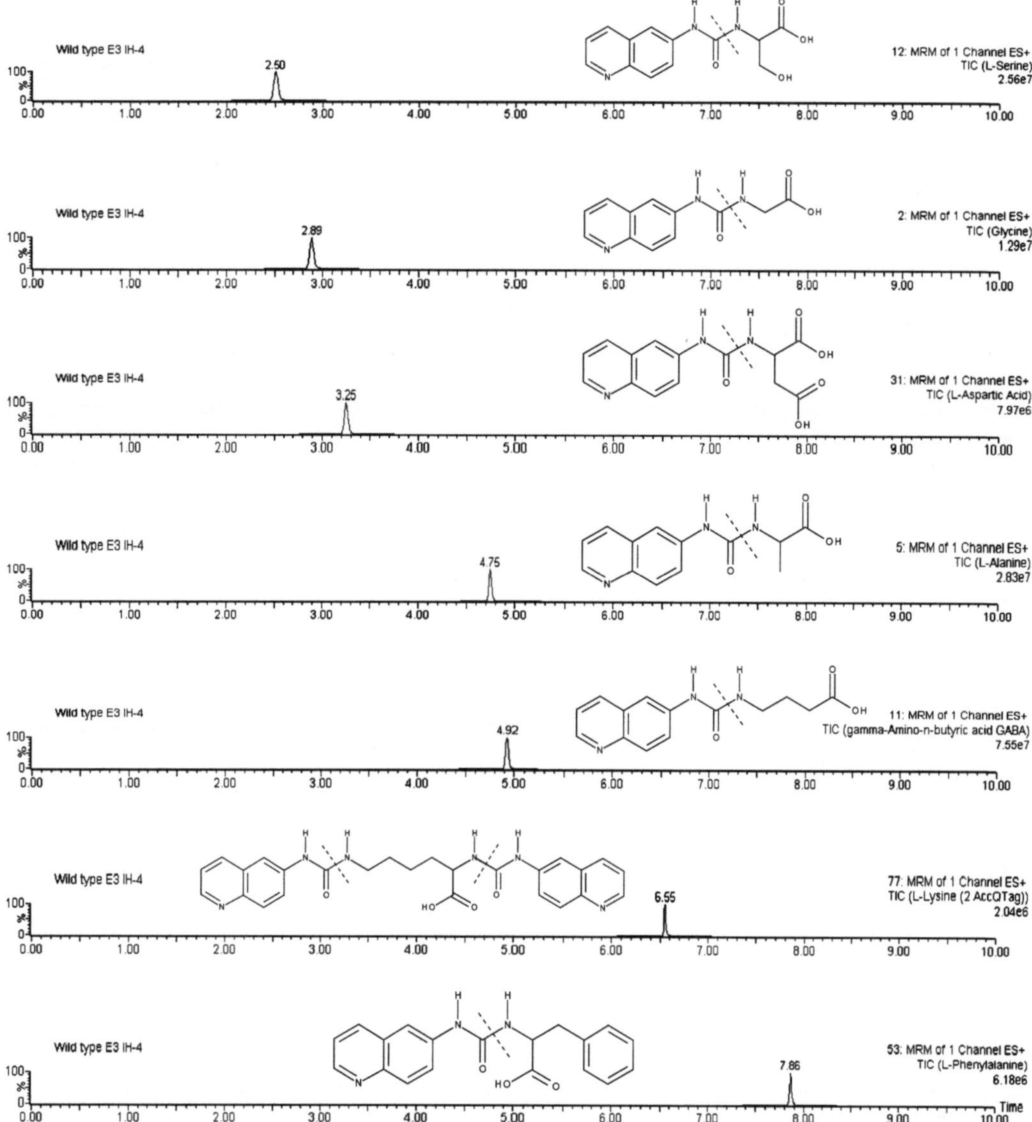

Fig. 1. Total ion chromatograms of selected AccQ·Tag-derivatized amino acids analyzed in wild-type Arabidopsis extracts by UPLC-ESI-MS/MS. Chemical structures show fragmentation site upon MS/MS.

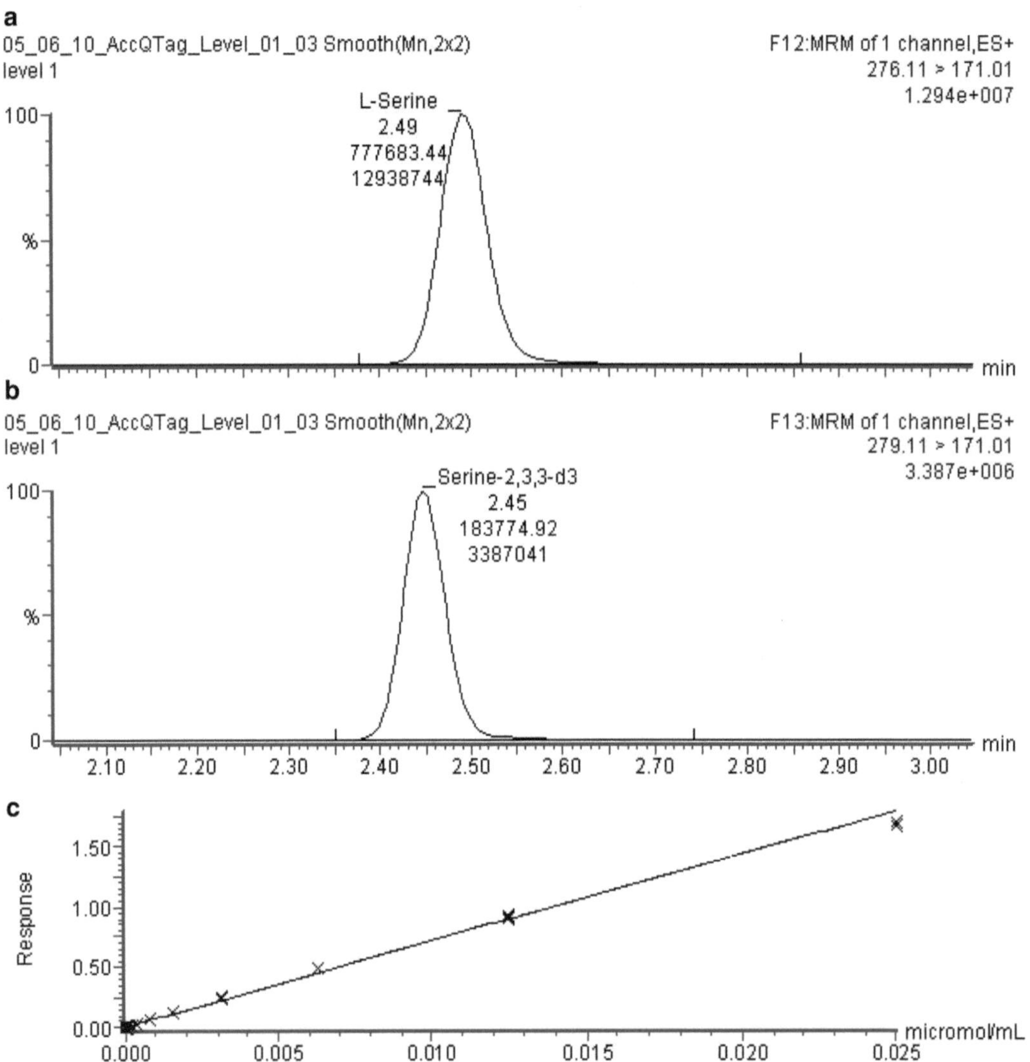

Fig. 2. Chromatogram (**a**) and internal calibration curve (**c**) for serine (AQC-derivatized serine at 0.025 μmol/mL; internal standard (**b**), serine-2,3,3-d_3, at 0.4 μg/mL).

analysis (calibration curves and amino acid quantitation) with the TargetLynx software (see Note 11). A typical chromatogram window for serine and its internal standard (serine-2,3,3-d_3), and the corresponding internal calibration curve is depicted in Fig. 2. A comparison of detection limits and dynamic ranges obtained by AccQ·Tag-UPLC-ESI-MS/MS amino acid analysis using the Xevo TQ and TQD mass spectrometric detectors is given in Table 2.

Table 2
Comparison of detection limits and dynamic ranges for AccQ·Tag-derivatized amino acids analyzed by Waters Acquity UPLC system interfaced with the Waters TQD MS and the Waters Xevo TQ MS

	UPLC-ESI-MS/MS			
	Xevo TQ		TQD	
Amino acid	Detection limit (M)	Dynamic range	Detection limit (M)	Dynamic range
Hydroxy-L-proline	4.86×10^{-11}	100,000	1.65×10^{-9}	100
Histidine	2.31×10^{-8}	1,000	1.33×10^{-8}	100
3-Methyl-histidine	2.16×10^{-11}	1,000	1.21×10^{-8}	1,000
Taurine	1.09×10^{-11}	10,000	3.95×10^{-9}	100
1-Methyl-histidine	1.02×10^{-11}	10,000	1.80×10^{-8}	100
Serine	1.10×10^{-8}	10,000	1.65×10^{-9}	1,000
Carnosine	4.13×10^{-11}	1,000	5.88×10^{-9}	1,000
Arginine	2.41×10^{-10}	10,000	1.19×10^{-8}	100
Glycine	3.20×10^{-9}	10,000	1.65×10^{-9}	10,000
Ethanolamine	3.02×10^{-9}	10,000	1.65×10^{-9}	1,000
Aspartic acid	3.17×10^{-9}	10,000	3.49×10^{-9}	100
Sarcosine	4.83×10^{-10}	10,000	1.65×10^{-9}	1,000
Glutamic acid	4.64×10^{-9}	10,000	5.40×10^{-9}	100
Citrulline	2.87×10^{-10}	10,000	3.98×10^{-9}	100
β-Alanine	1.43×10^{-9}	10,000	2.27×10^{-9}	1,000
Threonine	1.30×10^{-9}	10,000	2.63×10^{-9}	1,000
L-Alanine	1.06×10^{-9}	10,000	1.85×10^{-9}	1,000
γ-Amino-*n*-butyric acid	2.69×10^{-9}	10,000	1.65×10^{-9}	1,000
α-Amino adipic acid	9.27×10^{-11}	10,000	1.79×10^{-9}	1,000
β-Aminoisobutyric acid	9.13×10^{-11}	100,000	1.65×10^{-9}	1,000
Proline	1.55×10^{-9}	10,000	2.83×10^{-9}	1,000
α-Amino-*n*-butyric acid	1.43×10^{-9}	100,000	1.71×10^{-9}	1,000
Tyrosine	8.05×10^{-11}	10,000	3.85×10^{-9}	1,000
Methionine	1.60×10^{-9}	100,000	1.65×10^{-9}	1,000
Valine	8.24×10^{-10}	10,000	1.65×10^{-9}	1,000

(continued)

Table 2 (continued)

Amino acid	UPLC-ESI-MS/MS Xevo TQ Detection limit (M)	Xevo TQ Dynamic range	TQD Detection limit (M)	TQD Dynamic range
Leucine	2.74×10^{-10}	10,000	1.65×10^{-9}	1,000
Isoleucine	1.20×10^{-10}	10,000	1.65×10^{-9}	1,000
Phenylalanine	6.28×10^{-10}	10,000	4.01×10^{-9}	10,000
Tryptophan	6.18×10^{-10}	10,000	7.57×10^{-9}	1,000

4. Notes

1. Protect any unused reconstituted reagent solution from light and atmospheric water by wrapping the vial with aluminum foil, further sealing the cap with Parafilm, and storing the vial in a desiccator. Under these conditions, the reconstituted reagent last for up to 1 week.
2. The extraction buffer containing the labeled internal standards at 4 μg/mL should be defrosted only once for the extraction of amino acids.
3. The amino acid stock solution from Sigma-Aldrich contains physiological acidic, neutral and basic amino acids: β-alanine, L-alanine, L-α-aminoadipic acid, L-α-amino-*n*-butyric acid, D,L-β-aminoisobutyric acid, L-arginine, L-aspartic acid, L-carnosine, L-citrulline, creatinine, L-cystathionine, L-cystine, ethanolamine, L-glutamic acid, glycine, L-histidine, L-homocystine, δ-hydroxylysine, hydroxyl-L-proline, L-isoleucine, L-leucine, L-lysine, L-methionine, 1-methyl-L-histidine, 3-methyl-L-histidine, L-ornithine, L-phenylalanine, L-proline, L-sarcosine, L-serine, taurine, L-threonine, L-tryptophan, L-tyrosine, and L-valine.
4. Signal suppression is problematic during direct infusion of AccQ·Tag-derivatized amino acids into the Xevo TQ mass spectrometer when the borate buffer is used for optimum pH adjustment of the reaction solutions. This problem is attributed to the nonvolatile nature of the borate buffer. When this is the case, an alternative buffer for the derivatization protocol must be identified. A 50-mM ammonium acetate buffer (pH 9.3) demonstrated to be suitable for this purpose and was used for direct infusion of derivatized amino acid standards into the Xevo TQ mass spectrometer.

5. It is worth mentioning that signal suppression due to the nature of the borate buffer is only observed under direct infusion conditions in the Xevo TQ mass spectrometer. Such phenomenon is absent during UPLC-MS/MS analysis of AccQ·Tag-derivatized amino acids due to the dilution of the sample with the mobile phase. Although the ammonium acetate buffer is suitable for the AccQ·Tag derivatization, its use is limited to the derivatization of amino acids for infusion experiments in our protocol. The borate buffer is preferred for the UPLC-MS/MS analysis of AccQ·Tag-derivatized amino acids because larger peak areas are obtained with this buffer compare to those observed with ammonium acetate. Lower yields of amino acid derivatives are suspected to be produced when the ammonium acetate buffer is used.
6. The MRM transition used for each targeted derivatized amino acid in the final MS method corresponded to the intense parent–daughter transition m/z $(M+H)^{+}>171$. The m/z 171 diagnostic ion results from the collision-induced cleavage at the ureide bond of the AccQ·Tag adduct of each amino acid.
7. Since creating MRM-MS methods is costly and time consuming, the optimal cone voltage and collision energy parameters for each MRM transition were found only for a selected number of AccQ·Tag -amino acid adducts. Representative compounds from each amino acid group (i.e., polar, nonpolar, acidic, basic, and neutral) were selected. The average values of cone voltages and collision energies observed for the evaluated amino acids were applied for the remaining targeted compounds through all the experiments. In addition, the cone voltages and collision energies assigned to the stable isotope-labeled reference compounds corresponded to those observed for their corresponding non-labeled amino acid homologs.
8. This gradient is specific for the AccQ·Tag Ultra column, 2.1×100 mm, 1.7 μm particles. Optimum resolution of AccQ·Tag-derivatized amino acids is obtained with this column when operated under the original gradient conditions described by Waters. If a different column with different specifications is to be used, the gradient must be empirically modified to attain optimum separation of the analytes.
9. The adjustment of the time window and the number of MRM transitions monitored per function are optimized during the MS method development in order to obtain as many scans as possible within a MS peak and increase the reliability of the quantitation.
10. According to the manufacturer, derivatized amino acid samples are stable at room temperature up to 1 week.
11. A quantitation method is necessary in order to perform this step. In the TargetLynx method editor, enter the name of each

compound of the analysis and the parameters that describe their quantitation (acquisition function number, transition, retention time, internal reference, integration parameters, etc.). Once the entire quantitation method is built, set up a pre-acquired sample list of a set of calibration standards for quantitation (specify sample types and concentrations). Process the standards using the quantitation method. Review the integration, the calibration curves and make the necessary adjustments until acceptable results are obtained for the range of linear response. Save the data set and export it as a calibration curve file. Proceed to the analysis of the unknown samples specifying the quantitation method and calibration curve file. The concentration of amino acids in the unknown samples will be calculated based on the response indicated by the calibration curves. Report the final concentration of amino acids in the unknown samples as μmol/mg dry weight of Arabidopsis tissue.

References

1. Stein WH et al (1957) Observation on the amino acid composition of human hemoglobins. Biochim Biophys Acta 24: 640–642
2. Ullmer R, Plematl A, and Rizzi A (2006) Derivatization by 6-aminoquinolyl-N-hydroxysuccinimidyl carbamate for enhancing the ionization yield of small peptides and glycopeptides in matrix-assisted laser desorption/ionization and electrospray ionization mass spectrometry. Rapid Commun Mass Spectrom 20: 1469–1479
3. Badiou S et al (2004) Determination of plasma amino acids by fluorescent derivatization and reversed-phase liquid chromatographic separation. Clin Lab 50:153–158
4. Bernal JL et al (2005) A comparative study of several HPLC methods for determining free amino acid profiles in honey. J Sep Sci 28:1039–1047
5. Cohen SA (2000) Amino acid analysis using precolumn derivatization with 6-aminoquinolyl-N-hydroxysuccinimidyl carbamate. Methods Mol Biol 159:39–47
6. Cohen SA (2003) Amino acid analysis using pre-column derivatization with 6-aminoquinolyl-N-hydroxysuccinimidyl carbamate. Analysis of hydrolyzed proteins and electroblotted samples. Methods Mol Biol 211:143–154
7. Hou S et al (2009) Determination of soil amino acids by high performance liquid chromatography-electro spray ionization-mass spectrometry derivatized with 6-aminoquinolyl-N-hydroxysuccinimidyl carbamate. Talanta 80:440–447
8. Callahan DL et al (2007) Relationships of nicotianamine and other amino acids with nickel, zinc and iron in Thlaspi hyperaccumulators. New Phytol 176: 836–848
9. Armenta JM et al (2010) Sensitive and rapid method for amino acid quantitation in malaria biological samples using AccQ.Tag ultra performance liquid chromatography-electrospray ionization-MS/MS with multiple reaction monitoring. Anal Chem 82: 548–558
10. Bosch L, Alegria A, and Farre R (2006) Application of the 6-aminoquinolyl-N-hydroxysccinimidyl carbamate (AQC) reagent to the RP-HPLC determination of amino acids in infant foods. J Chromatogr B Analyt Technol Biomed Life Sci 831: 176–183
11. Pappa-Louisi A et al (2007) Optimization of separation and detection of 6-aminoquinolyl derivatives of amino acids by using reversed-phase liquid chromatography with on line UV, fluorescence and electrochemical detection, Anal Chim Acta 593: 92–97
12. Martínez-Girón AB et al (2009) Development of an in-capillary derivatization method by CE for the determination of chiral amino acids in dietary supplements and wines. Electrophoresis 30: 696–704
13. Cohen SA, and Michaud DP (1993) Synthesis of a fluorescent derivatizing reagent, 6-aminoquinolyl-N-hydroxysuccinimidyl carbamate, and its application for the analysis of hydrolysate amino acids via high-performance liquid chromatography. Anal Biochem 211: 279–287

Chapter 3

Isotope Dilution Liquid Chromatography-Tandem Mass Spectrometry for Quantitative Amino Acid Analysis

David M. Bunk and Mark S. Lowenthal

Abstract

The role of amino acid analysis in bioanalysis has changed from a qualitative to a quantitative technique. With the discovery of both electrospray ionization and matrix-assisted laser desorption ionization in the early 1990s, the use of amino acid analysis for qualitative analysis of proteins and peptides has been replaced by mass spectrometry. Accurate measurement of the relative molecular masses of proteins and peptides, peptide mapping, and sequencing by tandem mass spectrometry provide significantly better qualitative information than can be achieved from amino acid analysis. At NIST, amino acid analysis is used to assign concentration values to protein and peptide standard reference materials (SRMs) which, subsequently, will be used in the calibration of a wide variety of protein and peptide assays, such as those used in clinical diagnostics. It is critical that the amino acid analysis method used at NIST for SRM measurement deliver the highest accuracy and precision possible. Therefore, we have developed an amino acid analysis method that uses isotope dilution LC-MS/MS – the analytical technique routinely used at NIST to certify analyte concentrations in SRMs for a wide variety of analytes. Amino acid analysis by isotope dilution LC-MS/MS was first used to measure the concentration of bovine serum albumin in NIST SRM 927d ("bovine serum albumin, 7% solution"). We have recently refined our isotope dilution LC-MS/MS amino acid analysis method to certify the concentration of 17 amino acids in NIST SRM 2389a ("amino acids in 0.1 mol/L hydrochloric acid"). We present here our most recent method for the quantification of amino acids using isotope dilution LC-MS/MS.

Key words: Isotope dilution, Liquid chromatography, Tandem mass spectrometry, Quantification, Amino acid analysis

1. Introduction

The role of amino acid analysis has changed in bioanalysis from a qualitative to a quantitative technique. Amino acid analysis was originally used qualitatively to validate the identity of a purified protein or peptide. After hydrolysis of the protein or peptide into

Michail A. Alterman and Peter Hunziker (eds.), *Amino Acid Analysis: Methods and Protocols*, Methods in Molecular Biology, vol. 828, DOI 10.1007/978-1-61779-445-2_3, © Springer Science+Business Media, LLC 2012

constituent amino acids, the relative ratios of all amino acids were determined and compared to expected ratios. In this way, the pattern of amino acid ratios could be used as a fingerprint for identification. With the discovery of both electrospray ionization (ESI) and matrix-assisted laser desorption ionization (MALDI) in the early 1990s, the use of amino acid analysis for qualitative analysis of proteins and peptides has been replaced by mass spectrometry. Accurate measurement of the relative molecular masses of proteins and peptides, peptide mapping, and sequencing by tandem mass spectrometry provide significantly better qualitative information than can be achieved from amino acid analysis. While mass spectrometry has replaced amino acid analysis as a qualitative technique, the role of amino acid analysis has shifted to that of an important quantitative technique.

The analytical sensitivity and specificity of mass spectrometry makes it one of the most powerful techniques for the analysis of proteins and peptides, for both qualitative measurement and, most recently, quantification (1). Accuracy in quantitative mass spectrometry requires accuracy in the concentration of the peptide and protein calibrants used; yet, this is not always easy to achieve. Accuracy in the concentration of the calibration materials used in the analysis of other chemical analytes is typically achieved through gravimetry – the accurate weighing of solid analyte standards of known purity. Gravimetry is typically not feasible for protein and peptide analytes because often there is not a sufficient amount of solid protein or peptide standard available (if at all) for accurate weighing and, quite often, purity assessments of protein and peptide standards are crude. Because of the limitations of gravimetry to directly prepare accurate protein and peptide calibrants, amino acid analysis has become an important technique to assign concentration values to these calibrants.

Amino acid analysis used for quantitative analysis of protein and peptide solutions requires absolute quantification following hydrolysis. Amino acids are available in bulk with accurately assigned purity, allowing accurate preparation of calibrants to be produced using gravimetry for the amino acid analysis (i.e., solutions of amino acids). This allows protein and peptide quantification to be metrologically traceable (2) to gravimetrically prepared calibrants, producing the highest degree of accuracy (3).

In order for the accuracy and precision of gravimetrically prepared amino acid calibrants to results in accurate and precise peptide or protein calibration solutions, the quantitative amino acid analysis method should have high precision and be free of bias (or, have a known and constant bias). Of course, the required precision and accuracy of the amino acid analysis depend[s] on the accuracy and precision needed in the final measurement result of the protein or peptide analyte. At NIST, amino acid analysis is used to assign concentration values to protein and peptide standard reference materials (SRMs) which, subsequently, will be used in the calibration of a wide variety of protein and peptide assays, such as those used in clinical diagnostics. It is critical that the amino acid analysis method used at NIST for SRM measurement deliver the highest accuracy and precision possible. Therefore, we have developed an amino acid analysis method that uses isotope dilution LC-MS/MS (4) – the analytical technique routinely used at NIST to certify analyte concentrations in SRMs for a wide variety of analytes. Amino acid analysis by isotope dilution LC-MS/MS was first used to measure the concentration of bovine serum albumin in NIST SRM 927d ("bovine serum albumin, 7% solution"). We have recently refined our isotope dilution LC-MS/MS amino acid analysis method to certify the concentration of 17 amino acids in NIST SRM 2389a ("amino acids in 0.1 mol/L hydrochloric acid"). We present here our most recent method for the quantification of amino acids using isotope dilution LC-MS/MS.

2. Materials

Solvents used are all high purity "LC-MS" grade and reagents are analytical grade or better (see Note 1).

2.1. LC Components

1. LC mobile phase A: 0.5 mL/L trifluroacetic acid in 0.3 L/L aqueous acetonitrile. To 300 mL "LC-MS" grade acetonitrile, add "LC-MS" grade water to a total volume of 1 L. Add 0.5 mL of trifluoroacetic acid. Store at room temperature (see Note 2).
2. LC mobile phase B: 4.5 mL/L trifluroacetic acid in 0.3 L/L aqueous acetonitrile. To 300 mL "LC-MS" grade acetonitrile, add "LC-MS" grade water to a total volume of 1 L. Add 4.5 mL of trifluoroacetic acid. Store at room temperature (see Note 2).
3. LC column: Primesep 100, 2.1 mm × 250 mm, 5 μm particle size, 10-nm pore size (SIELC, Prospect Heights, IL) (see Note 3).
4. LC instrumentation: 1100 Series or 1200 Series liquid chromatography system (Agilent, Santa Clara, CA) consisting of a binary LC pump, well-plate autosampler with temperature control, and column oven.

2.2. MS Components

1. MS instrumentation: API 4000 or API 5000 triple quadrupole mass spectrometer (Applied Biosystems, Foster City, CA) equipped with an ESI source.

2.3. Labeled and Unlabeled Amino Acid Standards

1. Labeled amino acids: L-alanine (U-$^{13}C_3$ 98%, ^{15}N 98%), L-arginine (U-$^{13}C_6$ 98%), L-aspartic acid (U-$^{13}C_4$ 98%), L-glutamic acid (U-$^{13}C_5$ 98%), L-histidine (U-$^{13}C_6$ 98%), L-isoleucine (U-$^{13}C_6$ 98%),

L-leucine (U-$^{13}C_6$ 98%), L-lysine (U-$^{13}C_6$ 98%), L-methionine (U-$^{13}C_5$ 97–99%, ^{15}N 97–99%), L-phenylalanine (U-$^{13}C_9$ 97–99%, ^{15}N 97–99%), L-proline (U-$^{13}C_5$ 98%, ^{15}N 98%), L-serine (U-$^{13}C_3$ 98%, ^{15}N 98%), L-threonine (U-$^{13}C_4$ 97–99%), L-tyrosine (U-$^{13}C_9$ 98%, ^{15}N 98%), L-valine (U-$^{13}C_5$ 98%) (Cambridge Isotope Labs, Andover, MA) (see Note 4).

2. Unlabeled amino acids: all L-isoforms (Fluka/Sigma, St. Louis, MO) were used. Chemical purity of each amino acid standard was evaluated by the manufacturer and confirmed through in-house measurements (see Note 5).

3. Methods

3.1. Calibrant and Sample Preparation

1. Stock solutions of each isotopically labeled amino acid are prepared by weighing out the solid amino acid and dissolving in aqueous 0.1 mol/L hydrochloric acid. These stock solutions are stored at 5°C.
2. An internal standard solution is prepared by volumetrically blending the individual labeled amino acid stock solutions. The concentrations of the labeled amino acids in the internal standard solution should match as closely as possible the concentrations of amino acids in the hydrolysate of the protein or peptide solution being quantified. Aqueous 0.1 mol/L hydrochloric acid is used as the diluent, if necessary, to achieve the desired concentrations. The internal standard solution is stored at 5°C.
3. Stock solutions of each unlabeled amino acid are prepared by weighing out the solid amino acid and dissolving in aqueous 0.1 mol/L hydrochloric acid. To achieve the most accurate gravimetric weights, the solid amino acids should be stored in a dessicator for several days prior to use and should be equilibrated with the temperature of the room where will be weighed for at least several hours. These stock solutions are stored at 5°C.
4. Calibration solutions are prepared by gravimetrically (preferred, for highest accuracy and precision) or volumetrically blending the individual unlabeled amino acid stock solutions and the internal standard solution (containing the isotopically labeled amino acids). Several calibration solutions are prepared which contain varying molar concentration ratios of the unlabeled and labeled amino acids. For example, for preliminary measurements, calibration solutions with unlabeled-to-labeled amino acid molar concentrations of approximately 10, 3, 1, 0.3, and 0.1 are typically prepared. After preliminary measurement of the amino acid concentrations are made, higher accuracy and precision in the concentration measurement can be achieved using the technique of double exact matching isotope dilution

mass spectrometry (5, 6). Aqueous 0.1 mol/L hydrochloric acid is used as the diluent, if necessary, to achieve the desired concentrations. The internal standard solution is stored at 5°C.

5. Hydrolysate samples (see Note 6) are first lyophilized to dryness overnight (>16 h) without heating in a SPD 1010 SpeedVac (Savant/Thermo, Rockford, IL). The lyophilized hydrolysate is resolubilized with the internal standard solution and equilibrated overnight (≈16 h) at 5°C. If not analyzed immediately, samples are stored at 5°C.

3.2. LC-MS/MS Measurements

For several reasons (see Note 7), a single LC-MS/MS analysis method was not developed for all amino acids of interest. Instead, the amino acids are grouped into four sets with a LC-MS/MS method developed for each set. The amino acids are analyzed in the following sets: (1) proline, valine, isoleucine, leucine, phenylalanine; (2) aspartic acid, serine, tyrosine, lysine; (3) threonine, alanine, methionine, arginine; (4) glutamic acid, histidine.

1. For the chromatographic separation of most amino acids (groupings 2–4), gradient elution with mobile phases A and B is used. The elution starts at 100% mobile phase A and changes linearly to 50% mobile phase B in 30 min. Isocratic elution for the first 30 min using 100% mobile phase A used for the separation of amino acids in set 1. For both the gradient and isocratic elutions, after the first 30 min the mobile phase is changed to 95% mobile phase B for 5 min, to wash the column. Next, the column is re-equilibrated with the starting conditions (100% mobile phase A) for 25 min. A constant mobile phase flow rate of 200 μL/min is used for all separations. The Primesep 100 column is maintained at 30°C in the column oven. Samples are maintained at 10°C in the autosampler.
2. The triple quadrupole mass spectrometer is operated in positive ion mode using multiple reaction monitoring (MRM) scans. All analyses are performed with the following API 4000 MS instrumental parameters (if different, settings for measurements performed using the API 5000 are included in parenthesis): unit resolution in Q1 and Q3, collision gas = 41 kPa/6 psi, curtain gas (CUR) = 69 kPa/10 psi (275 kPa/40 psi), ion source gas 1 (GS1) = 552 kPa/80 psi (207 kPa/30 psi), ion source gas 2 (GS2) = 345 kPa/50 psi (276 kPa/40 psi), intensity threshold = 0, settling time = 10 ms, pause between mass ranges = 10 ms, *x*-axis spray position = 2 mm (0 mm), *y*-axis spray position = 5 mm (7 mm), ion spray voltage (IS) = 5,000 V, capillary temperature (TEM) = 500°C, interface heater = ON; dwell time = 200 ms.
3. The instrument parameters for the MRM scans used in the measurement of each set of labeled and unlabeled amino acids are listed in Table 1.

Table 1
Mass spectrometric instrument parameters for the MRM scans of amino acids measured using the Applied Biosystems API 4000 or API 5000 triple quadrupole mass spectrometer

Analyte	Period interval	MRM transition	DP	EP	CE	CXP
Set 1						
Pro	0–13.0 min	m/z 116.1 → m/z 70.0	46.0	10.0	17.0	8.0
		m/z 122.1 → m/z 75.2	46.0	10.0	17.0	8.0
Val	13.0–17.0 min	m/z 118.1 → m/z 72.0	46.0	10.0	17.0	8.0
		m/z 123.1 → m/z 76.1	46.0	10.0	17.0	8.0
Ile/Leu	17.0–28.0 min	m/z 132.2 → m/z 86.1	46.0	10.0	17.0	8.0
		m/z 138.2 → m/z 91.2	46.0	10.0	17.0	8.0
Phe	17.0–28.0 min	m/z 166.2 → m/z 120.0	46.0	10.0	17.0	8.0
		m/z 171.2 → m/z 125.0	46.0	10.0	17.0	8.0
Set 2						
Ser	0–12.0 min	m/z 106.1 → m/z 59.9	40.0	9.3	16.0	10.9
		m/z 110.3 → m/z 62.9	40.0	9.3	16.0	10.9
Asp	0–12.0 min	m/z 134.1 → m/z 73.9	40.0	9.0	19.0	13.4
		m/z 138.3 → m/z 75.9	40.0	9.0	19.0	13.4
Tyr	12.0–20.0 min	m/z 182.2 → m/z 136.0	45.0	8.2	20.0	13.2
		m/z 192.4 → m/z 145.1	45.0	8.2	20.0	13.2
Lys	20.0–35.0 min	m/z 147.2 → m/z 84.0	49.0	9.2	22.0	15.9
		m/z 153.3 → m/z 88.9	49.0	9.2	22.0	15.9
Set 3						
Ala	0–12.25 min	m/z 90.1 → m/z 44.0	34.5	9.3	18.8	7.5
		m/z 94.1 → m/z 47.0	34.5	9.3	18.8	7.5
Thr	0–12.25 min	m/z 120.1 → m/z 56.0	36.5	9.8	24.5	10.1
		m/z 124.1 → m/z 58.9	36.5	9.8	24.5	10.1
Met	12.25–18.25 min	m/z 150.2 → m/z 61.0	47.5	9.0	33.0	11.5
		m/z 156.2 → m/z 62.9	47.5	9.0	33.0	11.5
Arg	18.25–38.25 min	m/z 175.2 → m/z 70.0	53.0	9.4	33.6	13.1
		m/z 181.2 → m/z 74.0	53.0	9.4	33.6	13.1
Set 4						
Glu	0–14.0 min	m/z 148.2 → m/z 83.9	35.3	9.5	23.3	16.0
		m/z 153.2 → m/z 88.0	35.3	9.5	23.3	16.0
His	23.0–38.0 min	m/z 156.2 → m/z 110.0	44.3	9.0	20.4	20.3
		m/z 162.3 → m/z 115.0	44.3	9.0	20.4	20.3

DP declustering potential (V), *EP* entrance potential (V), *CE* collision energy (V), *CXP* collision cell exit potential (V)

4. Applied Biosystems Analyst software (version 1.4 or 1.5) is used for peak selection and integration (Fig. 1). Peaks are identified manually and peak areas were automatically integrated by Analyst using a bunching factor = 3, number of smooths = 1, and all other parameters set to default values.

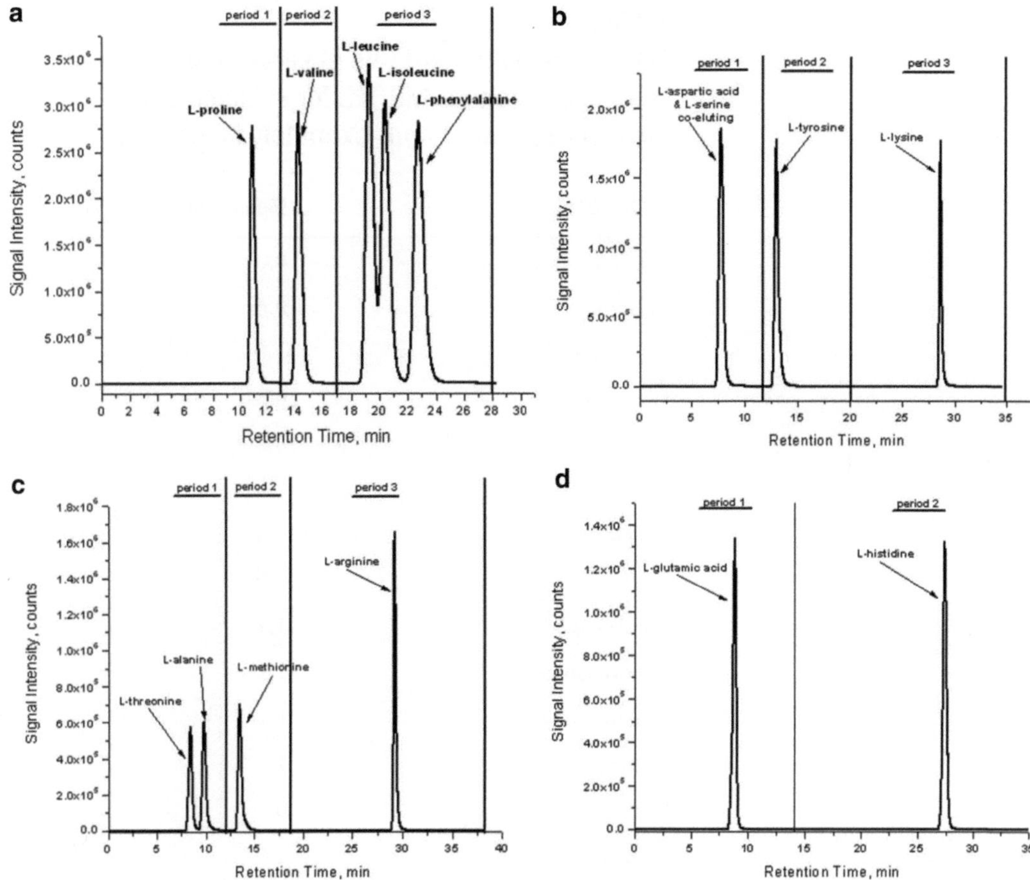

Fig. 1. Total ion chromatograms for the isotope dilution LC-MS/MS measurements of amino acids grouped into four sets: (**a**) L-proline, L-valine, L-leucine, L-isoleucine, and L-phenylalanine; (**b**) L-aspartic acid, L-serine, L-tyrosine, L-lysine; (**c**) L-theonine, L-alanine, L-methionine, L-arginine; (**d**) L-glutamic acid, L-histidine.

All peak integrations are visually inspected, and in some cases, manual integration is necessary (see Note 8). Peak area ratios are exported into Microsoft Excel for quantitative analysis. Unlabeled/labeled integrated peak area ratios are calculated from calibrant data and plotted against gravimetric mass ratios into calibration curves. For samples, molar mass ratios are extrapolated from the calibration curves according to the measured peak area ratios. From this data, amino acid concentrations are calculated and subjected to statistical evaluation. The measurement coefficients of variance (CV) achievable using this isotope dilution LC-MS/MS method to measure amino acids at concentrations of approximately 50 μmol/L range from 0.4% to 1.9% (Table 2).

Table 2
Typical measurement CVs observed using isotope dilution LC-MS/MS for selected amino acids measured at appr oximately 50 μmol/L

Amino acid	Measurement CV, %
Alanine	0.6
Arginine	0.6
Aspartic acid	1.7
Glutamic acid	0.5
Histidine	0.5
Isoleucine	2.0
Leucine	1.6
Lysine	0.9
Methionine	1.3
Phenylalanine	1.1
Proline	1.1
Serine	1.4
Threonine	0.4
Tyrosine	1.9
Valine	1.2

4. Notes

1. Certain commercial equipment, instruments, or materials are identified in this chapter to specify adequately the experimental procedure. Such identification does not imply recommendation or endorsement by the National Institute of Standards and Technology, nor does it imply that the materials or equipment identified are necessarily the best available for the purpose.
2. Solutions containing trifluoroacetic acid have limited stability and the breakdown products can increase the chemical background seen in mass spectrometry. Because of this, mobile phases containing trifluoroacetic acid are prepared each week.
3. Prior to using the Primesep 100 mixed-mode column (containing both anion exchange and reverse-phase column chemistry) for LC separations, we investigated using hydrophilic interaction chromatography (HILIC) using a polyhydroxyethyl aspartate column. We found that the HILIC separation

of amino acids often resulted in poor chromatographic peak shapes and the peak shape was often not consistent over time. The separations we achieve with the Primesep 100 column are of higher quality and more significant than what we observed with HILIC.

4. When choosing an isotopically labeled internal standard for isotope dilution mass spectrometry, we aim to have the labeled internal standard to be at least 3 Da greater in molecular mass than the unlabeled analyte. When there is at least a 3-Da mass difference, there will be minimal overlap of the isotopic distribution of the unlabeled analyte with the monoisotopic peak of the labeled internal standard; with greater mass difference, the overlap will be even less. Isotopically labeled internal standards containing ^{13}C and/or ^{15}N are preferable to those which are labeled with deuterium. Deuterium-labeled molecules may occasionally have slightly different chromatographic retention times than the unlabeled molecule which can impact quantification by isotope dilution.
5. The purity assessment performed by the manufacturer of small organic molecules, such as amino acids, is generally reliable. We have seen occasional discrepancies between the results of purity assessments performed at NIST and those of manufacturers which is what prompts us to perform in-house purity measurements when we are striving for high measurement accuracy. For the amino acids, the measurements performed at NIST for L-leucine and L-lysine indicated slightly lower purities than those reported by the manufacturer.
6. The success of the isotope dilution LC-MS/MS method discussed in this chapter to quantify proteins and peptides depends significantly on the successful hydrolysis of the protein or peptide into constituent amino acids. Incomplete hydrolysis or the degradation of amino acids during hydrolysis will adversely affect the accuracy of quantification by amino acid analysis. In our experience, the hydrolysis conditions must be validated for each protein or peptide analyte. It is important to note that for quantification of proteins or peptide by amino acid analysis, we choose to quantify only the amino acids that display stability under routine gas-phase hydrochloric acid hydrolysis conditions (110°C for 48 h). These amino acids are alanine, arginine, glycine, isoleucine, leucine, lysine, phenylalanine, proline, and valine.
7. We grouped amino acids into sets and developed unique LC-MS/MS methods for each sets in order to achieve maximal measurement accuracy and precision. It is our experience at NIST that the highest accuracy and precision can be achieved in isotope dilution LC-MS/MS through the careful design of the MRM scans. Although current triple quadrupole instruments

are capable of performing MRM scans with dwell times of 5 ms or less, fast scanning does not lead to higher quality measurements. In our experience, dwell times in the order of 200 ms produce the best quantitative results (although, for very narrow chromatographic peaks, shorter dwell times may be necessary to avoid distortion of peak shapes). Given a dwell time of 200 ms for each MRM transition, we aim to measure no more than four MRM scan functions in each time period of the overall MRM scan method. With these considerations, we found it necessary to group the amino acids into distinct sets. The inclusion of the amino acids in each set was based primarily on maintaining a range of retention times of the amino acids being measured so that time periods could be used in the MRM scan which contained no more than four MRM scan functions.

8. We could not achieve complete baseline chromatographic resolution for the isobaric amino acids, leucine and isoleucine (Fig. 1a). For samples in which there are equimolar or nearly equimolar concentrations of leucine and isoleucine, we have found that the chromatographic separation is sufficient to reliably use integrated peak areas for quantification. For samples that have significantly different molar concentrations of leucine and isoleucine, either using peak heights or using only the most abundant amino acid for quantification may yield the best results.

References

1. Brun V et al (2009) Isotope dilution strategies for absolute quantitative proteomics. J Proteomics 72:740–749
2. NIST Policy on Metrological Traceability. http://www.nist.gov/traceability/index.cfm. Accessed 1 December 2010
3. Burkitt WI et al (2008) Toward Système International d'Unité-traceable protein quantification: from amino acids to proteins. Anal Biochem 376:242–251
4. Sargent M, Harte R, Harrington C (2002) Guidelines for achieving high accuracy in isotope dilution mass spectrometry (IDMS). The Royal Society for Chemistry, London
5. Henrion A (1994) Reduction of systematic errors in quantitative analysis by isotope dilution mass spectrometry (IDMS): an iterative method. Fesenius J Anal Chem 350:657–658
6. Mackay LG et al (2003) High accuracy analysis by isotope dilution mass spectrometry using an iterative exact matching technique. Accred Qual Assur 8:191–194

Chapter 4

Analysis of Underivatized Amino Acids: Zwitterionic Hydrophilic Interaction Chromatography Combined with Triple Quadrupole Tandem Mass Spectrometry

Madeleine Dell'mour, Gunda Koellensperger, and Stephan Hann

Abstract

Analysis of underivatized amino acids is challenging regarding both separation and detection of this small, polar, and largely UV-inactive compounds. Additives for reversed phase chromatography such as ion-pairing reagents can hamper mass spectrometric detection. Zwitterionic hydrophilic interaction chromatography using MS compatible eluents together with tandem mass spectrometry in multiple reaction monitoring mode for selective detection is an attractive approach to overcome the abovementioned issues.

Key words: HILIC, Zwitterionic phase, Tandem mass spectrometry, MRM, Triple quadrupole mass spectrometry, Quantifier transition, Qualifier transition, Root exudation, Amino acid analysis

1. Introduction

The HILIC-MS/MS method described in the following has been successfully applied to the quantification of 16 amino acids in various aqueous root exudation samples from both soil and nutrient solution grown maize plants (highly differing in analyte concentration and soil matrix content) and has thereby proven to cope with the challenges that come along with the analysis of the mentioned samples and with the analysis of amino acids in general. Underivatized amino acids were separated via zwitterionic hydrophilic interaction chromatography (see Subheading 3.1) and were subsequently detected using a triple quadrupole mass analyzer in multiple reaction monitoring (MRM) mode after electrospray ionization (positive mode, see Subheading 3.2). This method allowed for fast and highly sensitive quantification of underivatized amino acids offering LODs in the nanomole per liter range without prior

Michail A. Alterman and Peter Hunziker (eds.), *Amino Acid Analysis: Methods and Protocols*, Methods in Molecular Biology, vol. 828, DOI 10.1007/978-1-61779-445-2_4,

Table 1
Retention times, efficiencies, calibration ranges, correlation coefficients, correlation types including weight, limits of detection, and limits of quantification of 16 individual amino acids after direct injection of amino acid standard mixes using isoleucine-$^{13}C_6$,^{15}N as internal standard

Amino acid	Retention time (min) (RSD/%; $n=12$)	Efficiency	Calibration range (μM)	Correlation coeff. R^2	Corr. type	Weight	$LOD_{conc.\ based}$ (nM)	$LOQ_{conc.\ based}$ (nM)
Alanine	8.76 (0.09)	65,000	0.1–5	0.9984	Linear	None	9	29
Serine	0.24 (0.10)	76,000	0.01–5	0.9979	Linear	None	10	34
Proline	8.33 (0.0007)	47,500	0.01–5	0.9969	Power	$1/x$	1	5
Valine	7.95 (0.07)	42,500	0.01–0.25	0.9966	Linear	$1/x$	2	5
Threonine	8.90 (0.11)	68,700	0.05–1	0.9915	Linear	None	15	50
Leucine	7.32 (0.11)	41,200	0.1–1	0.9946	Linear	$1/x$	5	15
Isoleucine	7.45 (0.12)	34,900	0.05–5	0.9987	Linear	None	1	3
Aspartic acid	9.04 (0.006)	98,000	0.1–1	0.9851	Linear	None	8	26
Lysine	11.00 (0.08)	82,600	0.01–1	0.9998	Quadratic	None	38	127
Glutamic acid	8.85 (0.13)	53,600	0.05–1	0.9944	Linear	None	28	93
Methionine	7.82 (0.11)	41,000	0.01–0.5	0.9943	Power	$1/x$	0.2	0.7
Histidine	10.66 (0.07)	87,200	0.01–1	0.9957	Linear	$1/x$	13	42
Phenylalanine	7.26 (0.07)	37,000	0.01–0.5	0.9964	Linear	$1/x$	0.9	3
Arginine	11.15 (0.08)	67,500	0.05–1	0.9901	Quadratic	None	4	12
Tyrosine	8.23 (0.01)	48,500	0.01–0.25	0.9917	Power	$1/x$	2	5
Cystine	9.99 (0.05)	95,700	0.01–1	0.9844	Linear	$1/x$	5	18

enrichment. Chromatographic runs with total analysis time of 19 min provided excellent separation efficiencies (N ranges from 35,000 to 100,000), peak shapes (FWHM < 0.1 min) and retention time repeatability (RSD < 0.13%; n = 12). Regarding mass spectrometric performance, simultaneous detection of both quantifier and qualifier transitions offered enhanced selectivity. Average precision of a microsuction cup sample – featuring the highest soil matrix content of the investigated root exudation sample types – spiked with 0.5 μM amino acid standard mix was 5% (n=6). Samples with amino acid concentrations ranging from 0.01 to 5 μM were directly injected, whereas low-concentrated samples (0.3–10 nM) were enriched using SPE prior to analysis. Following the SPE procedure described in Subheading 3.3, 11 amino acids could be enriched obtaining linear, quadratic or power correlation and allowing for an external calibration that included the enrichment step. Two isotope-labeled amino acids were added to the samples and the calibration standards: L-isoleucine-$^{13}C_6$,^{15}N for SPE control and $^{13}C_6$-tyrosine for internal standardization. SPE repeatability was 12.5% (n=30 samples). Quantification of directly injected samples was also carried out via external calibration; L-isoleucine-$^{13}C_6$,^{15}N was used for internal standardization in this case. Analytical figures of merit of the presented HILIC-MS/MS method for the individual amino acids of directly injected and SPE-enriched standard mixes are listed in Tables 1 and 2, respectively.

Table 2
Calibration ranges, correlation coefficients, and correlation types including weight of 11 individual amino acids after SPE enrichment and injection of amino acid standard mixes using $^{13}C_6$-tyrosine as internal standard

Amino acid	Calibration range (nM)	Correlation coeff. R^2	Corr. type	Weight
Proline	1–60	0.9967	Linear	1/x
Valine	20–80	0.9108	Linear	None
Leucine	1–40	0.9906	Linear	None
Isoleucine	1–40	0.9977	Linear	None
Lysine	1–40	0.9871	Linear	1/x
Methionine	20–100	0.9950	Power	None
Histidine	1–40	0.8664	Linear	None
Phenylalanine	1–60	0.9971	Linear	1/x
Arginine	1–60	0.9861	Quadratic	None
Tyrosine	1–80	0.9982	Quadratic	None
Cystine	1–40	0.9867	Linear	1/x

2. Materials

2.1. Chemicals

1. ACN B&J Brand LC-MS grade (Honeywell Specialty Chemicals Seelze GmbH, Seelze, Germany).
2. Methanol B&J Brand LC-MS grade (Honeywell Specialty Chemicals Seelze GmbH, Seelze, Germany).
3. Water was processed by reverse osmosis, filtration, and UV treatment (electric conductivity < 0.2 μS cm) and subsequently deionized using a Millipore high purification system (electric resistance > 18 MΩ).
4. Formic acid 98–100% v/v Suprapur (Merck KGaA, Darmstadt, Germany).
5. Hydrochloric acid fuming 37% v/v p.a. (Merck KGaA, Darmstadt, Germany).
6. Aqueous ammonium hydroxide 25% v/v p.a. (Carl Roth, Karlsruhe, Germany).
7. The amino acid standard mix from Agilent Technologies, Waldbronn, Germany, contained 1,000 μM of each L-aspartic acid, glycine, L-alanine, L-valine, L-leucine, L-isoleucine, L-threonine, L-tyrosine, L-proline, L-arginine, L-histidine, L-glutamic acid, L-cystine, L-phenylalanine, L-lysine, and L-methionine in 0.1 M hydrochloric acid.
8. Internal standards used – L-4-hydroxylphenyl-$^{13}C_6$-alanine (referred to as $^{13}C_6$-tyrosine in the text, min 99 atom % ^{13}C) and L-isoleucine-$^{13}C_6$,^{15}N (min 98 atom % ^{13}C, min 98 atom % ^{15}N) (Isotec, Miamisburg, OH, USA).

2.2. LC Components

1. The HPLC system consisted of an Agilent G1312A Binary Pump 1200 series, an Agilent G1367B high performance autosampler and an Agilent G1316A column compartment from Agilent Technologies.
2. A 2.1 × 20 mm ZIC-HILIC guard column (5 μm particle size) and a 4.6 × 150 mm ZIC-HILIC separation column (3.5 μm particle size) from SeQuant, Umeå, Sweden, were connected with polyether ether ketone (PEEK) tubings (O.D. 1/16 in., I.D. 0.005 in.) and 10–32 one-piece fingertight PEEK fittings (both from IDEX Health and Science, Oak Harbor, WA, USA) and were thermostated at 40°C. An inlet filter with PEEK frits (0.2 μm pore size) from IDEX Health and Science was installed in front of the columns.

2.3. Sample Enrichment Components

1. Gilson GX-271 Liquid Handler together with a Gilson 406 Syringe Pump (10 mL total volume) from Gilson International B.V. Deutschland, Limburg-Offenheim, Germany.

2. 1 mL Strata-X-C cartridges (Phenomenex, Torrance, CA, USA).
3. Savant ISS110 SpeedVac Concentrator (ThermoFisher Scientific/Thermo Electron LED GmbH, Vienna, Austria).
4. Glass inserts (VWR Österreich, Vienna, Austria).

3. Methods

The HILIC-MS/MS method as well as the enrichment procedure was recently published and has been modified for improvement (1).

3.1. Liquid Chromatography

Flow rate was set to 500 μL/min, autosampler temperature to 5°C, and injection volume to 10 μL. Gradient elution was performed using Eluent A (98% v/v water, 1% v/v ACN, 1% v/v formic acid) and Eluent B (98% v/v ACN, 1% v/v water, 1% formic acid) as follows: starting conditions of 90% B were held for 1 min, then B was decreased to 10% within 7 min and was constant for 1 min. Initial conditions were subsequently reconstituted within 0.1 min and the column was re-equilibrated with 90% B for 9.9 min (see Note 1). Hence, total analysis time amounted to 19 min.

3.2. MS Detection

1. MS detection was performed on Agilent 6410 Triple Quad LC/MS equipped with an ESI interface. The measurements were run in positive mode with the following source parameters: drying gas temperature 300°C, drying gas flow 8 L/min, nebulizer pressure 25 psi, and capillary voltage 4,000 V (see Note 2).
2. For the determination and optimization of the MRM transitions of 16 amino acids, after flow injection of a 20-μM amino acid standard mix (isocratic conditions: 1% v/v formic acid in ACN/H_2O 50:50 v/v), a four-stage Mass Hunter Optimizer Software from Agilent was applied: (a) variation of the fragmentor voltage in order to optimize the isolation of the selected precursor ion, (b) evaluation of the four most abundant product ions, (c) optimization of the collision energies for each of these product ions, and (d) evaluation of the exact *m/z* value of the product ions for optimum isolation (see Note 3).
3. For individual amino acid detection, the selection of the quantifier and qualifier transitions out of the four proposed transitions was based on the signal-to-noise ratios (highest and second highest ratio, respectively).
4. Dynamic MRM was carried out according to the particular retention times of the analytes with a time window of 2.0 min

and a cycle time of 1,000 ms. Including the ^{13}C-labeled internal standard, a total of 30 transitions resulted in a minimum dwell time of 41.95 ms (22 concurrent MRMs) and a maximum dwell time of 496.50 ms. For the precursor and product ions as well as the optimized values for fragmentor voltage and collision energy see Table 3.

3.3. Enrichment Procedure

The enrichment of low-concentrated samples was realized by SPE following a procedure adapted from Armstrong et al. proposing this method for cleanup of amino acids in plasma (2).

1. SPE was performed on an automated Gilson GX-271 Liquid Handler together with a Gilson 406 Syringe Pump (10 mL total volume) (see Note 4).
2. 1 mL Strata-X-C cartridges filled with 30 mg of a polymer-based mixed-mode stationary phase (strong cation exchange and reversed phase) were conditioned with 1 mL of methanol and subsequently equilibrated with 1 mL of 0.1 M hydrochloric acid in water.
3. After loading of 25 mL of sample and subsequent washing with 1 mL of methanol, amino acids were eluted into 1.5-mL test tubes using 1 mL of 5% v/v ammonium hydroxide in methanol.
4. The test tubes were then placed into a Savant ISS110 SpeedVac Concentrator at room temperature for solvent evaporation.
5. Thereafter, the samples were reconstituted with 100 μL of 1% v/v formic acid in water:ACN (50:50 v/v) and transferred into 200-μL glass inserts. Flow rates and equilibration times of the Liquid Handler were set as follows: conditioning and equilibration 5 mL/min (flow rate) and 20 s (equilibration time), sample loading 1 mL/min and 30 s, washing 1.3 mL/min and 20 s, elution 2 mL/min and 20 s. Elution was completed applying a N_2-push for 60 s.

3.4. Conclusion

The presented method is straight forward for quantification of amino acids in matrices of high ionic strength without prior sample preparation. To enhance sensitivity and decrease LODs the amino acids can be enriched via mixed mode SPE (reversed phase combined with cation exchange).

Table 3
Triple quadrupole operation settings for multiple reactions monitoring of 16 amino acids

Amino acid	MW	Precursor ion	Quantifier	Qualifier	FV (V)	$CE_{Quantifier}$ (V)	$CE_{Qualifier}$ (V)
Alanine	89.1	90.1	44.0	–	40	9	–
Serine	105.1	106.1	60.0	42.0	70	5	25
Proline	115.1	116.1	70.0	68.1	80	12	32
Valine	117.2	118.1	72.0	55.1	80	8	20
Threonine	119.1	120.1	101.9	74.1	140	4	8
Leucine	131.2	86.0	43.0	–	130	20	–
Isoleucine	131.2	86.0	57.0	–	130	20	–
Aspartic acid	133.1	134.0	115.9	–	80	4	–
Ile-$^{13}C_6$,^{15}N	138.2	139.0	92.0	74.0	80	5	15
Lysine	146.2	147.1	84.0	130.0	80	16	4
Glutamic acid	147.1	148.1	130.0	102.0	80	4	8
Methionine	149.2	150.1	132.9	104.0	80	4	4
Histidine	155.2	156.1	110.0	93.0	80	12	24
Phenylalanine	165.2	166.1	120.0	103.0	80	8	28
Arginine	174.2	175.1	116.0	130.0	110	12	12
Tyrosine	181.2	182.1	164.9	135.9	80	4	8
Ring-$^{13}C_6$-tyrosine	187.2	188.1	170.9	141.9	80	4	8
Cystine	240.3	241.0	151.9	119.9	80	8	16

4. Notes

1. HILIC separations usually require re-equilibration times of about 10 min in order to assure high retention time repeatability. Also, when applying this method to amino acid trace analysis, column care is essential in order to ensure optimum performance (i.e., installation of in-line filter and guard column, recommended flushing procedure and storage conditions).
2. Alternatively, negative ionization of amino acids may be carried out. However, the majority of the investigated amino acids showed higher abundance in positive ionization mode. Therefore, source parameters and fragmentation settings were only optimized for positively ionized amino acids.
3. If there is no such software available, the four described optimization steps may also be realized by manual variation of the settings. Since different instruments of different manufacturers may vary in their fragmentation behavior, it is recommended to optimize MRM transitions when transferring methods from one instrument to the other.
4. The operation mode of the used SPE automat is based on the generation of positive pressure in the cartridge (applying N_2-gas). However, the described SPE procedure can also be realized by the application of vacuum using a conventional vacuum SPE manifold as we have shown earlier (1).

Acknowledgment

We gratefully acknowledge the Austrian Science Fund for financial support (FWF project P20069).

References

1. Dell'mour M et al (2010) Hydrophilic interaction LC combined with electrospray MS for highly sensitive analysis of underivatized amino acids in rhizosphere research. J Sep Sci 33:911–922
2. Armstrong M, Jonscher K, Reisdorph NA (2007) Analysis of 25 underivatized amino acids in human plasma using ion-pairing reversed phase liquid chromatography/time-of-flight mass spectrometry. Rapid Commun Mass Sp 21:2717–2726

Chapter 5

Amino Acid Analysis via LC–MS Method After Derivatization with Quaternary Phosphonium

Shinsuke Inagaki and Toshimasa Toyo'oka

Abstract

(5-*N*-Succinimidoxy-5-oxopentyl)triphenylphosphonium bromide (SPTPP) is a highly sensitive and positively charged precolumn derivatization reagent for the analysis of amino acids in liquid chromatography–electrospray ionization-tandem mass spectrometry. The synthesis of this reagent and handling of the derivatization reaction are quite simple. It reacts with amino acids rapidly and with high efficiency. MS/MS analysis revealed that the SPTPP-derivatized amino acids formed strong product ions; thus, highly sensitive and selective detection is possible in the selected reaction monitoring mode. The limits of detection for the SPTPP-derivatized amino acids are in the sub-fmol range. The sensitivities of the derivatized amino acids increased about 500-fold, as compared to those of underivatized amino acids.

Key words: (5-*N*-Succinimidoxy-5-oxopentyl)triphenylphosphonium bromide, Liquid chromatography–electrospray ionization-tandem mass spectrometry, Charged derivatization, Quaternary phosphonium regent, Amino acid analysis, 3-Nitrotyrosine, GABA, Amino acid analysis

1. Introduction

High-performance liquid chromatography–electrospray ionization-tandem mass spectrometry (LC–ESI-MS/MS) is rapidly emerging as a powerful method for the highly sensitive and selective analysis of biological compounds, including amino acids. However, as a detection method, mass spectrometry (MS) is not as versatile as differential refractive index detection (RI) or charged aerosol detection (CAD) (1), and there are compounds for which this technique is not appropriate. The sensitivity of the technique is insufficient for the microanalysis of certain samples because the ionization efficiencies of the compounds in the samples are sometimes low; such compounds cannot be detected with high

Michail A. Alterman and Peter Hunziker (eds.), *Amino Acid Analysis: Methods and Protocols*, Methods in Molecular Biology, vol. 828, DOI 10.1007/978-1-61779-445-2_5,

sensitivity. Furthermore, in LC–ESI-MS/MS, a reversed-phase column such as an octadecylsilane (ODS) is commonly used. Highly polar analytes, including some amino acids, cannot be retained on such column. Therefore, it is difficult to separate and detect such compounds. To address these issues, chemical derivatization is performed to introduce charged or proton-affinitive species in a target functional group prior to analysis by LC–ESI-MS/MS to facilitate retention of analytes in the reversed-phase column and to promote ionization during ESI-MS. Chemical derivatization also improves the detection sensitivity. In particular, the introduction of permanently charged or polar substituents can improve the ionization of nonpolar compounds in the ESI mode (2–4).

The following conditions should be satisfied when derivatization reagents are used in LC–ESI-MS/MS: (1) the derivatized analytes should have high ionization efficiency, and they should be detected with high sensitivity; (2) reagents should react with the target functional group under mild conditions; (3) reagents should have a hydrophobicity that is appropriate for the separation of the derivatized analytes by a reversed-phase system; (4) reagents should have low susceptibility to ion suppression; (5) fragmentation of the target analytes should be accomplished easily by collision-induced dissociation (CID) to efficiently generate a particular production for sensitive MS/MS detection; (6) reagents should be inexpensive and easily obtainable, as compared to the numerous conventional derivatization reagents that are commercially available. Derivatization reagents for LC–ESI-MS/MS that meet these requirements have been highly anticipated.

A variety of techniques for converting amino acids into sensitive, analyzable fluorescent derivatives have been developed in the past three decades. The reagents used to produce the derivatives include *o*-phthalaldehyde (OPA), 5-(dimethylamino)-naphthalene-1-sulfonyl chloride (Dansyl-Cl), 9-fluorenylmethylchloroformate (Fmoc-Cl), 7-fluoro-4-nitrobenzo-2-oxa-1,3-diazole (NBD-F), and 6-aminoquinolyl-*N*-hydroxysuccinimidyl carbonate (AQC). Very recently, Shimbo et al. (5–7) developed derivatization reagents for the high-speed analysis of amino acids using LC–ESI-MS/MS, including *p*-*N*,*N*,*N*-trimethylammonioanilyl *N*′-hydroxysuccimidyl carbamate iodide (TAHS) and 3-aminopyridyl-*N*-hydroxysuccimidyl carbamate (APDS). These reagents have a reaction group for analytes, a chargeable group, a nonpolar region, and a readily cleavable group for the high-sensitivity detection using LC–ESI-MS/MS.

In this chapter, we describe the synthesis and application of a new derivatization reagent, (5-*N*-succinimidoxy-5-oxopentyl) triphenylphosphonium bromide (SPTPP), which has a permanently positively charged quaternary phosphonium functional group; it is a highly sensitive precolumn derivatization reagent for amino acids in LC–ESI-MS/MS (8). Amino acids are detected via selected reaction monitoring (SRM), whereby precursor ions are isolated in the first stage of the mass spectrometer followed by CID to generate fragment ions, which are detected after an additional stage of mass spectrometer isolation. SPTPP has been used for the analysis of a neurotransmitter, 4-aminobutanoic acid (GABA), and oxidative stress markers such as *o*-tyrosine (*o*-Tyr), *m*-tyrosine (*m*-Tyr), 3-nitrotyrosine (3NO_2-Tyr), and 3-chlorotyrosine (3Cl-Tyr), in biological samples.

In the near future, simultaneous and sensitive detection of D, L-amino acids can be potentially achieved by developing optically active derivatization reagents for LC–ESI-MS/MS.

2. Materials

2.1. Chemicals

1. Prepare all solutions using ultra pure water (prepared by purifying deionized water to attain a sensitivity of 18 MΩ cm at 25°C).
2. Acetonitrile is of LC–MS grade.
3. All other reagents were of analytical-reagent grade and were used without further purification.
4. Sodium borate buffer (pH 9.5) was prepared as follows. Boric acid was dissolved in distilled water, and the pH was adjusted to 9.5 by adding of 1 M NaOH and monitoring the pH using a pH meter.
5. Store all reagents at 5°C unless otherwise indicated.

2.2. Synthesis of (5-*N*-Succinimidoxy-5-Oxopentyl) Triphenylphosphonium Bromide

1. (4-Carboxybutyl)triphenylphosphine (120 mg, 0.25 mmol) was dissolved in 20 mL of acetonitrile.
2. *N*-Hydroxysuccinimide (29 mg, 0.25 mmol) and dicyclohexy-lcarbodiimide (52 mg, 0.25 mmol) were added to the solution.
3. After incubating for 16 h at room temperature, the precipitate was filtered off.
4. The filtrate was evaporated to dryness and washed with diethyl ether (15 mL×2).
5. The washed sample was evaporated to dryness again to yield SPTPP (see Notes 1 and 2).

2.3. LC–MS/MS Components

1. The LC–ESI-MS/MS apparatus comprised a 1100 series LC (Agilent Technologies, Santa Clara, CA, USA) and an API 3000 triple quadrupole mass spectrometer (AB SCIEX, Foster City, CA, USA) equipped with an ESI source.
2. Reversed-phase LC was performed using Symmetry C_{18} column (2.1 mm i.d.×100 mm, 3.5 μm; Waters). The column was maintained at 40°C. The flow rate of the mobile phase was 0.2 mL/min. The injection volume was fixed at 10 μL.

3. Methods

1. Carry out all procedures at room temperature unless otherwise specified.

3.1. Derivatization of Amino Acids with SPTPP

1. Fifty microliters of 10 mM SPTPP in acetonitrile and 100 μL of 100 mM borate buffer (pH 9.5, see Note 3) were added to 50 μL of the sample solution containing amino acids (0.010–5.0 μM, each).
2. The mixed solution was heated at 40°C for 10 min (Fig. 1), and then 800 μL of H_2O–acetonitrile (75:25) containing 0.1% formic acid (mobile phase) were added to the reaction mixture.
3. The solution was allowed to cool to ambient temperature with ice-water, and an aliquot (10 μL) was injected into the LC–ESI-MS/MS system (see Notes 3 and 4) (Figs. 2 and 3).

3.2. Analysis of Biological Samples Using SPTPP

1. Forty microliters of rat serum were deproteinized by the addition of 160 μL of acetonitrile containing an internal standard, and the sample was centrifuged at 2,000 × *g* for 5 min.
2. The resulting supernatant (40 μL) was transferred to another tube and evaporated to dryness under a gentle stream of nitrogen.

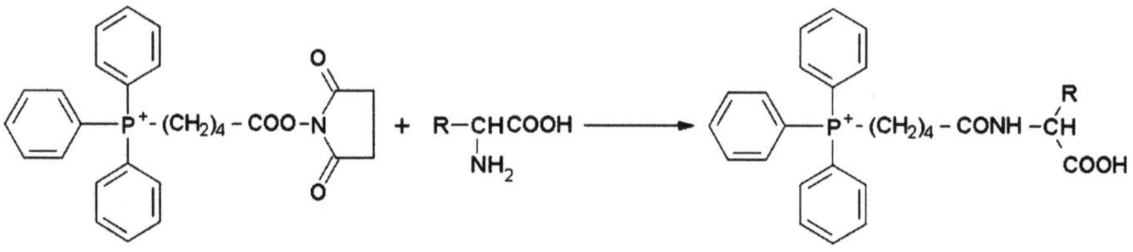

P+-(CH2)4-COO-N (O, O) + R-CHCOOH (NH2) → P+-(CH2)4-CONH-CH (R, COOH)

Fig. 1. Scheme for the derivatization reaction of amino acids with SPTPP.

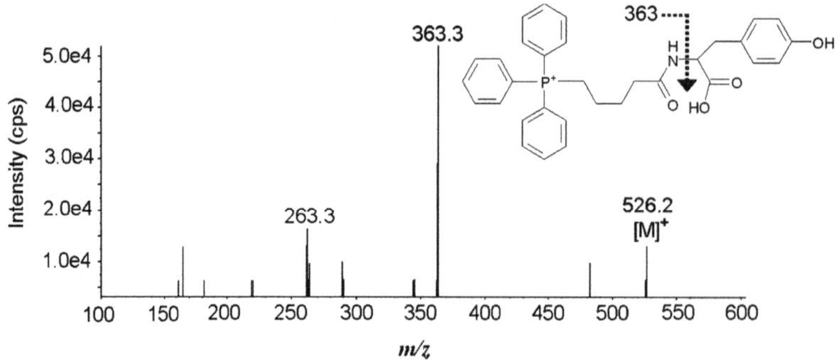

Fig. 2. Product ion mass spectrum of SPTPP-derivatized tyrosine (Precursor ion: *m/z* 526).

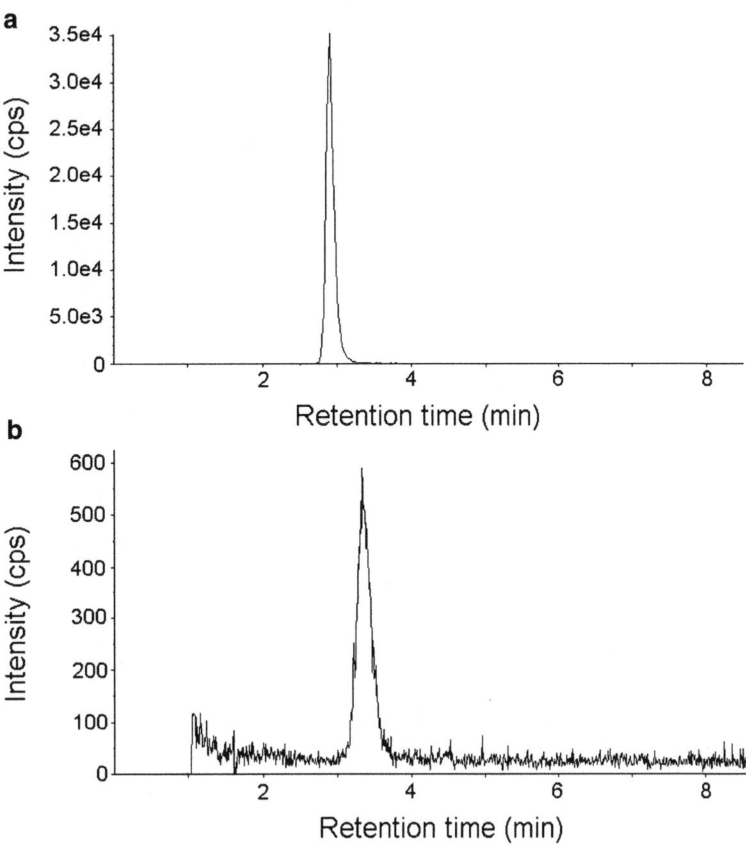

Fig. 3. SRM chromatograms of tyrosine derivatized with SPTPP (**a**, *m/z* 526 → 363) and underivatized tyrosine (**b**, *m/z* 182 → 165); the amounts of the two tyrosine samples were equal (1.0 pmol). The mobile phase was (**a**) H_2O–acetonitrile (75:25) containing 0.1% (v/v) formic acid and (**b**) H_2O–acetonitrile (98:2) containing 0.1% (v/v) formic acid.

3. Forty microliters of 100 mM borate buffer (pH 9.5) were added to the tube and reacted with 40 μL of 10 mM SPTPP.
4. The reaction mixture was heated at 40°C for 10 min, and 320 μL of the mobile phase were added.
5. The supernatant was filtered using a 0.20-μm filter, and an aliquot of the filtrate was injected into the LC–MS/MS system (see Note 5) (Figs. 4 and 5).

3.3. LC–ESI-MS/MS Conditions

1. The conditions for ESI-MS detection were optimized to obtain the highest signal intensity by using an optimizing program and were as follows: mode = positive-ion mode; ionspray voltage = 4,000 V; nebulizer gas pressure = 11 psi; curtain gas pressure = 10 psi; collision gas pressure = 6 psi; external turbo gas flow rate = 6 L/min; turbo gas temperature = 500°C; declustering potential = 76 V; focusing potential = 330 V; entrance potential = 10 V; collision cell exit potential = 15 V; and collision energy = 55 eV.

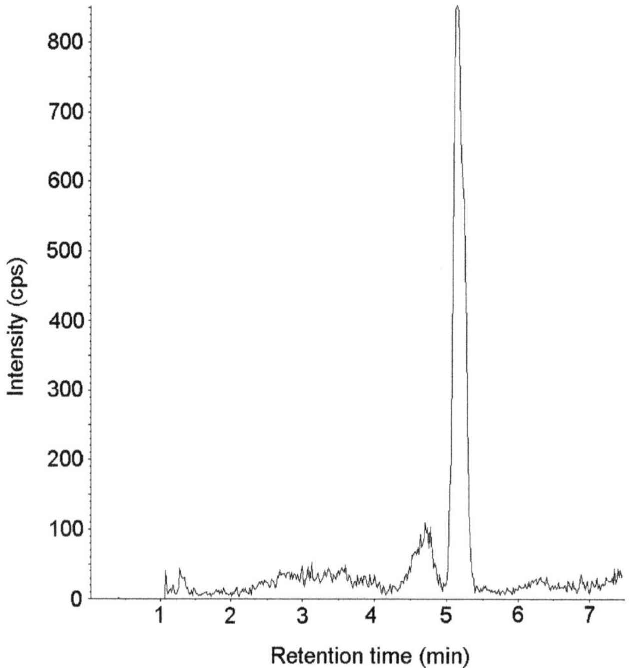

Fig. 4. SRM chromatograms of SPTPP-derivatized GABA obtained from rat serum. The mobile phase was H_2O–acetonitrile (80:20) containing 0.1% (v/v) formic acid.

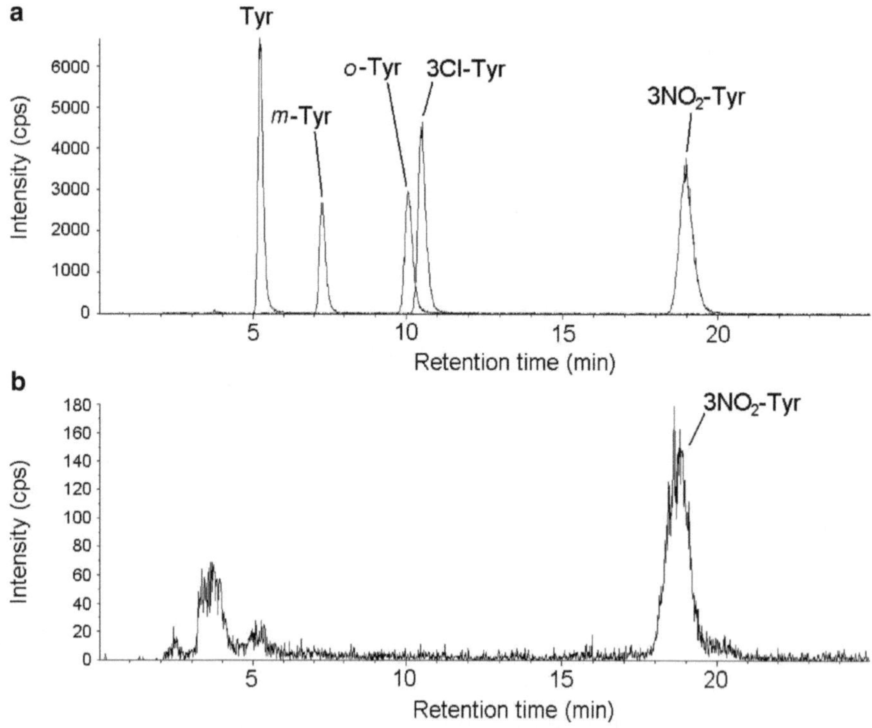

Fig. 5. SRM chromatograms of authentic SPTPP-derivatives of (**a**) tyrosine, *o*-tyrosine, *m*-tyrosine, 3-chlorotyrosine, and 3-nitrotyrosine and (**b**) analysis results for SPTPP-derivatized oxidative stress markers obtained from rat serum. The mobile phase was H_2O–acetonitrile (79:21) containing 0.1% (v/v) formic acid.

4. Notes

1. If the washed sample is oily, add a small amount of acetonitrile (~0.1 mL) to dissolve it, and add a large amount of diethyl ether (~5 mL). Then decant off the ether supernatant.
2. ESI-TOF-MS: *m/z* 460.164 ($[M]^+$). ^1H NMR in chloroform-*d* (ppm), 1.83 (m, 2 H), 2.23 (s, 2 H), 2.76 (t, 3 H), 2.81 (s, 4 H), 3.99 (m, 2 H), 7.80 (m, 15 H).
3. The borate buffer was observed to be effective at pH 9.5. SPTPP reacted with amino acids readily and with high efficiency without condensation reagents. The optimized reaction temperature was 40°C, and the reaction time was 10 min. Regular and intense product ions (*m/z* 363) were observed in the product ion mass spectra of the SPTPP-derivatized amino acids (precursor ion: $[M]^+$) (see Fig. 2). SPTPP enables the highly sensitive and selective detection of its derivatives in the SRM mode. For example, in the case of tyrosine, the signal-to-noise ratio of the SPTPP-derivatized tyrosine improved dramatically compared to the underivatized tyrosine. The limits of detection for intact tyrosine and the SPTPP-derivatized tyrosine were 170 fmol and 0.34 fmol, respectively (the amount of analyte per injection at a signal-to-noise ratio of 3). The signal-to-noise ratio of the former proved to be ~500 times greater than that of the latter (see Fig. 3).
4. Using 6-aminohexanoic acid as an internal standard, the calibration curve exhibited good linearity.
5. Using SPTPP, successful analysis of 4-aminobutanoic acid (GABA) in rat serum was achieved (see Fig. 4). GABA is a major inhibitory neurotransmitter in the central nervous system. Moreover, the analysis of 3-nitrotyrosine, which is an oxidative stress marker in rat serum, was also achieved (see Fig. 5).

Acknowledgments

The authors thank Mr. Yuma Tano and Mr. Yusuke Yamakata for their support with the experiments. This work was supported in part by an academic research grant from the Shimadzu Science Foundation and by the Global COE Program of the Ministry of Education, Culture, Sports, Science and Technology, Japan.

References

1. Dixon RW and Peterson DS (2002) Development and testing of a detection method for liquid chromatography based on aerosol charging. Anal Chem 74: 2930–2937
2. Barry SJ et al (2003) Derivatisation for liquid chromatography/electrospray mass spectrometry: synthesis of pyridinium compounds and their amine and carboxylic acid derivatives. Rapid Commun Mass Spectrom 17:603–620
3. Zaikin VG and Halket JM (2006) Derivatization in mass spectrometry. 8. Soft ionization mass spectrometry of small molecules. Eur J Mass Spectrom 12:79–115
4. Santa T et al (2007) Derivatization reagents in liquid chromatography/electrospray ionization tandem mass spectrometry for biomedical analysis. Drug Discov Ther 1:108–118
5. Shimbo K et al (2009) Multifunctional and highly sensitive precolumn reagents for amino acids in liquid chromatography/tandem mass spectrometry. Anal Chem 81:5172–5179
6. Shimbo K et al (2009) Precolumn derivatization reagents for high-speed analysis of amines and amino acids in biological fluid using liquid chromatography/electrospray ionization tandem mass spectrometry. Rapid Commun. Mass Spectrom 23:1483–1492
7. Shimbo K et al (2010) Automated precolumn derivatization system for analyzing physiological amino acids by liquid chromatography/mass spectrometry. Biomed Chromatogr 24:683–691
8. Inagaki S et al (2010) Highly sensitive and positively charged precolumn derivatization reagent for amines and amino acids in liquid chromatography/electrospray ionization tandem mass spectrometry. Rapid Commun Mass Spectrom 24:1358–1364

Chapter 6

Amino Acid Analysis by Hydrophilic Interaction Chromatography Coupled with Isotope Dilution Mass Spectrometry

Megumi Kato and Akiko Takatsu

Abstract

Here, we describe an amino acid analysis that is based on the use of hydrophilic interaction liquid chromatography coupled with isotope dilution mass spectrometry for the accurate quantification of underivatized amino acids from hydrolyzed peptide/protein. Twelve underivatized amino acids were separated and detected during an 88-min runtime. The absolute limits of detection and limits of quantification (on column) of the four amino acids (isoleucine, phenylalanine, proline, and valine) were in the range of 6–80 and 20–200 fmol, respectively. As little as 25 pmol of peptide or protein hydrolysates is sufficient for determining absolute content.

Key words: Amino acid analysis, Hydrophilic interaction liquid chromatography, Isotope dilution mass spectrometry, Peptide/protein quantification, Accuracy

1. Introduction

Amino acid analysis is a classical and well-documented method in common use. As almost all amino acids have no chromophore groups, many analytical techniques have been developed centering on derivatization chemistry and the subsequent separation of those derivatives. The most general derivatization reagents are ninhydrin (1, 2), *o*-phthalaldehyde (OPA) (3), phenylisothiocyanate (PTC) (4, 5), and aminoquinolylhydroxysuccinimidyl carbamate (AQC) (6, 7).

Mass spectrometry has offered a new methodology to detect underivatized amino acids or to analyze several analytes that show the same retention time if they have a different m/z value (8).

Michail A. Alterman and Peter Hunziker (eds.), *Amino Acid Analysis: Methods and Protocols*, Methods in Molecular Biology, vol. 828, DOI 10.1007/978-1-61779-445-2_6, © Springer Science+Business Media, LLC 2012

It also allows for isotope dilution mass spectrometric (IDMS) amino acid analysis that determines accurately the amounts of peptides and proteins (9–12). According to the fundamental principle of IDMS, isotope-labeled amino acids (internal standards) behave in the same manner as the corresponding target (natural) amino acids throughout hydrolysis and analytical processes.

Here, we demonstrate an application of high-performance liquid chromatography (HPLC)-electrospray ionization (ESI)-MS coupled with IDMS of the underivatized amino acids for peptide and protein quantification (10). The quantitative results from the IDMS method were compared with those of the commercially available precolumn AQC method, and better recovery and more precise data were obtained with the IDMS method (10). We adopted the hydrophilic interaction liquid chromatography (HILIC) separation mode, which can retain underivatized amino acids without any mobile phase modifier(s). Moreover, the MS sensitivity is expected to increase because of a lack of interference by easy-ionized mobile phase modifier(s) and the high content of organic solvent in the mobile phase. Twelve underivatized amino acids were separated and detected during an 88-min runtime. The absolute limits of detections (LODs) and limits of quantification (LOQ) (on column) of the four amino acids (isoleucine, phenylalanine, proline, and valine) were in the range of 6–80 and 20–200 fmol, respectively. As little as 25 pmol of peptide or protein hydrolysates is sufficient for determining absolute content.

2. Materials

2.1. Preparation of Calibration and Sample Blends (See Note 1)

1. HCl (Amino acid analysis grade).
2. Working HCl, 0.1 M concentrated HCl in water.
3. Each natural amino acid solution, gravimetrically prepared with 0.1 M HCl (see Note 2).
4. Working natural amino acid mixture, gravimetrically prepared with each natural amino acid solution (see Note 3).
5. Each labeled amino acid solution, prepared with 0.1 M HCl (see Note 4).
6. Working labeled amino acid mixture, prepared with each labeled amino acid solution (see Note 3).
7. Calibration mixtures, gravimetrically prepared by mixing natural and labeled amino acid mixtures (see Note 5).
8. Peptide/protein sample solution, gravimetrically prepared with proper solvent. Use directly if sample is an aqueous solution.
9. The sample mixture, gravimetrically prepared by mixing peptide/protein sample solution and labeled amino acid mixture (see Notes 5 and 6).

2.2. Hydrolysis

1. 5 mm × 50 mm sampling glass tubes, pyrolyzed by heating at 500°C for 4 h (see Notes 7 and 8).
2. Pico-Tag vacuum vials (Waters, Milford, MA, USA). The polyethylene-modified cap should be cleaned by sonication in 50% methanol/water. The glass vessel should be pyrolyzed by heating at 500°C for 4 h (see Note 8).
3. Pico-Tag Workstation and hydrolysis chamber (Waters, Milford, MA, USA) (see Note 9).
4. Dry ice (see Note 10).
5. Nitrogen gas (more than 99.99% grade).
6. HCl (Amino acid analysis grade).
7. Phenol (crystalline or liquefied, Amino acid analysis grade).
8. Working 6 M HCl with 1% phenol.

2.3. Pretreatment

1. Anion-exchange diethylaminoethyl cellulose (DEAE) resin, DE52 (Whatman, KENT, UK).
2. PVDF membrane unit, Ultrafree-MC (Millipore, Bilerica, MA, USA).

2.4. Chromatographic Analysis

1. Acetic acid (LC-MS grade).
2. Mobile Phase A: Working acetic acid, 10 mM concentrated acetic acid in water.
3. Mobile Phase B: Acetonitrile (LC-MS grade).
4. Zwitterion chromatography (ZIC)-HILIC column, 5 μm × 2.1 mm × 250 mm (Merck Sequant, Sweden) (see Note 11).
5. HPLC system consisting of two pumps, an in-line degasser, an autosampler, a column heater, and a switching valve (see Note 12).
6. Single-quadrupole mass spectrometer (see Note 13).

3. Methods

3.1. Hydrolysis (Carefully Read Pico-Tag Workstation Manual)

1. Use a pipet to place each calibration and sample mixture into a 5 mm × 50 mm glass tube (see Note 1).
2. Place the tubes in a Pico-Tag vacuum vial. Use the Pico-Tag Workstation to dry each blend completely.
3. Add 200 μL of 6 M HCl with 1% phenol to the bottom of the vacuum vial.
4. Seal the vial after alternate steps of vacuum and nitrogen purging, finishing with a vacuum step.

5. Put the vial in a hydrolysis chamber. The standard hydrolysis condition is 110°C for 24 h (see Note 14).
6. After hydrolysis, open the vacuum vial and remove the tubes with a Teflon-coated forceps. After wiping the outside of the tubes with a Kimwipe, insert the tubes into another clean vacuum vial, and then evacuate individual tubes to dryness (see Note 1).

3.2. Pretreatment (See Notes 1 and 15)

1. Wash the anion-exchange DEAE cellulose resin, DE-52 with Milli-Q water to remove the preservative (see Note 16).
2. Add a known quantity of the prewashed resin into each tube (see Note 17).
3. After removing extra Milli-Q water, add the hydrolysates onto the resin.
4. Stir the mixture of hydrolysates and resin for several minutes until the chloride ion to be separated is adsorbed.
5. Centrifuge the mixture of hydrolysates and resin and transfer the supernatant to a fresh PVDF membrane unit (see Note 18).
6. Filtrate the supernatant with the PVDF membrane unit.
7. Subject the sample to chromatographic analysis (see Note 19).

3.3. Chromatographic Analysis

1. Set the column heater to 30°C.
2. Set the m/z value for each amino acid to measure samples in the selective ion monitoring (SIM) mode.
3. Equilibrate the column in 25:75, 10 mM acetic acid:acetonitrile, at a flow rate of 0.1 mL/min.
4. Analyze the samples using the gradient program: initial = 25% A, 75% B, 25 min = 90% A, 10% B, 55 min = 25% A, 75% B (all segments linear) followed by re-equilibration for 33 min at 25% A, 75% B.

3.4. Calibration

1. Calculate the amount of each amino acid based on the mass of each solution provided at the time of preparation and each peak area provided at the time of measurement (see Fig. 1) (see Note 20).
2. Calculate the amount of the peptide/protein sample based on that of each amino acid.

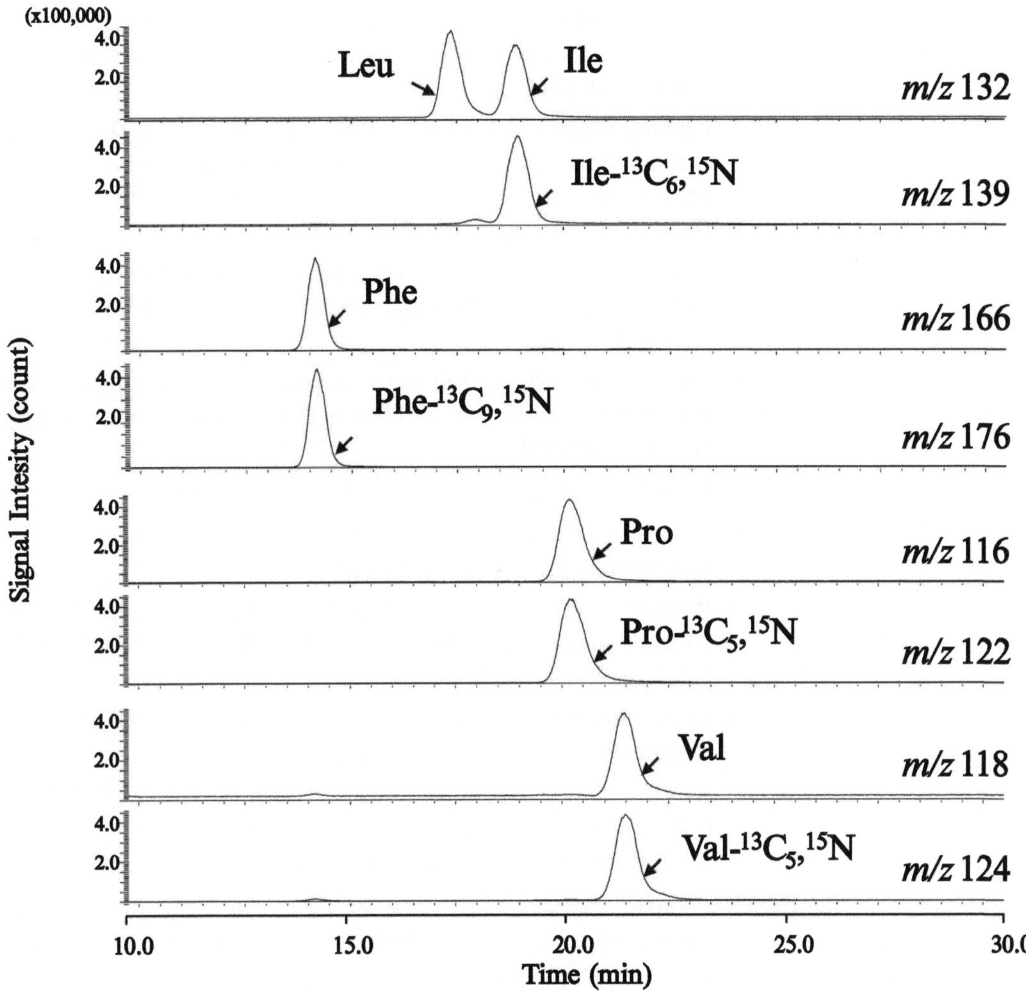

Fig. 1. Selected ion chromatograms of the natural and isotope-labeled amino acids of interest. The injected volume was 2 μL with a total of 50 pmol of each amino acid injected. The chromatographic conditions described in Subheading 3.3 (reproduced from ref. 10 with permission from Elsevier).

4. Notes

1. To avoid the contamination by organic compounds such as amino acids and proteins, wear gloves, use clean apparatus, and work in a clean area.
2. Using highly purified natural amino acids is recommended. For an experienced user who wants to establish the traceability and calculate the uncertainty, amino acid certified reference materials (CRMs) are available from the National Metrology Institute of Japan (NMIJ) (13).

3. Prepare the mixture with the ratio of each component being almost the same molar ratio as that of the target peptide/protein sample.
4. Choose the labeled amino acid in which more than three atoms are substituted by “heavy” stable analogs if possible, so as to prevent the interference by the natural isotope derived from the corresponding natural amino acid.
5. The final concentration of each natural and labeled amino acid in the calibration and sample mixtures should be adjusted to be almost the same.
6. To avoid the adsorption of the peptide/protein sample before hydrolysis, use of tubes and tips with low adsorption properties is recommended.
7. Heat-resistant glass such as Pyrex is recommended.
8. Heating at 500°C for 4 h is sufficient to decompose all contaminating organic compound.
9. Waters no longer produces Pico-Tag Workstation, one can buy a refurbished one or a new one from Eldex, name Hydrolysis/Derivatization (H/D) WorkStation.
10. Liquid nitrogen can be substituted for dry ice to cool the cold trap of the Pico-Tag Workstation.
11. Among several HILIC mode columns, the ZIC-HILIC column showed favorable separation between isomeric compounds L-leucine and L-isoleucine.
12. A switching valve is useful to avoid the infusion of buffer constituent from the peptide/protein sample solution into the mass spectrometer.
13. A triple quadrupole mass spectrometer can be substituted for a single quadrupole mass spectrometer (8).
14. Other variations include 130°C for 48 h or 150°C for 1 h. The hydrolysis condition of each target peptide/protein sample should be optimized individually. Peptide or protein CRMs, which are available from NMIJ, are useful for validating analytical methods (14).
15. For ZIC-HILIC column separation, the chloride ion in the hydrolysate breaks the water-enriched liquid layer on the ZIC-HILIC stationary phase and diminishes the retention of amino acids. DE-52 resin, a weak anion-exchanger, effectively removes only the chloride ion from the hydrolysate.
16. DE-52 is preswollen type resin which can be used without the initial precycling. However, it should never be dried by any method, at any stage of pretreatment, or in subsequent use.
17. The double resin of the exchange capacity was quantitatively sufficient to remove the chloride ion from the sample solution.

Too much resin will remove not only the chloride ions but also the amino acids.

18. Residual resin in the supernatant will decrease the MS sensitivity. Commercially available Ultrafree Centrifugal Filter units are convenient for fast filtration.
19. For HILIC separation, injecting too much hydrophilic solvent, such as 100% water, will result in lower retention, poorer efficiency, and inferior separation. It is therefore preferable to dilute the aqueous sample with an organic solvent to at least 50%.
20. One-point calibration can be applied where the linearity is good enough in the range of the solutions tested prior to the sample analyses. Insert each value into equation.

$$C_x = C_s \times \frac{m_z}{m_{yc}} \times \frac{m_y}{m_x} \times \frac{R'_B}{R'_{BC}}$$

In the equation, C_x is the amount of amino acid in the sample, C_s is the amount of amino acid in a standard solution, m_z is the mass of the standard amino acid solution added to the calibration blend, m_x is the mass of the sample solution added to the sample blend, m_{yc} is the mass of the isotopically labeled standard solution added to the calibration blend, m_y is the mass of the isotopically labeled standard solution added to the sample blend, R'_B is the measured ratio (peak area of the natural amino acid/peak area of the isotopically labeled amino acid) of the sample blend, and R'_{BC} is the average measured ratio (peak area of the natural amino acid/peak area of the isotopically labeled amino acid) of the calibration blend injected before and after the sample.

References

1. Moore S, and Stein WH (1948) Photometric ninhydrin method for use in the chromatography of amino acids. J Biol Chem 176:367–388
2. Moore S, Spackman DH, and Stein WH (1958) Chromatography of Amino Acids on Sulfonated Polystyrene Resins. An Improved System. Anal. Chem. 30:1185–1190
3. Benson JR, and Hare PE (1975) O-phthalaldehyde: fluorogenic detection of primary amines in the picomole range. Comparison with fluorescamine and ninhydrin, Proc Natl Acad Sci USA 72:619–622
4. Cohen SA, and Strydom DJ (1988) Amino acid analysis utilizing phenylisothiocyanate derivatives. Anal Biochem 174:1–16
5. Molnár-Perl I (1994) Advances in the high-performance liquid chromatographic determination of phenylthiocarbamyl amino acids. J Chromatog A 661:43–50
6. Cohen SA, and Michaud DP (1993) Synthesis of a fluorescent derivatizing reagent, 6-aminoquinolyl-N-hydroxysuccinimidyl carbamate, and its application for the analysis of hydrolysate amino acids via high-performance liquid chromatography. Anal Biochem 211: 279–287
7. Strydom DJ, and Cohen SA (1994) Comparison of amino acid analyses by phenylisothiocyanate and 6-aminoquinolyl-N-hydroxysuccinimidyl carbamate precolumn derivatization. Anal Biochem 222:19–28

8. Dell'mour M et al (2010) Hydrophilic interaction LC combined with electrospray MS for highly sensitive analysis of underivatized amino acids in rhizosphere research. J Sep Sci 33:911–922
9. Burkit, WI et al (2008) Toward Systeme International d'Unite-traceable protein quantification: from amino acids to proteins. Anal Biochem 376:242–251
10. Kato M et al (2009) Application of amino acid analysis using hydrophilic interaction liquid chromatography coupled with isotope dilution mass spectrometry for peptide and protein quantification. J Chromatogr 877:3059–3064
11. Woolfitt AR et al (2009) Amino acid analysis of peptides using isobaric-tagged isotope dilution LC-MS/MS. Anal Chem 81:3979–3985
12. Munoz A, Kral R, and Schimmel H (2010) Quantification of protein calibrants by amino acid analysis using isotope dilution mass spectrometry. Anal Biochem 408:124–131
13. National Metrology Institute of Japan (2009) Reference Material Certificate, NMIJ CRM 6013–6016
14. National Metrology Institute of Japan (2008) Reference material certificate, NMIJ CRM 6201

Chapter 7

A Universal HPLC-MS Method to Determine the Stereochemistry of Common and Unusual Amino Acids

Sonja Hess

Abstract

The determination of the stereochemistry of common and unusual amino acids is important in food chemistry, archeology, medicine, and life sciences, including such diverse areas as marine biology and extraterrestrial chemistry and has greatly contributed to our current knowledge in these fields.

To determine the stereochemistry of amino acids, many chromatographic methods have been developed and refined over the last decades. Here, we describe a state-of-the-art indirect chromatography-based LC-MS method. Diastereomers were formed from amino acids that were reacted with chiral derivatizing agents, such as Marfey's reagent (FDAA), GITC, S-NIFE, and OPA-IBLC and separated on a reversed phase column using mass spectrometry compatible buffers.

Key words: Amino acid derivatization, Enantiomer, Marfey's reagent, FDAA, GITC, S-NIFE, OPA-IBLC, Amino acid analysis

1. Introduction

The 20 (or if we include pyrrolysine and selenocysteine 22 (1)) proteinogenic amino acids and nonproteinogenic amino acids, such as γ-aminobutyric acid, selenomethionine, *N*-methylthreonine are commonly found in their L-configuration in nature. However, there is also a small but considerable amount of D-amino acids present in plants and food, bacteria, marine organisms, and even higher organisms, such as frogs, snails, spiders, rats, chickens, and humans (2–11). D-Amino acids are usually synthesized through different routes than their L-amino acid counterparts, but some D-amino acids are the result of racemization of L-amino acids. Particularly in humans, the occurrence of D-amino acids has been associated with racemization during ageing processes, when protective D-amino acid oxidases become less active (6).

Michail A. Alterman and Peter Hunziker (eds.), *Amino Acid Analysis: Methods and Protocols*, Methods in Molecular Biology, vol. 828, DOI 10.1007/978-1-61779-445-2_7,

D-Amino acids are present as free amino acids, or are incorporated in proteins and peptides. Their biological functions are diverse (e.g., as part of the cell wall peptidoglycans or antibiotics in bacteria, toxins in spiders) and, quite often, currently unknown (e.g., free amino acids in human tissue and plasma). In addition to their occurrence on earth, unusual amino acids have even been found in meteorites of the outer space, both in their D- and L- enantiomeric configuration (12, 13). Although D-amino acids are underrepresented when compared to L-amino acids, one can thus conclude that both, D- and L-amino acids play an important role in our universe. In fact, it is generally believed that research on the occurrence of D- and L-amino acids increases our fundamental understanding of life and universe.

Conventionally, the determination of D- and L-amino acids is carried out via composition analysis of proteins or peptides using acid hydrolysis followed by direct or indirect liquid chromatography with either UV/VIS or fluorescence detection. While direct chromatography uses chiral columns, indirect chromatography relies on the formation of diastereomers by reacting the amino acids with chiral derivatizing agents (CDAs). The most commonly used CDA by far is Marfey's reagent (FDAA, *N*-α-(2,4-dinitro-5-fluorophenyl)-L-alaninamide; Scheme 1) (14). In addition, 2,3,4,6-tetra-*O*-acetyl-β-D-glucopryanosyl isothiocyanate (GITC) (15), (*S*)-*N*-(4-nitrophenosycarbonyl)phenylalanine methoxyethyl ester (*S*-NIFE) (16), *o*-phthalaldehyde/isobutyryl-L-cysteine (OPA-IBLC) (17) have been used successfully (Scheme 1).

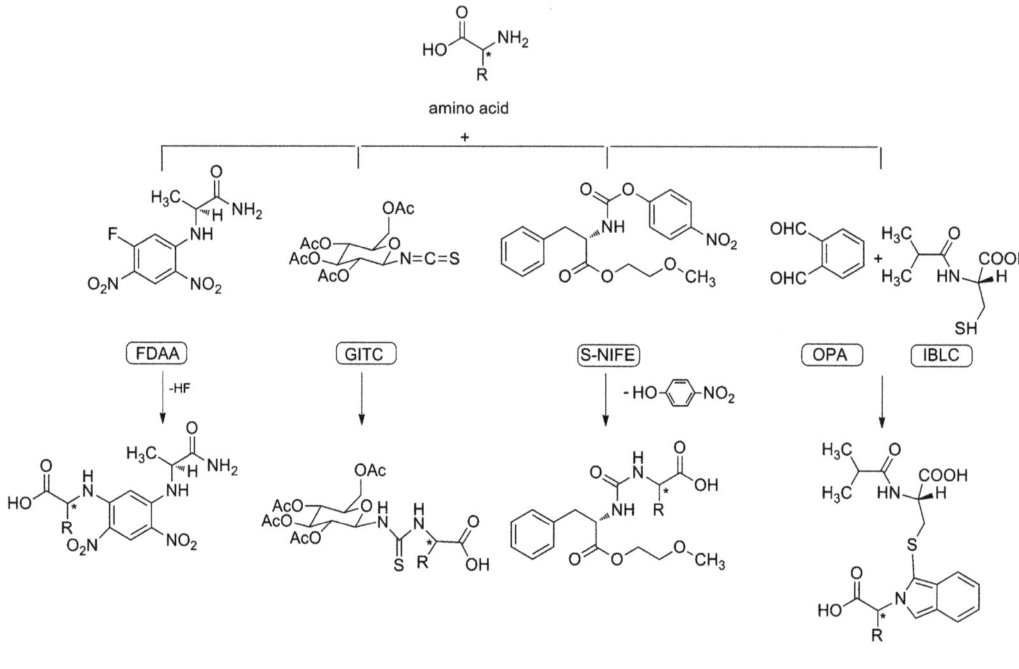

Scheme 1. Common and uncommon amino acids are reacted with FDAA, GITC, S-NIFE or OPA-IBLC to form diastereomers prior to indirect chromatography.

UV/VIS and fluorescence detection are traditionally established methods, but mass spectrometry offers specific advantages in terms of sensitivity and selectivity (see Note 1). A requirement for a successful UV/VIS or fluorescence detection is essentially baseline separation of all amino acids. Due to the additional mass spectrometric dimension where selected ions (for selected amino acids) can be extracted from the total ion chromatograms, baseline separation of all amino acids is not essential and overlapping derivatized amino acids can often be separated based on their mass. In addition, mass spectrometry allows the interrogation of isotopic contributions. This can be important when the isotopic distribution carries additional information, for instance in current stable isotope labeling by amino acids in cell culture (SILAC) experiments, where the incorporation of stable isotopes is used for quantitative information (18–21) (see Note 2).

For a successful transfer of methods based on traditional UV/VIS or fluorescence detection to mass spectrometry detection, the originally used mobile phases (e.g., potassium phosphate or borate buffers) needed to be replaced and optimized for mass spectrometry. While this may change the elution profile and sometimes even resolution of the separated diastereomers, the additional information gain from the mass spectrometry dimension is advantageous (7, 20).

Amino acids were reacted with several CDAs (FDAA, GITC, S-NIFE, OPA-IBLC) and separated by RP-HPLC-ESI-MS. The stereochemical assignment was accomplished with CDA-derivatized D- and L-amino acid standards that were separated under identical conditions. This method is applicable to both, common and uncommon amino acids, such as *N*-methylthreonine or beta-methoxytryosine.

2. Materials

2.1. Chemicals

1. Acetonitrile (HPLC gradient grade quality).
2. Water (HPLC gradient grade quality).
3. All other chemicals and reagents should be of the highest purity available.
4. Triethylamine buffer (6%; pH = 11.9) is used as a basic buffer. To prepare it add 6.1 μl of triethylamine (99%) to 93.9 μl of water.
5. Acetic acid (5%; pH = 2.6): Add 50 ml of acetic acid (99.7%) to 950 ml of water.

2.2. Hydrolysis of Proteins or Peptides

1. Protein or peptide of which the amino acid composition should be determined.
2. 6 N (constant boiling) HCl.
3. Vacuum hydrolysis tube (Kontes, Vineland, NJ).

4. 3-way valve in vacuum apparatus to remove oxygen from hydrolysis tube and replace it with nitrogen.
5. Dry ice to freeze sample.
6. Leather gloves to handle dry ice and frozen samples.
7. Heat block with thermometer.
8. Timer.
9. Lyophylizer.
10. pH-paper.

2.3. Preparation of Standard Amino Acids

1. D- and L-amino acids (Sigma Aldrich, St. Louis, MO; and/or BACHEM Bioscience Inc. King of Prussia, PA) (see Note 3).
2. Prepare aqueous solutions of amino acid standards at a concentration of 1 mg/ml using HPLC water. If necessary (e.g., for hydrophobic amino acids), add methanol to dissolve otherwise precipitating amino acids.
3. Scale.
4. Eppendorf tubes; pipettes.
5. Water (HPLC gradient grade quality).
6. Methanol (HPLC gradient grade quality), if necessary to dissolve amino acid.
7. DMSO (99.9%) if necessary to dissolve amino acid.

2.4. Derivatization with FDAA

1. Eppendorf tubes; autosampler vials.
2. *N*-α-(2,4-dinitro-5-fluorophenyl)-L-alaninamide (FDAA; Marfey's reagent; Sigma Aldrich, St. Louis, MO).
3. Acetone (HPLC grade).
4. 6% triethylamine (pH = 11.9).
5. pH meter, pH paper.
6. Heat block.
7. Timer.
8. 5% acetic acid.

2.5. Derivatization with GITC

1. Eppendorf tubes; autosampler vials.
2. 2,3,4,6-tetra-*O*-acetyl-β-D-glucopyranosyl isothiocyanate (GITC; Sigma Aldrich, St. Louis, MO).
3. Acetone (HPLC grade).
4. 6% triethylamine.
5. pH paper.
6. Heat block.
7. Timer.
8. 5% acetic acid.

2.6. Derivatization with S-NIFE

1. Eppendorf tubes; autosampler vials.
2. (*S*)-*N*-(4-nitrophenoxycarbonyl)phenylalanine methoxyethyl ester (*S*-NIFE; Peptisyntha; Brussels, Belgium).
3. Acetone (HPLC grade).
4. 6% triethylamine.
5. pH paper.
6. Heat block.
7. Timer.
8. 5% acetic acid.

2.7. Derivatization with OPA-IBLC

1. Eppendorf tubes; autosampler vials.
2. *o*-Phthalaldehyde (OPA; Sigma Aldrich, St. Louis, MO).
3. Methanol (HPLC gradient grade quality).
4. Isobutyryl-L-cysteine (IBLC; Calbiochem-Novabiochem; San Diego, CA).
5. Water (HPLC gradient grade quality).
6. 6% triethylamine.
7. pH paper.
8. Heat block.
9. Timer.
10. 5% acetic acid.

2.8. Liquid Chromatography–Mass Spectrometry

1. Agilent 1100 LC-MSD (Agilent Technologies, Palo Alto, CA) with a binary pump, a degasser, an autosampler, a DAD detector, ChemStation.
2. Agilent C18 ZorbaxSB column (150 mm × 2.1 mm).
3. C18 Opti-Guard column (15 mm × 1 mm, Bodman, Baltimore, MD).
4. Mobile phase A: 5% acetic acid, pH 2.6. Add 50 ml of acetic acid to 950 ml of water. Check pH.
5. Mobile phase B: Acetonitrile containing 10% methanol. Add 100 ml of methanol to 900 ml of acetonitrile (see Note 4).

2.9. Data Analysis

ChemStation software.

3. Methods

3.1. Hydrolysis of Proteins or Peptides

1. To ca. 200 μg of protein or peptide, add 500 μl of 6 N (constant boiling) HCl in a vacuum hydrolysis tube. Using a three-way valve, flush repeatedly with nitrogen to remove any remaining oxygen (see Note 5).

2. Heat the closed tube at 108°C for 24 h.
3. Transfer the sample in an Eppendorf tube using a pipette. Wash the hydrolysis tube with 200 μL water twice to remove all samples.
4. Remove the solvent of the combined sample in vacuo.
5. To remove residual HCl, dissolve residue in 200 μl of water and lyophilize repeatedly (see Note 6).

3.2. Preparation of Standard Amino Acids

Mix 10 μl each of D- and L-amino acids in an Eppendorf tube. Remove the solvent in vacuo. Redissolve in 10 μl of water (see Note 7).

3.3. Derivatization with FDAA

1. Dissolve 1 mg of *N*-α-(2,4-dinitro-5-fluorophenyl)-L-alaninamide (FDAA; Marfey's reagent) in 100 μl of acetone.
2. To 10 μl of peptide/protein hydrolysate or standard amino acid solution, add 10 μl of freshly prepared 6% triethylamine (see Note 6) in an Eppendorf tube.
3. Add 10 μl of FDAA solution, close the Eppendorf tube and heat to 50°C for 1 h (see Note 8).
4. Add 10 μl of 5% acetic acid, transfer the sample into an autosampler vial and analyze 20 μl by liquid chromatography–mass spectrometry (LC-MS) (see Notes 9 and 10).

3.4. Derivatization with GITC

1. Dissolve 1 mg of GITC in 100 μl of acetone.
2. To 10 μl of peptide/protein hydrolysate or standard amino acid solution, add 10 μl of freshly prepared 6% triethylamine (see Note 6).
3. Add 10 μl of GITC solution and react at RT for 10 min.
4. Add 10 μl of 5% acetic acid, transfer the sample into an autosampler vial, and analyze 20 μl by LC-MS (see Notes 9 and 11).

3.5. Derivatization with S-NIFE

1. Dissolve 1 mg of *S*-NIFE in 100 μl of acetone.
2. To 10 μl of peptide/protein hydrolysate or standard amino acid solution, add 10 μl of 6% triethylamine.
3. Add 10 μl of S-NIFE solution and react at RT for 20 min.
4. Add 10 μl of 5% acetic acid, transfer the sample into an autosampler vial, and analyze 20 μl by LC-MS (see Notes 9 and 12).

3.6. Derivatization with OPA-IBLC

1. For a 30 mM OPA solution, dissolve 402 μg of OPA in 100 μl of methanol.
2. For a 90 mM IBLC solution, dissolve 1719 μg of IBLC in 100 μl of water.
3. To 10 μl of peptide/protein hydrolysate or standard amino acid solution, add 10 μl of freshly prepared 6% triethylamine (see Note 6).

4. Add 10 μl of 30 mM OPA solution and react at RT for 20 min.
5. Add 10 μl of the 90 mM IBLC solution.
6. Add 10 μl of 5% acetic acid, transfer the sample into an autosampler vial, and analyze 20 μl by LC-MS (see Notes 9 and 13).

3.7. Liquid Chromatography–Mass Spectrometry

1. For the separation and analysis of the derivatized amino acids, an Agilent 1100 LC-MSD can be used (see Note 4).
2. The derivatized amino acids are separated using a C18 ZorbaxSB column (150 mm × 2.1 mm) equipped with a C18 Opti-Guard column (15 mm × 1 mm) at a column temperature of 50°C and a flow rate of 250 μl/min.
3. Linear gradients from 5 to 50% B are applied within 50 min, followed by 100% B for 10 min.
4. The mass spectra are recorded in positive ion mode acquiring data from *m/z* 300 to *m/z* 750 every 4 s.
5. If an additional diode array detector is used, it is set to 340 nm for FDAA derivatives, 254 nm for GITC and S-NIFE derivatives, and 280 nm for OPA-IBLC derivatives (see Note 14).

3.8. Data Analysis

To display total ion chromatograms (TIC) and extract the specific ions of interest, ChemStation software was used (see Note 15).

4. Notes

1. Sensitivity in mass spectrometry is largely dependent on the setup, with nanoLC and nanoelectrospray setups being the most sensitive. While we describe here a system using 250 μl/min flow rate, sensitivity can be further improved when the method is transferred to a nanoflow system.
2. There are different ways how one can determine the incorporation of stable isotopes in proteomes, either on the protein, peptide, or amino acid level. The amino acid determination can be used to determine the latter. Figure 1 shows a 1:1 contribution of unlabeled and ^{13}C-labeled Gly derivatized with FDAA. For comparison, the inset shows the unlabeled Gly derivatized with FDAA.
3. Common and unusual amino acids can be studied with this method including but not limited to D- and L-threonine, D- and L-*allo*-threonine, D- and L-*N*-methylthreonine, D- and L-*allo*-*N*-methyl-threonine, D- and L-homoproline, D- and L-proline, D- and L-leucine, D- and L-isoleucine, D- and L-*allo*-isoleucine,

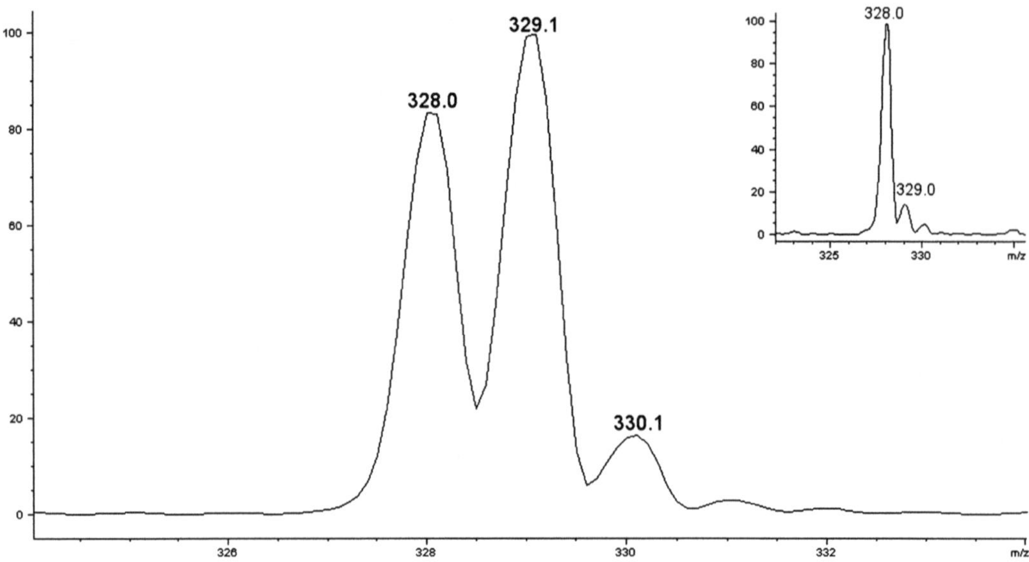

Fig. 1. Monoisotopic peaks of FDAA labeled Gly spiked with 50% ^{13}C Gly. The *inset* shows the unspiked Gly.

D- and L-norleucine, D- and L-*tert*-leucine, D- and L-*N*-methyl-leucine, D- and L-*N*-methyl-isoleucine, D- and L-*allo*-*N*-methyl-isoleucine, D- and L-*N*-methyl-phenylalanine, D- and L-*threo*-β-methyoxytyrosine, D- and L-*erythro*-β-methyoxytyrosine. Amino acids, such as *N*-methylthreonines and β-methyoxytyrosines may need to be synthesized as previously described (22, 23).

4. The use of the Agilent 1100 LC-MSD is not mandatory. The protocol can be adapted to HPLCs and mass spectrometers from other vendors.

5. We usually freeze the sample before placing a vacuum on the tube as shown in Fig. 2. This way, one prevents the sample to splash out of the tube. While a vacuum is applied, slowly warming the tube between your thumb and index finger helps to remove small air bubbles as soon as the sample thaws.

   During the hydrolysis step, asparagines and glutamines are converted to aspartic acid and glutamic acid. Tryptophan is generally completely hydrolyzed under standard conditions, but can be rescued when reducing conditions are used (24).

6. For the subsequent derivatization, it is important that no HCl remains in the sample. It may be necessary to repeat the washing step more than once to remove all HCl. This is sample-dependent. Check pH.

7. Depending on the solubility of the investigated amino acids, it may be necessary to add methanol (or DMSO) to the sample rather than water.

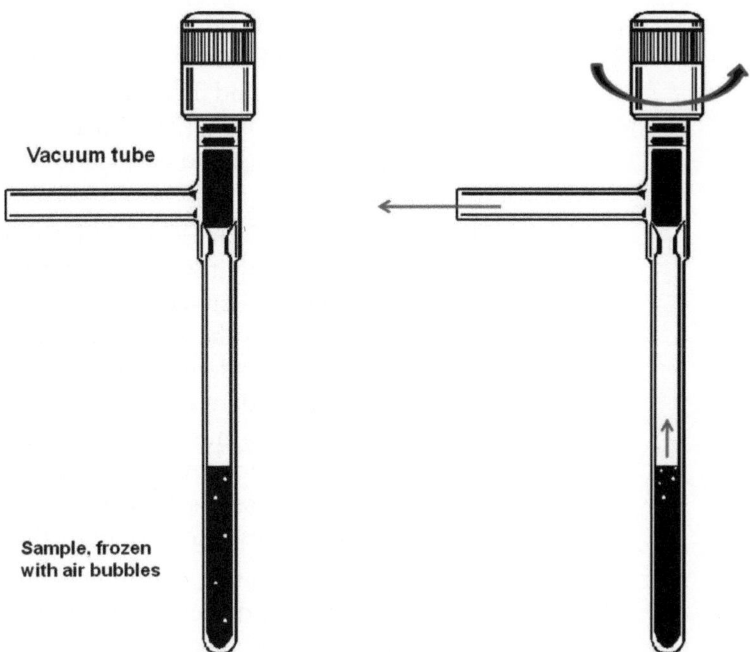

Fig. 2. Hydrolysis tubes are used for the acid hydrolysis of proteins and peptides. Samples are frozen and a vacuum is applied to the tube. Gently warm up the sample so that the air bubbles are removed.

8. The final product should have an orange color. If the sample turns dark red, most likely the pH was still too acidic for the reaction to occur.
9. Acetic acid quenches the reaction.
10. FDAA-derivatized (also referred to as 2,4-dinitrophenyl5-L-alanine amide = DNPA) amino acids are stable and highly enantioselective. Nucleophiles, such as primary amines, thiols, aromatic alcohols lead to single and double substitution in case of Lys, His, Cys, and Tyr (25–30). Excess of FDAA reduces the amount of singly substituted amino acids. As internal standard, an FDAA derivative with *m/z* 348 with an elution time of approximately 38 min is used to determine the unadjusted relative retention $r_G = t_{Ri}/t_{R(St)}$, where t_{Ri} is the retention time of the analyte and $t_{R(St)}$ is the retention time of the internal standard.
11. GITC-derivatized amino acids are not as stable as FDAA and should be analyzed within 24 h of the reaction. GITC reacts specifically with primary amines, so double substitutions are only observed for Lys. GITC reacts with Pro, but not with HomoPro, presumably due to steric hindrance (15, 31). As internal standard, a GITC derivative with *m/z* 463 with an elution time of approximately 38 min can be used to determine the unadjusted relative retention $r_G = t_{Ri}/t_{R(St)}$.

12. Upon addition of S-NIFE, the sample turns immediately yellow. S-NIFE derivatives are stable. S-NIFE can react twice with Lys, Cys, and Tyr (16). As internal standard, an S-NIFE derivative with m/z 473 with an elution time of approximately 45 min can be used to determine the unadjusted relative retention $r_G = t_{Ri}/t_{R(St)}$.
13. Traditionally, OPA-IBLC derivatives have been analyzed by fluorescence, and it was noticed that the fluorescence intensity of Lys and Gly derivatives rapidly decreased (17). This could be attributed to the transformation of the kinetically formed mono-substitute derivatives into the thermodynamically stable di-substituted derivatives (32, 33). OPA-IBLC derivatized amino acids are thus not recommended for determination of Gly and Lys derivatives. As internal standard, an OPA-IBLC derivative with m/z 481 with an elution time of approximately 48 min can be used to determine the unadjusted relative retention $r_G = t_{Ri}/t_{R(St)}$.
14. The additional detection can sometimes be useful, but is not necessary.
15. Each of the CDAs has advantages and disadvantages. In general, using the described procedures, amino acids reacted with FDAA exhibit higher enantioselectivity and more stability, but slightly less sensitivity than the GITC derivatives. With the exception of hydrophilic hydroxyl and acidic amino acids, (25) TFA reportedly improves resolution of the FDAA-derivatized amino acids. However, it also decreases the MS signal intensity significantly (by a factor of 4) (7). In contrast, using 5% acidic acid in the water phase and adding 10% methanol to the acetonitrile phase generally improves signal intensity. If sensitivity is not a concern, FDAA is the CDA of choice for non-aromatic amino acids.

    While GITC is not as stable as FDAA and generally shows lower enantioselectivity, it provides higher sensitivity. Enantioselectivity of S-NIFE is higher than that of GITC, but lower than that of FDAA. A general order of sensitivity is: GITC > S-NIFE ≈ FDAA > OPA-IBLC. The limit of detection is in the low picomolar range using the described procedures and can be further improved with a nanoflow and nanoelectrospray setup.

    The reported high enantioselectivity for OPA-IBLC could not be maintained with the mass spectrometry compatible solvents. OPA-IBLC showed the lowest enantioselectivity for most amino acids except for aromatic amino acids. A general order of enantioselectivity was FDAA > S-NIFE > GITC > OPA-IBLC, but for aromatic amino acids, the order was S-NIFE > OPA-IBLC > GITC > FDAA.

The use of the unadjusted relative retention ($r_G = t_{Ri}/t_{R(St)}$) allows for an easy transfer of this method on other instrument platforms. Table 1 shows the molecular masses, elution order, retention times, unadjusted relative retention (r_G), and separation factors of selected amino acids derivatized with FDAA using the conditions described here.

Table 1
Retention times ($t_R i$), unadjusted relative retention (r_G), and separation factors of amino acids derivatized with FDAA

Amino acid	$(M+H)^+$	Elution order	t_{Ri} (L) (min)	t_{Ri} (D) (min)	r_G (L)[a]	r_G (D)[a]	r_G (L)/r_G (D)
Thr	372	L<D	19.41	24.85	0.511	0.654	0.781
allo-Thr	372	L<D	19.01	22.38	0.506	0.588	0.861
N-MeThr	386	L<D	21.17	22.93	0.556	0.602	0.924
allo- *N*-MeThr	386	L<D	25.60	26.81	0.673	0.705	0.955
homoPro	382	D<L	35.10	33.20	0.929	0.875	1.062
Pro	368	L<D	26.00	27.93	0.683	0.734	0.931
Leu	384	L<D	38.18	43.80	1.008	1.150	0.877
Ile	384	L<D	37.15	43.18	0.976	1.135	0.860
allo-Ile	384	L<D	37.31	43.32	0.980	1.138	0.861
NorLeu	384	L<D	38.37	44.23	0.985	1.165	0.845
t-Leu	384	L<D	36.93	42.68	0.970	1.121	0.865
N-MeLeu	398	L<D	41.03	44.70	1.079	1.175	0.918
N-MeIle	398	L<D	40.81	42.81	1.074	1.127	0.953
allo- *N*-MeIle	398	L<D	41.32	44.73	1.086	1.173	0.926
N-MePhe	432	L<D	38.93	39.42	1.024	1.037	0.987
erythro-β-MeOTyr	463	D<L	24.81	23.75	0.653	0.627	1.041
threo-β-MeOTyr	463	L<D	23.97	24.83	0.631	0.655	0.963
Di-*erythro*-β-MeOTyr	716	D<L	45.36	45.05	1.194	1.189	1.004
Di-*threo*-β-MeOTyr	716	D<L	51.05	47.19	1.344	1.246	1.079

As internal standard, a DNP derivative with m/z 348 was chosen, which eluted around 38 min. Reprinted from ref. 7 with permission from Elsevier

[a] The unadjusted relative retention r_G was calculated as $t_{Ri}/t_{R(St)}$

Acknowledgments

This work was supported by the Intramural Program of the National Institute of Diabetes, Digestive and Kidney Diseases, Z01 DK070004-04, the Beckman Institute and the Betty and Gordon Moore Foundation.

References

1. Atkins JF, and Gesteland R (2002) Biochemistry – The 22nd amino acid. Science 296:1409–1410
2. Bruckner H, and Westhauser T (2003) Chromatographic determination of L- and D-amino acids in plants. Amino Acids 24:43–55
3. Patzold R, and Bruckner H (2006) Gas chromatographic determination and mechanism of formation of D-amino acids occurring in fermented and roasted cocoa beans, cocoa powder, chocolate and cocoa shell. Amino Acids 31:63–72
4. Nagata Y et al (1998) Occurrence of peptidyl D-amino acids in soluble fractions of several eubacteria, archaea and eukaryotes. BBA-Gen Subjects 1379:76–82
5. Mor A, Delfour A, and Nicolas P (1991) Identification of a D-Alanine-Containing Polypeptide Precursor for the Peptide Opioid, Dermorphin. J Biol Chem 266:6264-6270
6. Fujii N (2002) D-amino acids in living higher organisms. Origins of Life and Evolution of the Biosphere 32:103–127
7. Hess S et al (2004) Chirality determination of unusual amino acids using precolumn derivatization and liquid chromatography-electrospray ionization mass spectrometry. J Chromatogr A 1035:211–219
8. Oku N et al (2004) Neamphamide A, a new HIV-inhibitory depsipeptide from the Papua New Guinea marine sponge Neamphius huxleyi. J Nat Prod 67:1407–1411
9. Davies-Coleman M T et al (2003) Isolation of homodolastatin 16, a new cyclic depsipeptide from a Kenyan collection of Lyngbya majuscula. J Nat Prod 66:712–715
10. Hashimoto A et al (1993) Embryonic-Development and Postnatal Changes in Free D-Aspartate and D-Serine in the Human Prefrontal Cortex. J Neurochem 61:348–351
11. Ratnayake A S et al (2006) Theopapuamide, a cyclic depsipeptide from a Papua New Guinea lithistid sponge Theonella swinhoei. J Nat Prod 69:1582–1586
12. Engel M H, and Macko S A (1997) Isotopic evidence for extraterrestrial non-racemic amino acids in the Murchison meteorite. Nature 389:265–268
13. Cronin J R, and Pizzarello S (1997) Enantiomeric excesses in meteoritic amino acids. Science 275:951–955
14. Marfey P (1984) Determination of D-Amino Acids. 2. Use of a Bifunctional Reagent, 1,5-Difluoro-2,4-Dinitrobenzene. Carlsberg Research Communications 49:591–596
15. Nimura N, Toyama A, and Kinoshita T (1984) Optical resolution of amino acid enantiomers by high-performance liquid chromatography. J Chromatogr A 316:547–552
16. Peter A, Vekes E, and Torok G (2000) Application of (S)-N-(4-nitrophenoxycarbonyl) phenylalanine methoxyethyl ester as a new chiral derivatizing agent for proteinogenic amino acid analysis by high-performance liquid chromatography. Chromatographia 52:821–826
17. Brückner H et al (1994) Liquid chromatographic determination of d- and l-amino acids by derivatization with o-phthaldialdehyde and chiral thiols: Applications with reference to biosciences. J Chromatogr A 666:259–273
18. Ong S E et al (2002) Stable isotope labeling by amino acids in cell culture, SILAC, as a simple and accurate approach to expression proteomics. Mol Cell Proteomics 1:376–386
19. Blanco FJ et al (2001) Solid-state NMR data support a helix-loop-helix structural model for the N-terminal half of HIV-1 Rev in fibrillar form. J Mol Biol 313:845–859
20. Hess S, van Beek J, and Pannell L K (2002) Acid hydrolysis of silk fibroins and determination of the enrichment of isotopically labeled amino acids using precolumn derivatization and high-performance liquid chromatography-electrospray ionization-mass spectrometry. Anal Biochem 311:19–26
21. Graham R L J, and Hess S (2010) Mass Spectrometry in the Elucidation of the Glycoproteome of Bacterial Pathogens Curr Proteomics 7:57–81
22. Ford P W et al (1999) Papuamides A–D, HIV-Inhibitory and Cytotoxic Depsipeptides from

the Sponges Theonella mirabilis and Theonella swinhoei Collected in Papua New Guinea. J Am Chem Soc 121:5899–5909

23. Cranfill D C, and Lipton M A (2007) Enantio- and Diastereoselective Synthesis of (R,R)-β-Methoxytyrosine. Org Let 9: 3511–3513
24. Ng LT, Pascaud A, and Pascaud M (1987) Hydrochloric acid hydrolysis of proteins and determination of tryptophan by reversed-phase high-performance liquid chromatography. Anal Biochem 167:47–52
25. Fujii K et al (1997) A nonempirical method using LC/MS for determination of the absolute configuration of constituent amino acids in a peptide: Combination of Marfey's method with mass spectrometry and its practical application. Anal Chem 69:5146–5151
26. Brückner H, and Gah C (1991) High-performance liquid chromatographic separation of dl-amino acids derivatized with chiral variants of Sanger's reagent. J Chromatogr A 555:81–95
27. Fujii K et al (1997) A nonempirical method using LC/MS for determination of the absolute configuration of constituent amino acids in a peptide: Elucidation of limitations of Marfey's method and of its separation mechanism. Anal Chem 69:3346–3352
28. Harada K et al (2001) Abnormal elution behavior of ornitine derivatized with 1-fluoro-2,4-dinitrophenyl-5-leucinamide in advanced Marfey's method. J Chromatogr A 921: 187–195
29. Kochhar S, and Christen P (1989) Amino acid analysis by high-performance liquid chromatography after derivatization with 1-fluoro-2,4-dinitrophenyl-5-L-alanine amide. Anal Biochem 178:17–21
30. B'Hymer C, Montes-Bayon M, and Caruso J A (2003) Marfey's reagent: Past, present, and future uses of 1-fluoro-2,4-dinitrophenyl-5-L-alanine amide. J Sep Sci 26:7–19
31. Tian ZP et al (1991) Resolution of 2,3,4,6-Tetra-O-Acetyl-Beta-D-Glucopyranosylisothiocyanate Derivatives of Alpha-Methyl Amino-Acid Enantiomers by High-Performance Liquid-Chromatography. J Chromatogr 541:297–302
32. Mengerink Y et al (2002) Advances in the evaluation of the stability and characteristics of the amino acid and amine derivatives obtained with the o-phthaldialdehyde/3-mercaptopropionic acid and o-phthaldialdehyde/N-acetyl-L-cysteine reagents – High-performance liquid chromatography-mass spectrometry study. J Chromatogr A 949:99–124
33. Molnar-Perl I, and Vasanits A (1999) Stability and characteristics of the o-phthaldialdehyde/3-mercaptopropionic acid and o-phthaldialdehyde/N-acetyl-L-cysteine reagents and their amino acid derivatives measured by high-performance liquid chromatography. J Chromatogr A 835:73–91

Chapter 8

Amino Acid Analysis by Capillary Electrophoresis-Mass Spectrometry

Akiyoshi Hirayama and Tomoyoshi Soga

Abstract

A method for the determination of underivatized amino acids based on capillary electrophoresis-mass spectrometry is described. To analyze free amino acids simultaneously, a low acidic pH electrolyte was used to confer positive charge on whole amino acids. All protonated amino acids migrated toward the cathode in CE and then were detected by electrospray ionization mass spectrometry with high selectivity and sensitivity. This method is simple, rapid, and selective and could be readily applied to the analysis of free amino acids in various samples.

Key words: Capillary electrophoresis, Mass spectrometry, Quadrupole mass spectrometry, Triple-quadrupole tandem mass spectrometry, Time-of-flight mass spectrometry, Tumor tissue, Amino acid analysis

1. Introduction

Quantitative amino acid analysis is an important component in many medical, scientific and industrial applications. Since amino acids are highly polar compounds and most of them have minimal UV absorbance, they have been commonly analyzed by high-performance liquid chromatography (HPLC) methods with pre- (1, 2) or postcolumn (3, 4) derivatization using UV chromophore or fluorophore reagents. However, these methods have some drawbacks including that they require manual derivatization procedures or a complicated analytical system are necessary or that they lack selectivity and sensitivity. In addition, the analysis of amino acids in biological, physiological, and food samples is rather complicated, mainly due to the presence of heterogeneous matrices that interfere with the separation and detection of amino acids.

Michail A. Alterman and Peter Hunziker (eds.), *Amino Acid Analysis: Methods and Protocols*, Methods in Molecular Biology, vol. 828, DOI 10.1007/978-1-61779-445-2_8,

Another approach to amino acid analysis is liquid chromatography coupled to mass spectrometry (5). Although it generally provides outstanding performance, its quantification accuracy can be affected by ion suppression effects.

Recently, capillary electrophoresis-mass spectrometry (CE-MS) has emerged as a powerful analytical tool for charged species. While CE provides rapid analysis time and efficient resolution, MS provides high selectivity and sensitivity. In CE, charged species are separated on the basis of their charge and size, therefore, basic and acidic amino acids migrate in opposite directions at a medium pH range. Recently, our group has developed global and quantitative amino acid analysis methods using quadrupole (6), triple-quadrupole tandem- (7) or time-of-flight mass spectrometers (8). To achieve the simultaneous analysis of amino acids, 1 M formic acid (pH lower than 2) was employed. Since the isoelectric points (p*I*) of most amino acids range from 2.77 to 10.76, every amino acid was positively charged below a pH value of 2.77 and thus migrated toward the cathode which was coupled to the mass spectrometer and then detected with high selectivity and sensitivity. The choice of the sheath liquid parameters was also very important in developing a CE-MS method. The optimized conditions are described in Subheading 3. This chapter describes the technical details of the CE-MS method for amino acid analysis using human tumor tissues (9).

2. Materials

2.1. Chemicals

1. All chemicals are of analytical or reagent grade.
2. All solutions should be prepared using Milli-Q water by purifying deionized water to attain a resistivity of 18 MΩ cm at 25°C.

2.2. Amino Acid Standard Mixture

1. Individual stock solutions of amino acids and internal standard (methionine sulfone), at a concentration of 10 or 100 mM was prepared in water, 0.1 N HCl, or 0.1 N NaOH, depending on the nature of the compound (see Note 1). All stock solutions were stored at 4°C.
2. The working standard mixture was prepared by diluting these stock solutions with water immediately prior to use.

2.3. CE-MS Components

1. Capillary conditioning and running buffer: 1 M formic acid. Weigh 4.6 g formic acid and transfer to a 100-mL measuring flask. Add water to a marked line. Mix well and store at 4°C until use.
2. Calibration standard solution: 100 μM hexakis (2,2-difluorothoxy) phosphazene (SynQuest Laboratories). Weigh 6.2 mg hexakis (2,2-difluorothoxy) phosphazene and transfer to a 100-mL measuring flask. Add methanol to a marked line. Mix well and store at 4°C until use.

3. Sheath liquid: 50% (v/v) methanol–water containing 0.1 μM hexakis (2,2-difluorothoxy) phosphazene. Mix 25 mL each of methanol and water and add 50 μL of calibration standard solution. Store at 4°C until use.

2.4. Instrumentation

1. We routinely perform CE-TOFMS experiments using the Agilent CE capillary electrophoresis system (Agilent Technologies, Waldbronn, Germany), the Agilent G3250AA LC/MSD TOF system (Agilent Technologies, Palo Alto, CA), the Agilent 1100 series isocratic HPLC pump, the G1603A Agilent CE-MS adapter kit, and the G1607A Agilent CE-ESI-MS sprayer kit (see Note 2).
2. The CE system control was performed using the G2201 AA Agilent ChemStation software, while TOFMS control, data acquisition, and evaluation were performed with the MassHunter software (Agilent Technologies).
3. Cell disrupter (MS-100R; TOMY).

3. Methods

3.1. Sample Preparation

1. To extract amino acids from tissues, preweighed deep-frozen samples (approximately 50 mg each) were completely homogenized by a cell disrupter at 2°C after adding 500 μL of methanol containing 20 μM methionine sulfone as an internal standard.
2. The homogenate was then mixed with 200 μL of water and 500 μL of chloroform and centrifuged at 9,000 × *g* for 15 min at 4°C.
3. The 600 μL of aqueous solution was centrifugally filtered through a 5-kDa cutoff filter to remove proteins. The 480 μL of filtrate was centrifugally concentrated and dissolved in 50 μL water immediately before CE-TOFMS analysis.

3.2. CE-TOFMS Conditions

1. Prior to first use, a new capillary should be rinsed with the capillary conditioning buffer for 20 min (see Note 3). In addition, before each injection, the capillary should be equilibrated for 5 min by flushing with the capillary conditioning buffer.
2. Separations are carried out on a fused-silica capillary (50 μm I.D. × 100 cm total length) (see Note 3).
3. The sample is hydrodynamically injected with a pressure of 50 mbar for 3 s (ca. 3 nL) (see Notes 5 and 6). The applied voltage was set at 30 kV and the capillary temperature was maintained at 20°C during analysis.
4. The sheath liquid, 50% (v/v) methanol–water that contained 0.1 μM hexakis (2,2-difluorothoxy) phosphazene, was delivered

at 10 μL/min using the HPLC pump equipped with a 1:100 splitter.

5. ESI-TOFMS was conducted in the positive ion mode and the capillary voltage was set at 4,000 V.
6. The flow of heated dry nitrogen gas (heater temperature of 300°C) and nebulizer gas flow were maintained at 10 L/min and 10 psig, respectively.
7. The fragmentor voltage, skimmer voltage, and octopole radio frequency voltage (Oct RFV) were set at 75, 50, and 125 V, respectively.
8. An automatic recalibration function was performed by using two reference masses of reference standards; (^{13}C isotope ion of protonated methanol dimer $[2MeOH+H]^+$, *m/z* 66.0632) and ($[$hexakis (2,2-difluorothoxy)phosphazene$+H]^+$, *m/z* 622.0290). Exact mass data were acquired at the rate of 1.5 cycles/s over a 50–1,000 *m/z* range.
9. Figure 1 shows typical electropherogram and Table 1 shows the reproducibility, linearity, and sensitivity of individual amino acids.

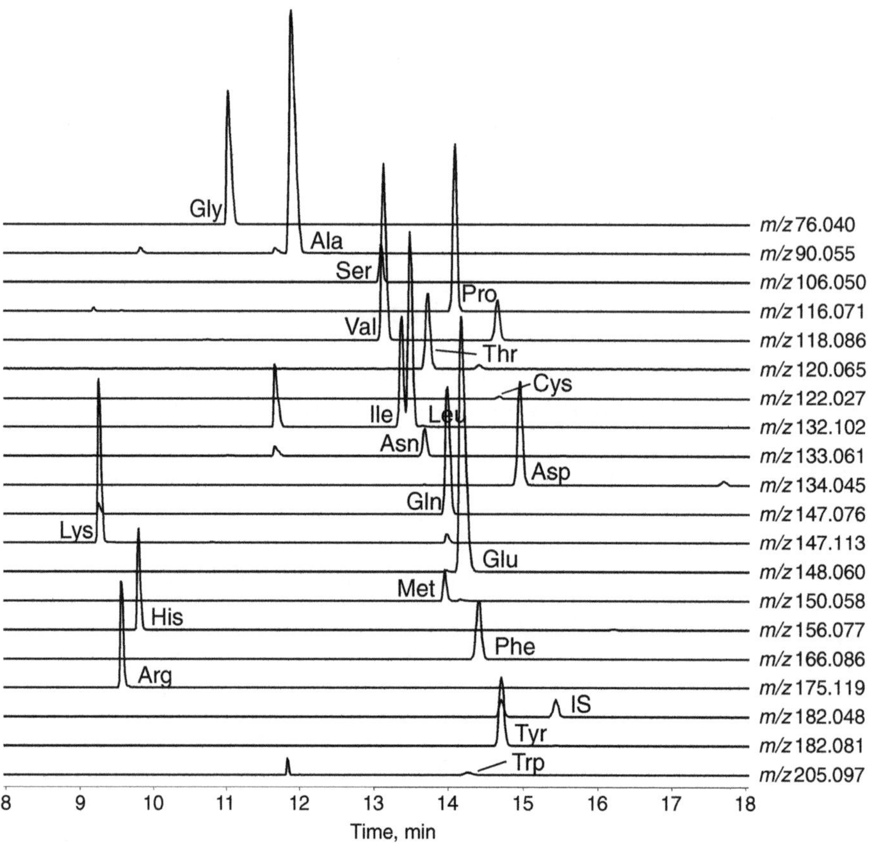

Fig. 1. Selected CE-TOFMS electropherograms for amino acids in human stomach tumor tissue. The numbers on the *right* indicate the exact mass of each amino acid obtained in this analysis.

Table 1
Reproducibility, linearity, and sensitivity of amino acid detection by CE-TOF-MS

	RSD ($n=6$)(%)			
Amino acid	Migration time	Peak area	Linearity correlation	Detection limit (μmol/L)
Gly	0.6	2.3	0.999	0.4
Ala	0.6	1.9	0.999	0.4
Ser	0.7	2.5	0.999	0.5
Pro	0.7	2.7	0.999	0.2
Val	0.7	3.2	0.999	0.6
Thr	0.7	3.0	0.999	0.3
Cys	0.8	6.4	0.998	0.6
Ile	0.7	6.0	0.996	0.1
Leu	0.7	4.7	0.999	0.1
Asn	0.7	2.7	0.999	0.3
Asp	0.8	2.9	0.997	0.3
Gln	0.7	3.1	0.997	1.4
Lys	0.6	1.9	0.999	0.1
Glu	0.8	2.1	0.999	0.3
Met	0.7	2.8	0.994	0.1
His	0.6	5.6	0.998	0.1
Phe	0.8	6.1	0.997	0.1
Arg	0.6	4.2	0.999	0.1
Tyr	0.8	3.2	0.999	0.2
Trp	0.8	6.3	0.997	0.1

4. Notes

1. The concentration and solvent for individual stock solutions of amino acids are as follows: Ala: 100 mM, water; Arg: 100 mM, 0.1 N HCl; Asn: 100 mM, 0.1 N HCl; Asp: 100 mM, 0.1 N NaOH; Cys: 10 mM, 0.1 N HCl; Gln: 100 mM, 0.1 N NaOH; Glu: 100 mM, 0.1 N HCl; Gly: 100 mM, water; His: 100 mM, 0.1 N HCl; Ile: 100 mM, 0.1 N HCl; Leu: 100 mM, water; Lys: 100 mM, water; Met: 100 mM, 0.1 N HCl; Phe: 100 mM, 0.1 N HCl; Pro: 100 mM, 0.1 N HCl; Ser: 100 mM, water;

Thr: 100 mM, water; Trp: 100 mM, 0.1 N HCl, Tyr: 10 mM, 0.1 N NaOH; Val: 100 mM, 0.1 N HCl; methionine sulfone (internal standard): 10 mM, water.

2. To avoid a siphoning effect, the CE inlet vial should be at the same height as the sprayer tip of the mass spectrometer.
3. We recommend avoiding capillary conditioning with sodium hydroxide because this degrades performance for this application.
4. The electrospray performance depends on the quality of the capillary cut. Jagged edges prevent the formation of a uniform spray and can also act as adsorption sites for sample components. We recommend using a diamond blade cutter (e.g., 5183–4669 CE column cutter; Agilent) or a fiber cleaving tool (INFOCUT cleaving tool; SEDI Fibres Optiques, France).
5. We recommend using polypropylene sample vials rather than glass vials.
6. If the current suddenly drops or broadened peaks are observed, the sample should be further diluted with water to reduce its conductivity and then reanalyzed.

References

1. Heinrikson RL, Meredith SC (1984) Amino acid analysis by reverse-phase high-performance liquid chromatography:precolumn derivatization with phenylisothiocyanate. Anal Biochem 136: 65–74.
2. Einarsson S et al (1983) Determination of amino acids with 9-fluorenylmethyl chloroformate and reversed-phase high-performance liquid chromatography. J Chromatogr A 282: 609–618.
3. Moore S, Stein WH (1951) Chromatography of amino acids on sulfonated polystyrene resins. J Biol Chem 192: 663–681.
4. Spackman DH et al (1958) Automatic Recording Apparatus for Use in Chromatography of Amino Acids. Anal Chem 30: 1190–1206.
5. Yoshida H et al (2007) Comprehensive analytical method for the determination of hydrophilic metabolites by high-performance liquid chromatography and mass spectrometry. J Agric Food Chem 55: 551–560.
6. Soga T, Heiger DN (2000) Amino acid analysis by capillary electrophoresis electrospray ionization mass spectrometry. Anal Chem 72: 1236–1241.
7. Soga T et al (2004) Qualitative and quantitative analysis of amino acids by capillary electrophoresis-electrospray ionization-tandem mass spectrometry. Electrophoresis 25: 1964–1972.
8. Soga T et al (2006) Differential metabolomics reveals ophthalmic acid as an oxidative stress biomarker indicating hepatic glutathione consumption. J Biol Chem 281: 16768–16776.
9. Hirayama A et al (2009) Quantitative metabolome profiling of colon and stomach cancer microenvironment by capillary electrophoresis time-of-flight mass spectrometry. Cancer Res 69: 4918–4925.

Chapter 9

New Advances in Amino Acid Profiling by Capillary Electrophoresis-Electrospray Ionization-Mass Spectrometry

Philip Britz-McKibbin

Abstract

Capillary electrophoresis-electrospray ionization-mass spectrometry (CE-ESI-MS) offers a selective, sensitive yet robust approach for amino acid profiling in complex biological samples with minimal sample pretreatment. Direct analysis of amino acids and their analogs is routinely performed using strongly acidic buffer conditions under positive-ion mode ESI-MS with a coaxial sheath liquid interface. New advances in online sample preconcentration, chemical derivatization, and/or ESI interface designs can further improve assay performance allowing for resolution of amino acid stereoisomers and labile aminothiols with low nanomolar detection limits. Accurate prediction of the electromigration behavior of amino acids offers a convenient approach for their qualitative identification complementary to ESI-MS. Simultaneous analysis of amino acids together with other classes of cationic metabolites can be realized by CE-ESI-MS for comprehensive metabolite profiling applications relevant to disease prognosis, drug efficacy, and food safety/quality control.

Key words: Amino acids, CE-ESI-MS, Chemical derivatization, Online sample preconcentration, Aminothiols, Amino acid analysis

1. Introduction

High efficiency separation of amino acids by capillary electrophoresis (CE) was first demonstrated by Jorgenson and Lukacs (1) when using an open-tubular fused-silica capillary format. The selectivity of zwitter-ionic amino acid separations in CE is determined by differences in apparent electrophoretic mobility ($\mu_{ep}{}^{A}$), which can be accurately predicted based on their characteristic weak acid dissociation constant(s) (pK_a) and absolute mobility (μ_o) under defined background electrolyte (BGE) conditions (2). CE offers a low cost yet high resolution microseparation platform for ultrasensitive analysis of amino acids when using laser-induced

Michail A. Alterman and Peter Hunziker (eds.), *Amino Acid Analysis: Methods and Protocols*, Methods in Molecular Biology, vol. 828, DOI 10.1007/978-1-61779-445-2_9, © Springer Science+Business Media, LLC 2012

fluorescence detection, including D/L-enantiomers (3) and posttranslational modifications (4). Moreover, sample preconcentration and chemical derivatization steps can be integrated directly in-capillary during electromigration without complicated off-line sample handling procedures (5). However, unambiguous identification and reliable quantification of amino acids in complex biological samples is limited by CE when using photometric or electrochemical detection (6)

CE-electrospray ionization-mass spectrometry (CE-ESI-MS) offers a powerful hyphenated technique for amino acid analysis without chemical labeling when using a strongly acidic BGE with positive-ion mode detection (7). The coaxial sheath liquid interface remains the most robust design for coupling CE to ESI-MS that generates a stable ion spray while allowing for independent optimization of separation and ionization conditions (8). Recent introduction of low flow/sheathless interfaces in CE-ESI-MS can further improve concentration sensitivity for detection of nanomolar levels of amino acids, peptides and protein by reducing postcapillary dilution effects (9, 10). However, the ionization efficiency of native amino acids have wide disparity in ESI-MS that can often vary over two orders of magnitude due to their different physicochemical properties influencing ion desorption (11). Given the poor ion responses of low molecular weight/hydrophilic aminothiols that are also susceptible to auto-oxidation and/or thiol-disulfide exchange, online sample preconcentration together with thiol-selective maleimide labeling offers a strategy for comprehensive thiol speciation in human plasma, including reduced thiols and intact oxidized disulfides (12). Similarly, CE-ESI-MS offers a specific method for confirmatory testing in expanded newborn screening programs as compared to LC-MS since resolution of amino acids, acylcarnitines and their major stereoisomers and/or isobars (e.g., *allo*-Ile, Leu, Ile, Hyp) that differ widely in their overall polarity can be achieved under a single elution condition without complicated sample workup (13). A recent inter-laboratory comparison of methods used for amino acid profiling in plant cell extracts demonstrated that CE-ESI-MS offers greater coverage of various classes of amino acids while allowing for higher sample throughput, lower operating costs, and reduced sample handling relative to GC-MS (14). Moreover, accurate prediction of analyte migration behavior by CE can facilitate the identification of novel amino acid analogs and/or peptides that supports accurate mass and isotopic pattern information by MS (15, 16). Herein, standardized protocols are described for reliable profiling of amino acids by CE-ESI-MS that are applicable to complex biological samples, including dried blood spots (DBS), plasma, cell lysates, and urine.

2. Materials

2.1. Chemicals

1. Barnstead EASYpure II LF ultrapure water system (Dubuque, USA).
2. Acetonitrile (ACN), methanol (MeOH), ammonium acetate, sodium chloride (NaCl), sodium hydroxide (NaOH), sodium dihydrogen phosphate (NaH_2PO_4), concentrated formic acid (>95% purity, Sigma-Aldrich Inc.).
3. *O*-acetyl-L-carnitine (C2), *N*-acetyl-L-cysteine (NAC), *S*-adenosyl-L-methionine (SAM), L-alanyl-L-alanine (diAla), L-arginine (Arg), L-aspartic acid (Asp), asymmetric dimethyl-L-arginine (ADMA), L-ascorbic acid (AA), Betaine (Bet), L-carnitine (C0), L-cysteine (Cys), L-cystine (CySS), L-cysteinyl-cysteinyl-glycine disulfide (CySSCysGly), L-cysteinyl-glutathione disulfide (CySSG), L-cysteinyl-L-glutamylcysteine disulfide (CySSGluCys), L-cysteinyl-glycine (CysGly), L-cysteinyl-glycine disulfide (CysGlySS), L-citrulline (Cit), dimethylglycine (DMG), L-glutamine (Gln), L-glutamic acid (Glu), L-glutamyl-cysteine (GluCys), glutathionine – oxidized (GSSG), glutathionine – reduced (GSH), L-histidine (His), L-homocysteine (Hcy), L-homocystine (HcySS), L-leucine (Leu), L-isoleucine (Ile), L-*allo*-isoleucine (*allo*-Ile), L-lysine (Lys), L-methionine (Met), 3-methyl-L-histidine (MeHis), *N*-methylmaleimide (NMM), 2-mercaptoethane-sulfonate (MESNA), monomethyl-L-arginine (MMA), *O*-myristoyl-L-carnitine (C14), L-ornithine (Orn), *O*-palmitoyl-L-carnitine (C16), 5-oxo-L-proline or pyroglutamic acid (5-oxo-Pro), L-phenylalanine (Phe), L-proline (Pro), *trans*-4-hydoxy-L-proline (4-OH-Pro), L-serine (Ser), symmetric dimethyl-L-arginine (SDMA), *N*-*tert*-butylmaleimide (TBM), L-threonine (Thr), L-tryptophan (Trp), L-tyrosine (Tyr) L-valine (Val) (>95% purity, Sigma-Aldrich Inc.).
4. *N*-(2-(trimethylammonium)ethyl)maleimide chloride (NTAM), L-cysteinyl-glutathione disulfide (CysSSG) (>95% purity, Toronto Research Chemicals Inc., Toronto, Canada).
5. Individual stock solutions of amino acids (10 mM) were diluted in 1:1 methanol/water, whereas all reduced thiol standards (e.g., GSH) were prepared fresh daily in 0.1% formic acid using degassed deionized water due to their susceptibility to auto-oxidation.
6. All solutions/buffers were degassed by sonication and amino acid stock solutions were stored at 4°C prior to use.
7. A quality control mixture consisting of ten standard amino acids was prepared for daily runs.

8. Polypropylene centrifuge tubes 0.5 mL and 1.5 mL (VWR International Inc., Toronto, Canada).
9. 3 kDa Nanosep® centrifugal filters (Pall Life Sciences Inc., Michigan, USA).
10. Unistik 3 disposable lancets (Owen Munford Ltd., Georgia, USA).
11. Grade 903 Protein Saver Card filter paper and 1/8th inch diameter hole puncher (Whatman Inc., New Jersey, USA).
12. Excel 2007 (Microsoft Inc. WA, USA) and Igor 5.0 (Wavemetrics Inc., OR, USA) software were used for all data processing and statistical analyses.

2.2. CE-ESI-MS System with Coaxial Sheath Liquid Interface

1. Agilent 7100 CE system (see Note 1) equipped with an XCT 3D ion trap or 6224 series TOF-MS, an Agilent 1100 series isocratic pump, and a G16107 CE-ESI-MS coaxial sheath-liquid sprayer interface (Agilent Technologies Inc., Santa Clara, USA).
2. Flexible open-tubular fused-silica capillaries (Polymicro Technologies Inc., Phoenix, AZ) with total length of 80 cm and internal diameter of 50 μm were prepared (see Note 2) using a ShortixTM diamond capillary cutter (Sigma-Aldrich Inc., St. Louis, USA).

3. Methods

Amino acid profiling by CE-ESI-MS is optimally performed using a volatile acidic BGE, 1 M formic acid, pH 1.8 under suppressed electroosmotic (EOF) conditions in conjunction with positive-ion mode detection. This not only ensures that zwitter-ionic amino acids are adequately ionized (i.e., positive mobility) given the strong acidity of α-carboxylic acid moieties ($pK_a \approx 1.8$–2.4), but also avoids the pH range at which the EOF tends to be more unstable (pH ≈ 4–8). Various dynamic coating procedures have been introduced to improve migration time reproducibility in CE-MS (17); however, normalization of amino acid migration times using one or more internal standards provides excellent long-term precision under 2% when using inexpensive bare fused-silica capillaries (15, 16). In addition, online sample preconcentration with desalting can be readily integrated during electromigration in order to enhance concentration sensitivity of submicromolar levels of amino acids in complex biological samples prior to ESI-MS without matrix-induced ion suppression (11, 15, 18). To date, several different types of mass analyzers have been used with a coaxial sheath liquid interface, including tandem MS (7), 3D ion trap (11), and Fourier-transform ion cyclotron MS (19); however, the high mass accuracy, moderate resolution and fast duty cycle of time-of-flight (TOF)-MS offers an ideal configuration for reliable quantification and qualitative identification of amino acids that migrate as narrow zones in CE (16).

In some cases, modifications to the CE-ESI-MS methodology are required when analyzing certain classes of labile/low abundance amino acids and their stereoisomers. For example, better selectivity and/or improved sensitivity can be achieved when using alkaline BGE conditions in CE with negative-ion mode ESI-MS for acid-labile amino acids analogs, such as sulfotyrosine (20). Also, in-capillary oxidation of labile aminothiols at the capillary inlet (i.e., anode) can introduce bias when quantifying oxidized disulfides using CE-ESI-MS given the high abundance of reducd glutathione (GSH) in cell lysates; this artifact is corrected by displacing the sample plug from the electrode prior to voltage application (21). Indeed, rigorous validation of sample collection, pretreatment and storage procedures are critical for accurate determination of amino acids in complex biological samples (12, 21). Resolution of amino acid enantiomers by CE require the addition of a chiral selector in the BGE, which often contributes to detrimental ion suppression and reduced sensitivity when using ESI-MS (20). In addition, since most amino acids are highly hydrophilic, chemical derivatization is often required to enhance their affinity with chiral selectors for optimal resolution, such as modified cyclodextrins (22). Alternatively, 18-crown tetracarboxylic acid has a promising chiral selector that functions as both the BGE and complexing agent for native amino acids with enhanced ionization efficiency since ions are monitored as their crown ether-amino acid complex (23). In this report, robust methodologies for routine achiral analysis of amino acids, aminothiols, and their major diastereomers by CE-ESI-MS are described, including simple yet validated protocols for sample workup of cells and biofluids.

3.1. Sample Pretreatment of Dried Blood Spots

1. Collect human blood samples using a finger-prick method via disposable lancets and spot them on a Grade 903 Protein Saver Card and dry overnight.
2. Punch out a 3.2 mm (1/8 in.) disk (≈3.4 μL) manually from each dried blood spot (DBS) with a hole puncher into a 0.5 mL centrifuge tube that contained 100 μL of ice-cold 1:1 MeOH:H_2O with the internal standard dialanine (diAla, 100 μM).
3. Extract the DBS under sonication for 10 min.
4. Filter the resulting extract solution through a 3 kDa Nanosep centrifugal filter at 150 × *g* at 4°C for 10 min prior to analysis (see Note 3).
5. Dilute the filtrate 1:1 with an aqueous ammonium acetate solution (400 mM, pH 7.0) to produce the final sample

solution used for analysis (200 mM ammonium acetate, 25% MeOH, 50 μM dialanine).

6. In most cases, amino acids derived from filtered DBS extracts are analyzed directly by CE-ESI-MS without sample pretreatment steps, such as chemical derivatization, lyophilization, and/or solid-phase extraction (see Note 3).

3.2. Sample Pretreatment of Human Plasma and Red Blood Cell Lysates

1. Collect human blood samples using a finger-prick method (see above) or via a venous catheter inserted in the ante-cubital vein that is kept patent using a saline (0.9% w/v) solution.
2. Immediately place each blood sample on ice and subsequently centrifuge at 20 × *g* at 4°C for 5 min to fractionate plasma from erythrocytes.
3. Transfer the plasma supernatant and remove plasma proteins using a 3 kDa Nanosep centrifugal filter at 150 × *g* for 10 min prior to dilution and subsequent analysis (see Note 4).
4. Wash the red blood cells (RBCs) with phosphate buffered saline (10 mM NaH_2PO_4, 150 mM NaCl, pH 7.4, 4°C), then vortex and centrifuge at 70 × *g* at 4°C for 1 min to isolate RBCs.
5. Repeat the washing steps three times until the supernatant is clear without any evidence of premature hemolysis.
6. Hemolyze 200 μL of RBCs by adding 600 μL of prechilled deionized water, then centrifuge at 70 × *g* at 4°C for 1 min to sediment cell debris.
7. Next, filter 100 μL of RBC lysate with a 3 kDa filter 150 × *g* at 4°C for 10 min to remove excess hemoglobin.
8. Dilute all filtered protein-free RBC lysates twofold in BGE containing 50 μM diAla as internal standard for positive-mode ESI-MS and store frozen at −80°C. Thaw only once prior to analysis (see Note 5).

3.3. Sample Pretreatment of Single-Spot Urine Collection

1. Collect human urine mid-stream as a first void morning sample and store at 4°C, where creatinine is used to correct for urine dilution effects (see Note 6).
2. Centrifuge the urine samples at 70 × *g* at 4°C for 2 min to remove particulate matter, and then dilute the samples tenfold using deionized water containing 50 μM diAla as internal standard when using CE-ESI-MS under positive ion mode.

3.4. Capillary Conditioning, Buffer Preparation, and CE-ESI-MS Operation

1. Prior to first use, condition fused-silica capillaries installed in the coaxial sheath liquid interface (see Note 7) for 15 min each with MeOH, 1 M NaOH, deionized H_2O followed by rinsing with 1 M formic acid for 60 min prior to usage.

2. Perform several trial runs with the newly conditioned capillary using a quality control test mixture as the sample and the desired BGE until stable currents and reliable ion signals are attained (see Note 8).
3. Once conditioning is complete, perform a 10 min prerinse/flush of the capillary with BGE prior to each separation.
4. High purity (>98%) volatile buffer reagents used for CE separations were prepared on a weekly basis, filtered, and sonicated prior to use.
5. In most cases, 1.0 M formic acid, pH 1.8 (diluted from concentrated formic acid) can be used as the aqueous acidic BGE for amino acids, aminothiols, and other classes of cationic metabolites (see Note 9).
6. Perform CE separations at 20°C with an applied voltage of 30 kV unless otherwise indicated (see Note 10).
7. The sheath liquid consisting of 1:1 MeOH:H_2O with 0.1% v/v formic acid is supplied by the 1100 series isocratic pump at a flow rate of 10 μL/min.
8. Nitrogen serves as both a nebulizing and a drying gas supplied at 6 psi and 10 L/min, respectively.
9. Perform all ESI-MS analyses using a +4 kV cone voltage in positive-ion mode at a temperature of 300°C unless otherwise stated.
10. Record full-scan MS data within a range of 50–750 *m/z* (see Note 11).

3.5. Comprehensive Amino Acid Profiling by CE-ESI-MS

1. Introduce all samples to the capillary after a two- or tenfold dilution in 100 mM ammonium acetate, pH 7.0 using a low-pressure hydrodynamic injection by first injecting a sample for 75 s at 50 mbar (see Note 12) followed by a 60 s injection of the acidic BGE (1 M formic acid, pH 1.8) at 50 mbar prior to voltage application (see Note 13).
2. Perform all separations for amino acids using bare fused-silica capillaries under normal polarity by CE with positive-ion mode ESI-MS (see Note 14).
3. In most cases, all analyses should be performed in triplicate using a fresh BGE reservoir (inlet/anode) for each run. Data should be processed (e.g., peak integration) using extracted ion electropherograms (EIE) based on the *m/z* for the protonated molecular ion ($M+H^+$) of each amino acid relative to an internal standard (see Note 15).
4. Perform a prerinse with acidic BGE for 10 min after each run in order to recondition the capillary.

4. Notes

1. Alternative CE systems can also be coupled with other mass analyzers using recent low flow/sheathless interface designs.
2. The polyimide coating at the distal end of the capillary (≈2–3 cm) that is directed toward the ion source was removed by burning it with a match and subsequent cleaning using a Kimwipe wetted with a few drops of methanol (MeOH). This capillary was then installed into the coaxial sheath liquid interface to generate a bare fused-silica tip that served as the effective emitter in the ion source.
3. The solvent of the DBS extract can also be evaporated under a gentle stream of N_2 and reconstituted in 20 μL of sample solution in order to improve concentration sensitivity, which can provide additional tenfold sample enrichment prior to CE-ESI-MS (13). Reduced thiols (e.g., Cys, GSH) are not reliably analyzed when using conventional DBS collection methods due to auto-oxidation when exposed to ambient air unless maleimide labeling is used.
4. Ultrafiltration is the deproteinization method of choice when processing whole blood or cell extracts without oxidation artifacts, which are induced by protein denaturation (e.g., oxygenases, hemoglobin) with acid or organic solvent precipitation (21).
5. Blood collection and sample workup should be processed within about 2 h while blood specimens are kept on ice (4°C) at all times using degassed/prechilled solutions in order to avoid premature hemolysis and oxidation artifacts when measuring unlabeled reduced thiols (e.g., GSH).
6. The creatinine concentration is used to normalize urinary amino acid concentration levels (i.e., μmol/mmol creatinine) that corrects for dilution effects when using single-spot morning urine samples assuming no underlying kidney dysfunctions.
7. Reproducible alignment of the capillary emitter into the ion source was facilitated by the Agilent coaxial sheath liquid ESI-MS interface, where the distal end of the bare fused-silica capillary was allowed to protrude from the sprayer by about 0.1 mm in order to minimize postcapillary dilution effects.
8. The average capillary lifespan is about 3 weeks when analyzing complex biological samples by CE-ESI-MS under continuous operation. In most cases, capillary replacement is made evident by an unstable capillary current and/or increasing noise in the total ion electropherogram that cannot be resolved by subsequent conditioning or rinsing with BGE. The incorporation of amino acid quality control test runs randomly throughout the

day is strongly recommended to confirm the lack of instrumental drift. Also, periodic cleaning of sprayer assembly and ion source is recommended to improve long-term performance.

9. In some cases, a small fraction of organic solvent may be added to the BGE (e.g., 15% v/v acetonitrile or methanol) in order to improve the solubility of hydrophobic analytes or suppress micellar formation of surface-active analytes, such as long-chain acylcarnitines (13).

10. In all cases, the current generated during voltage application in CE should be kept under 50 μA in order to minimize Joule heating effects and improve ion signal stability in ESI-MS.

11. All full-scan MS data were acquired at a rate of 10 spectra/s. The 3D ion trap was operated using an ultrascan mode of 26,000 *m/z* per second with helium at 6×10^{-6} mbar as the damping gas. TOF-MS was performed with the fragmentor, skimmer and Oct RFV voltage set at 75 V, 50 V, and 125 V, respectively. The methanol adduct ion $(2MeOH+H_2O+H)^+$, *m/z* 83.0703) and reserpine (0.5 μM, $(M+H)^+$, *m/z* 609.2806) provided the lock mass for exact mass measurements which were delivered by the sheath liquid at 10 μL/min.

12. Hydrodynamic injection of a long sample plug (≈10% of capillary length) allows for online sample preconcentration of amino acids directly in-capillary during electromigration prior to ionization based on differences in co-ion electrolyte mobility and pH at the sample and BGE interface. Up to a 50-fold improvement in concentration sensitivity can be realized by this method without compromising separation efficiency or resolution when using conventional instrumentation (15).

13. The second injection sequence is used to displace the original sample plug within the capillary past the electrode interface at the inlet (anode), which is required to avoid CE-induced oxidation artifacts when analyzing low micromolar levels of oxidized glutathione (GSSG) in the presence of excess reduced GSH derived from filtered cell lysates when not using maleimide labeling (21, 24).

14. Under these conditions, neutral metabolites (e.g., urea) co-migrate with the suppressed EOF (>20 min), whereas strongly acidic anions (e.g., chloride) migrate out of the capillary at the inlet upon voltage application. The relative migration time (RMT) of an ion can be accurately predicted in CE based on its characteristic μ_o and thermodynamic pK_a as derived from its putative chemical structure (15), which provides a novel strategy for identification of amino acids complementary to ESI-MS.

15. In most cases, amino acids are detected as their singly charged protonated molecular ion $(M+H^+)$ with the exception of pep-

tides such as GSSG, which predominately forms a divalent molecular ion ($M+2\ H^{2+}$) in the gas-phase. Optimization of ionization conditions can reduce the extent of in-source fragmentation that can impact sensitivity.

16. Stereoselective analysis of *allo*-Ile from DBS extracts allows for specific diagnosis of Maple Syrup Urine Disease without false-positives associated with total Leu, whereas resolution of 4-OH-Pro prevents misdiagnosis of hydroxyprolinemia, a benign disorder. Similarly, the two positional isomers of symmetric and asymmetric dimethylarginine can also be resolved by CE prior to ESI-MS (14).
17. A variety of classes of cationic metabolites can be analyzed by this method, including amino acids, (biogenic) amines, peptides, acylcarnitines and nucleosides, where the relative response factor of ions can be predicted based on their fundamental physicochemical properties in CE-ESI-MS (11).
18. As an alternative procedure to Subheading 3.2, filtered human plasma were processed promptly from whole blood specimens followed by thiol-selective maleimide labeling at 4°C in 200 mM ammonium acetate, pH 5.0. This chemical derivatization strategy is both rapid (<2 min) and selective under weakly acidic conditions even at 4°C resulting in generation of a stable thioether adduct that is needed for quantification of nanomolar levels of labile reduced thiols in plasma, such as free Hcy (12). Greater sensitivity and faster analysis times was achieved when using NTAM (1 mM) as a cationic maleimide label for reduced thiols. Excess NTAM was subsequently quenched with MESNA (1.3 mM, an acidic synthetic thiol) to form a neutral adduct that does not interfere with NTAM-labeled thiols. This was then followed by NMM (0.5 mM) addition in order to avoid thiol-disulfide exchange reactions of endogenous oxidized disulfides with residual MESNA. All three steps of the reaction were performed in series at 4°C in the same diluted plasma sample after mixing for about 2 min at each stage. Alternatively, a neutral maleimide analog (TBM) offers a simpler approach without subsequent quenching steps since excess maleimide does not interfere with thiol-maleimide adducts as it co-migrates with the EOF.
19. Figure 1 depicts the coaxial sheath liquid interface configuration used in CE with an acidic BGE (pH 1.8) under positive-ion mode ESI-MS for the analysis of amino acids and other cationic metabolites (e.g., amines, acylcarnitines, nucleosides, etc.), which migrate in order of their apparent charge density (see Note 14).

    Figure 2 illustrates the mechanism of online preconcentration and desalting by CE when injecting long sample plugs of

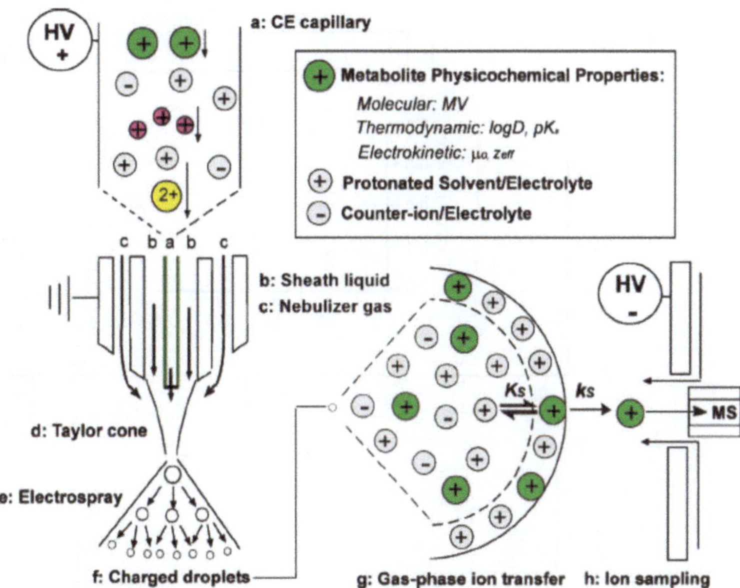

Fig. 1. High efficiency separation of amino acids by CE with positive-ion mode ESI-MS using a coaxial sheath liquid interface. Decoupling of separation and ionization conditions allows for a stable electrospray formation, where cationic amino acids are desorbed into the gas-phase depending on their intrinsic physicochemical properties, such as molecular volume (*MV*), octanol-water partition coefficient (*logD*) and absolute mobility (μ_0) or effective charge state (z_{eff}). (Reproduced from ref. 11 with permission from ACS).

zwitter-ionic amino acids (e.g., Trp) in the presence of major involatile salts (e.g., Na^+) prior to ESI-MS (15). This method is readily compatible with standard instrumentation by preparation of the sample in a neutral/weakly acidic ammonium acetate buffer (pH 5–7), which boosts sensitivity for submicromolar detection of amino acids in complex biological matrices without ion suppression (see Note 12).

Figure 3 depicts the simultaneous analysis of polar amino acids (e.g., Arg) and surface-active acylcarnitines (e.g., palmitoyl-L-carnitine, C16) under a single elution condition by CE-ESI-MS relevant to expanded newborn screening of inborn errors of metabolism (13), which also allows for the resolution of isomeric (e.g., Ile, *allo*-Ile) and isobaric (e.g., 4-OH-Pro) co-ion interferences (see Note 16). Ion suppression from major co-ions from cell lysates is also minimized since excess sodium ions (Na^+) and reduced GSH are fully resolved from other amino acids (see Note 17) by CE prior to ESI-MS (11).

Figure 4a depicts the altered injection sequence in CE required to avoid oxidation artifacts of unlabeled reduced aminothiols for unbiased determination of intracellular half-cell reduction potential based on the GSH/GSSG redox couple (21). Figure 4b demonstrates that this validated method can

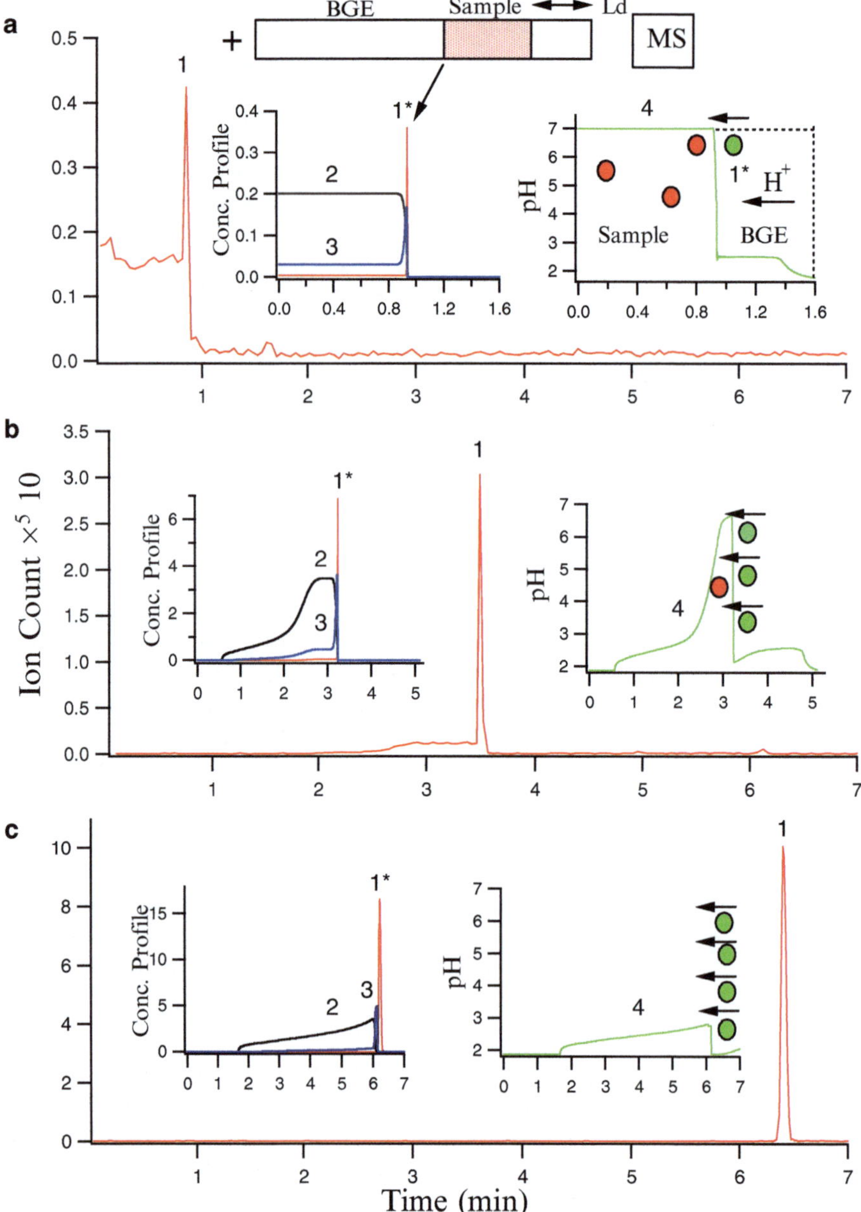

Fig. 2. Time-resolved electropherograms highlighting the dynamics of online sample preconcentration/desalting of Trp via dynamic pH junction/transient isotachophoresis based on experimental data using CE-ESI-MS and computer simulations (*insets*). A long sample plug of 12.5 cm was placed at increasing distance from the distal end of the capillary (L_d) from about (**a**) 7.5, (**b**) 25, and (**c**) 40 cm. Direct sample enrichment and desalting of complex biological samples can be realized during electromigration without changes in instrumental setup or complicated sample pretreatment. Conditions: BGE: 1 M formic acid, pH 1.8, sample: 200 mM ammonium acetate, pH 7.0, 15 mM NaCl. Analyte peak numbering corresponds to the concentration profiles of 1-Trp (experimental), 1^* – Trp (simulated), 2 – NH_4^+, 3 – Na^+, and 4 – pH. (Reproduced from ref. 15 with permission from ACS).

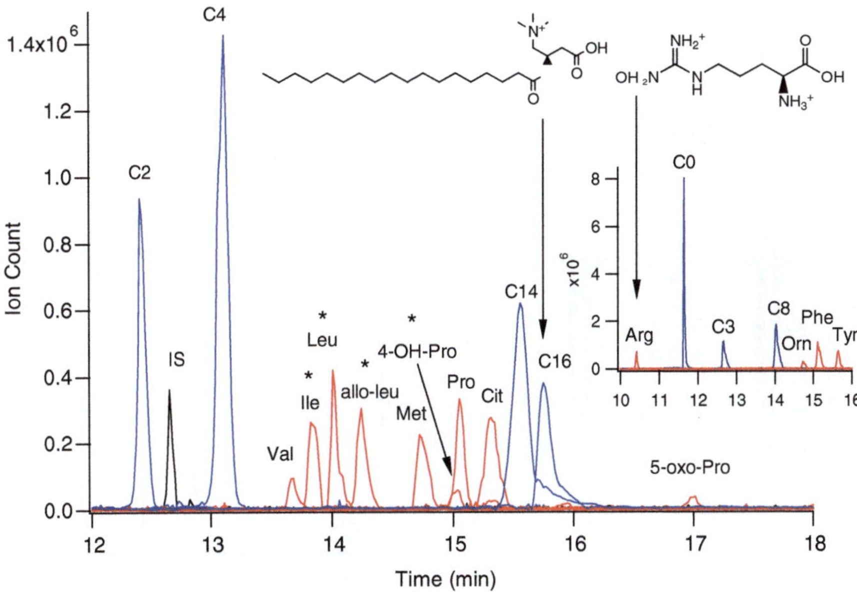

Fig. 3. Extracted ion electropherograms depicting recovery studies for a filtered dried blood spot extract spiked at 12.5% maximum calibration level with 20 different amino acid and acylcarnitine metabolites commonly targeted in expanded newborn screening programs. Note that metabolites of widely different polarity can be analyzed simultaneously by CE-ESI-MS (e.g., Arg and C16) in free solution while providing resolution of (*) isomeric and isobaric ions, including Leu, Ile, *allo*-Ile, and 4-OHPro. (Reproduced from ref. 13 with permission from ACS).

accurately measure the dynamics of strenuous exercise and subsequent recovery on a cellular level as a way to evaluate the efficacy of high-dose antioxidant pretreatment for attenuation of exercise-induced oxidative stress on a human volunteer (24).

Figure 5a highlights comprehensive thiol speciation in human plasma by CE-ESI-MS over a wide dynamic range with unprecedented selectivity and sensitivity, including low abundance reduced thiols (e.g., Hcy, GSH) as well as intact oxidized thiols, namely symmetric (e.g., CysSS, GSSG) and mixed (e.g., CysSSCysGly, CysSSG) disulfides. Thiol-selective maleimide labeling is performed on filtered plasma samples as a way to enhance ionization efficiency and improve robustness for quantification of nanomolar levels of labile reduced thiols without complex sample handling (see Note 18). Figure 5b also demonstrates that amino acids and various other classes of cationic metabolites can also be measured simultaneously by this approach that is useful for untargeted metabolite profiling applications (12).

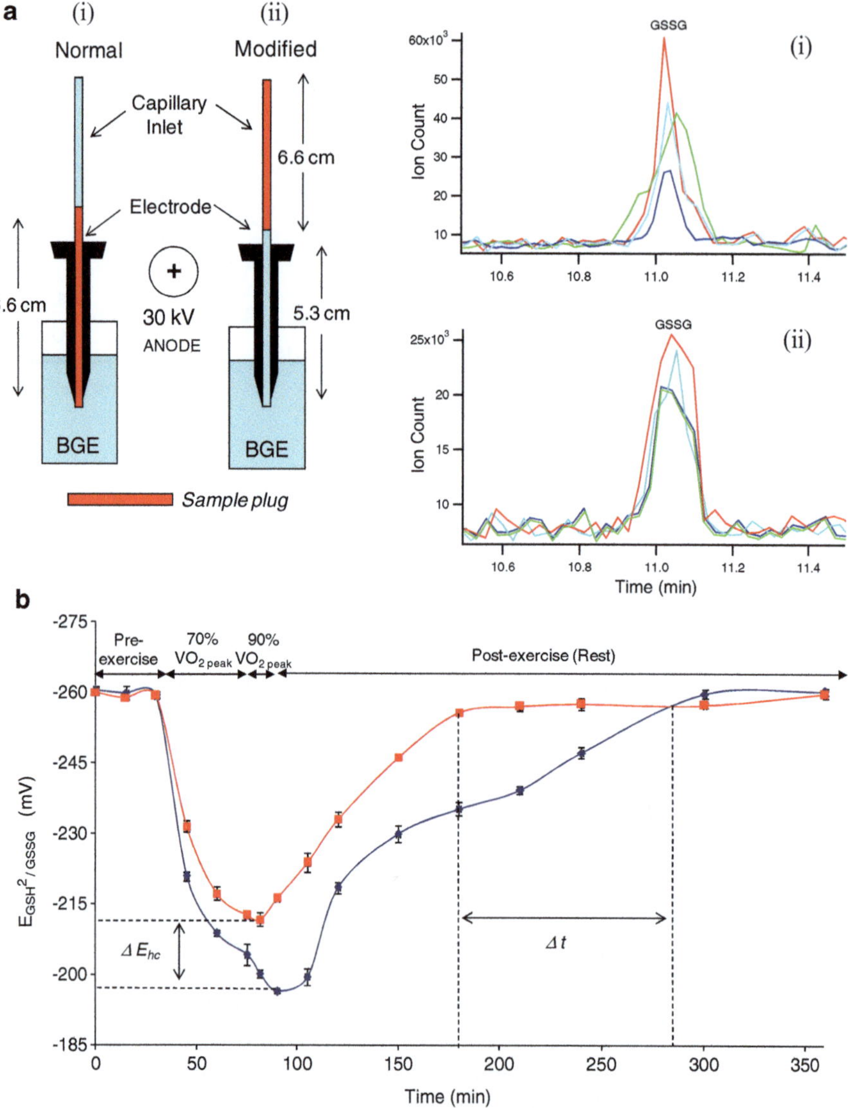

Fig. 4. (**a**) A comparison of (i) normal and (ii) modified hydrodynamic sample injection sequences for avoiding oxidation artifacts at the anodic electrode interface for reliable analysis of low micromolar levels of GSSG ($MH_2^{2+} = 307$ *m/z*) in the presence of a 500-fold excess of GSH by CE-ESI-MS. (**b**) Overlay traces of the in vivo half-cell reduction potential of intracellular glutathione or $E_{GSSG}/2_{GSH}$ derived from filtered red blood cell lysates as a function of time for control and NAC trials, where a greater $E_{GSSG}/2_{GSH}$ at peak oxidation ($\Delta E_{peak} \approx +15$ mV) and a shorter recovery time to redox homeostasis ($\Delta t_{rec} \approx 100$ min) was measured with NAC pretreatment relative to the control for the same subject measured at four key time intervals during standardized exhaustive cycling regimes, namely pre-exercise (0–30 min), 70% $0_{2\ peak}$ (30–75 min), 90% $0_{2\ peak}$ (75–81.5 min) and post-exercise (81.5–360 min) while at rest. (Reproduced from refs. 21 and 24 with permission from ACS).

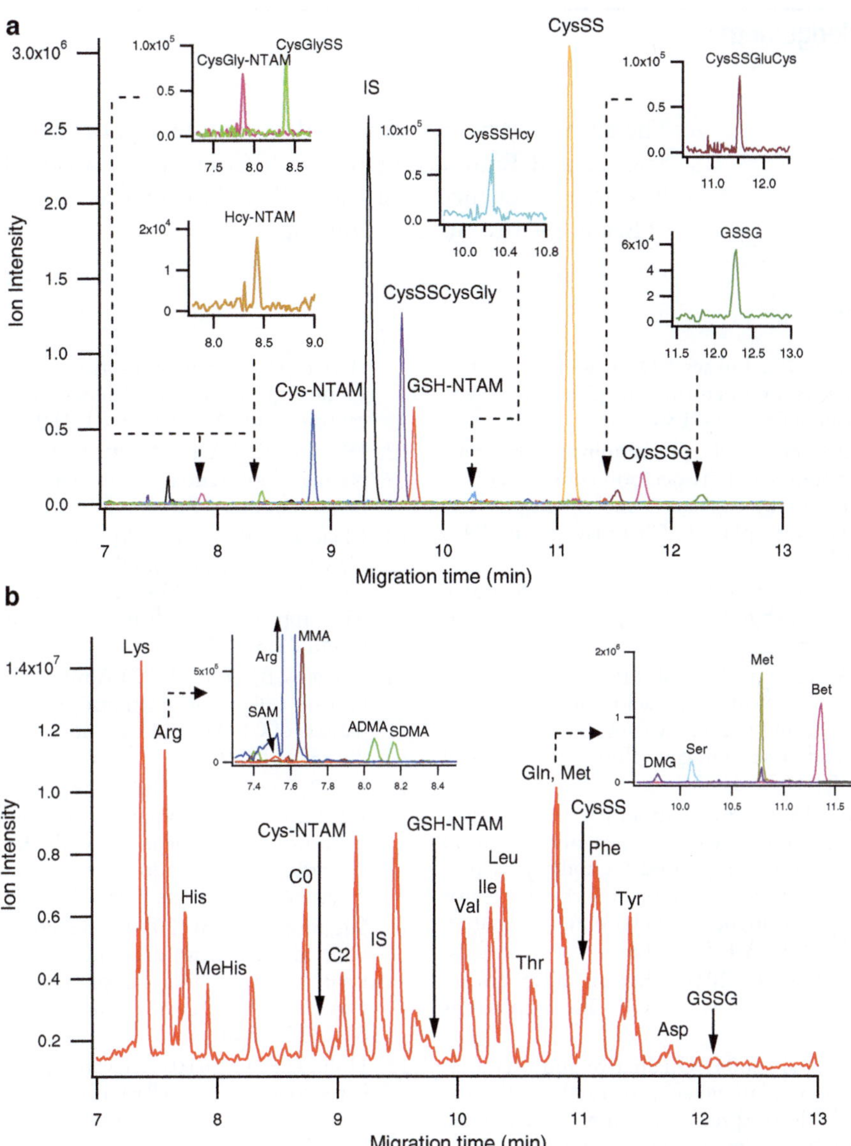

Fig. 5. (**a**) Series of extracted ion electropherograms (EIEs) demonstrating comprehensive assessment of plasma thiol redox status by CE-ESI-MS. Filtered plasma samples were diluted threefold prior to NTAM derivatization in conjunction with online sample preconcentration by CE-MS for simultaneous analysis of low abundance thiols (e.g., Hcy, GSH), as well as intact symmetric (e.g., CysSS, GSSG) and mixed oxidized disulfides (e.g., CysSSCysGly, CysSSG) over a wide dynamic range. (**b**) Overlay of the total ion electropherogram for the same plasma specimen that highlights integration of untargeted metabolite profiling for deeper insight into metabolic pathways modulated by thiol dysregulation, including of various classes of polar metabolites/cofactors related to thiol-amino acid metabolism, including methionine (*Met*), betaine (*Bet*), dimethylglycine (*DMG*), *S*-adenosylmethionine (*SAM*), and asymmetric dimethylarginine (*ADMA*). (Reproduced from ref. 12 with permission from ACS).

Acknowledgements

The author wishes to acknowledge funding support from National Science and Engineering Research Council of Canada, Premier's Research Excellence Award, as well as a Japan Society for Promotion of Science – Invitation Fellowship.

References

1. Jorgenson JW, Lukacs KD (1998) Zone electrophoresis in open-tubular glass capillaries. Anal Chem 53:1298–1302
2. Zuskova I et al (2006) Determination of limiting mobilities and dissociation constants of 21 amino acids by capillary zone electrophoresis at very low pH. J Chromatogr B 841: 129–134
3. Sanchez-Hernandez L et al (2008) Sensitive chiral analysis by CE: An update. Electrophoresis 29:237–251
4. Britz-McKibbin P et al (1999) Analysis of gamma-carboxyglutamic acid content of protein, urine and plasma by capillary electrophoresis and laser-induced fluorescence. Anal Chem 71:1633–1637
5. Ptolemy AS, Britz-McKibbin P (2006) Sample preconcentration with chemical derivatization by capillary electrophoresis: Capillary as preconcentrator, microreactor and chiral selector for high-throughput metabolite screening. J Chromatogr A 1106:7–18
6. Poinsot V et al (2010) Recent advances in amino acid analysis by CE. Electrophoresis 31:105–121
7. Soga T et al (2004) Qualitative and quantitative analysis of amino acids by capillary electrophoresis-electrospray ionization-tandem mass spectrometry. Electrophoresis 25:1964–1972
8. Maxwell JE, Chen DDY (2008) Twenty years of interface development for capillary electrophoresis-electrospray ionization-mass spectrometry. Anal Chim Acta 627:25–33
9. Maxwell EJ, Zhong X, Chen DDY (2010) Asymmetrical emitter geometries for increased range of stable electrospray flow rates. Anal Chem 82:8377–8381
10. Haselberg R et al (2010) Performance of a sheathless porous tip sprayer for capillary electrophoresis–electrospray ionization-mass spectrometry of intact proteins. J Chromatogr A 1217:7605–7611
11. Chalcraft KR et al (2009) Virtual quantification of metabolites by capillary electrophoresis-electrospray ionization-mass spectrometry: Predicting ionization efficiency without chemical standards. Anal Chem 81:2506–2515
12. D'Agostino LA et al (2011) Comprehensive plasma thiol redox status determination for metabolomics. J Proteome Res 10:592–603
13. Chalcraft KR, Britz-McKibbin P (2009) Newborn screening of inborn errors of metabolism by CE-ESI-MS: A second-tier method with improved specificity and sensitivity. Anal Chem 81:307–314
14. Williams BJ et al (2007) Amino acid profiling in plant cell cultures: An inter-laboratory comparison of CE-MS and GC-MS. Electrophoresis 28:1371–1379
15. Lee R, Niewzcas L, Britz-McKibbin P (2007) Integrative metabolomics for characterizing unknown low-abundance metabolites by capillary electrophoresis-mass spectrometry with computer simulations. Anal Chem 79:403–415
16. Sugimoto M et al (2010) Prediction of metabolite identity from accurate mass, migration time prediction and isotopic pattern information in CE-TOF/MS data. Electrophoresis 31:2311–2318
17. Huhn C et al (2010) Relevance and use of capillary coatings in capillary electrophoresis-mass spectrometry. Anal Bioanal Chem 396: 297–314
18. Mayboroda OA et al (2007) Amino acid profiling in urine by capillary zone electrophoresis-mass spectrometry. J Chromatogr A 1159: 149–153
19. Baidoo EE et al. (2008) Capillary electrophoresis-fourier transform ion cyclotron resonance mass spectrometry for the identification of cationic metabolites via a pH-mediated stacking-transient isotachophoretic method. Anal Chem 80:3112–3122
20. Desiderio C et al (2010) Capillary electrophoresis-mass spectrometry for the analysis of amino acids. J Sep Sci 33:2385–2393
21. Lee R, Britz-McKibbin P (2009) Differential rates of glutathione oxidation for assessment of

cellular redox status and antioxidant capacity by capillary electrophoresis-mass spectrometry: An elusive biomarker of oxidative stress. Anal Chem 81:7047–7056

22. Giuffrida A et al (2009) Modified cyclodextrins for fast and sensitive chiral-capillary electrophoresis-mass spectrometry. Electrophoresis 30:1734–1742

23. Moini M, Schultz CL, Mahmood H (2003) CE/electrospray ionization-MS analysis of underivatized D/L-amino acids and several small neurotransmitters at attomole levels through the use of 18-crown-6-tetracarboxylic acid as a complexation reagent/background electrolyte. Anal Chem 75:6282–6287

24. Lee R et al (2010) Differential metabolomics for quantitative assessment of oxidative stress and nutritional intervention with strenuous exercise: Thiol-specific regulation of cellular metabolism with *N*-acetyl-L-cysteine pretreatment. Anal Chem 82:2959–2968

Chapter 10

Optimal Conditions for the Direct RP-HPLC Determination of Underivatized Amino Acids with Online Multiple Detection

A. Pappa-Louisi, P. Agrafiotou, and S. Sotiropoulos

Abstract

The combined use of a dual-UV detector as well as a fluorimetric and a multielectrode electrochemical detector (equipped with a dual electrode, consisting of a conventional size 3 mm diameter glassy carbon electrode (GCE) and of a pair of 30 μm thick carbon microfibers) is proposed for the detection of the following 15 underivatized amino acids: L-histidine (his), L-cysteine (cys), creatine (crn), S-methyl-L-cysteine (me-cys), DL-homocysteine (hcy), L-methionine (met), beta-(3,4-dihydroxyphenyl)-L-alanine (dopa), L-tyrosine (tyr), DL-m-tyrosine (m-tyr), L-a-methyl-dopa (me-dopa), L-phenylanine (phe), DL-alpha-methyltyrosine (me-tyr), 5-hydroxy-tryptophan (5htp), 3-nitro-L-tyrosine (NO_2Tyr) and L-tryptophan (trp), as well as of two dipeptides: L-cystathionine (cysta), L-carnosine (car), and of creatinine (cre). A multilinear solvent (acetonitrile) gradient elution program, determined by a simple optimization algorithm, is required for the efficient reversed phase separation of the above mixture of 18 solutes within 27 min at a flow rate of 1.0 mL/min and at 25°C.

Key words: Underivatized amino acids and related compounds, Multiple UV-fluorescence-electrochemical detection, Dual electrode, Microfiber electrode, Computer-aided multilinear gradient reversed phase separation optimization, Amino acid analysis

1. Introduction

Analysis of amino acids (AAs) is of continued interest due to the important role they play in several biological processes, such as protein synthesis and metabolic pathways. Up to now, a variety of methods to separate and detect specific AAs in samples of interest have been developed. However, since most AAs lack large hydrophobic sides and natural strong chromophore, fluorophore, or electroactive groups for photometric, fluorometric, or amperometric detection, respectively, the majority of the present methods employ

Michail A. Alterman and Peter Hunziker (eds.), *Amino Acid Analysis: Methods and Protocols*, Methods in Molecular Biology, vol. 828, DOI 10.1007/978-1-61779-445-2_10,

some form of pre- or post-column derivatization procedures in order to enhance detection or chromatographic separation.

Direct detection of AAs without performing derivatization is preferred not only for convenience, flexibility, simplicity, and accuracy, but also for avoiding problems introduced by derivatizing procedures, such as repeatability, derivative stability, side reactions, and reagent interferences. Several HPLC methods for the detection of underivatized AAs have been described. These methods include, direct detection by UV absorbance at low wavelengths when high sensitivity is not necessary (1, 2), conductivity detection (2, 3), electrospray mass spectrometry (MS) methods (2, 4–10), detection via native fluorescence (FL) (11), and some other detection techniques, e.g., evaporative light scattering (ELS) (2, 12), refractive index (RI) (2), chemiluminescent nitrogen (CLN) (2), or nuclear magnetic resonance (NMR) detection (2) as well as a combination of two different methods of detection, such as NMR-ELS (13) or UV-ELS detection (14). The detector that has been investigated the most for underivatized AA analysis is the electrochemical (EC) detector (15–21). The number of AAs that can be detected at conventional carbon-based electrodes is limited (15, 16), but as an alternative several transition metal-based electrodes, such as copper (17) or nickel-titanium (18), nickel-gold (19) as well as chemical modifications of some of the above electrodes (20, 21) have been proposed as EC sensors for the direct determination of underivatized AAs. However, no detection method proposed for the direct analysis of AAs is superior to all the others and free of drawbacks. As a result, in a recent paper we have tried to overcome the problems of assaying underivatized AAs by using a multiple online detection system along with reversed-phase HPLC (22). A laboratory-made computer-controlled system permits the simultaneous monitoring of five signals in a single run: two UV responses at two different wavelengths, a native FL signal and two EC current signals obtained at a dual-electrode arrangement consisting of a glassy macro-disk electrode and a couple of cylindrical carbon microfiber electrodes, in a series configuration. Obviously, the use of UV, FL, and EC detectors is limited to AAs with a chromophore, fluorophore, or electroactive groups, respectively.

The high polarity of most of AAs makes their separation by the reverse-phase HPLC difficult without adding an ion-paring reagent to the mobile phase (23–26). Thus, a multilinear solvent (acetonitrile) gradient elution program was proposed for the efficient separation of the mixture of 18 AAs and relative compounds tested in (22). The best gradient profile used in this publication, found by a simple optimization algorithm, was started at 0% of the organic modifier.

2. Materials

2.1. Chemicals

1. Use acetonitrile (MeCN) of HPLC grade.
2. Prepare buffer solutions using analytical grade KH_2PO_4 and 85% H_3PO_4.
3. Prepare all solutions using double distilled water.
4. Filter mobile phases through 0.45 μm regenerated cellulose membranes and sonicate them by ultrasound for about 20 min (see Note 1).

2.2. Buffers and Mobile Phases

1. 0.2 M aqueous phosphate buffer with pH 2.5: In a 1.0 L volumetric flask, dissolve 27.2 g of KH_2PO_4 and 3.95 mL 85% H_3PO_4 with 900 mL double distilled water. Incubate it at 25°C for about 15 min and complete the volume to the mark with double distilled water.
2. Mobile phase 1 (MP_1) consists of 0.02 M aqueous phosphate without organic modifier: In a 1.0 L volumetric flask, add 100 mL of the 0.2 M phosphate buffer (pH 2.5) and complete to the mark with double distilled water after incubation at 25°C for about 15 min.
3. Mobile phase 2 (MP_2) consists of 0.02 M aqueous phosphate and 15% MeCN: In a 1.0 L volumetric flask add 100 mL of the 0.2 M phosphate buffer (pH 2.5), 150 mL of MeCN and complete the volume to the mark with double distilled water after incubation at 25°C for about 15 min.
4. Mobile phase 3 (MP_3) consists of 0.02 M aqueous phosphate and 6% MeCN: In a 1.0 L volumetric flask add 100 mL of the 0.2 M phosphate buffer (pH 2.5), 60 mL of MeCN and complete the volume to the mark with double distilled water after incubation at 25°C for about 15 min.

2.3. HPLC System and Conditions

1. The liquid chromatography system consisted of a Shimadzu LC-20AD pump, a Shimadzu online degasser (DGU-20A3), a model 7125 syringe loading sample injector fitted with a 20 μL loop, a 250×4.6 mm MZ-Analytical column (MZ-Aqua Perfect C18 5 μm, see Note 2), a Shimadzu dual UV-Visible spectrophotometric detector (Model SPD-10A), a Shimadzu spectrofluorometric detector RF-10AXL and a laboratory-made multiple electrochemical detector (see Note 3) equipped with a laboratory-modified dual electrode (see details for the construction of the dual electrode in the following Subheading 2.4).
2. The upstream electrode was a glassy carbon macro-disk electrode and the downstream electrode was a couple of carbon microfibers. The two electrodes were in a series configuration and the

signal could be monitored simultaneously at both electrodes. The same potential (1.1 V vs. the Ag/AgCl reference electrode) was applied at both electrodes.

3. The UV, FL, and EC detectors were connected in series so that the analytes separated on the HPLC column eluted through the dual UV detector first, then through the FL detector and finally through the EC system previously described. This allows multiple measurements of analytes by UV absorbance, FL detection and oxidation response at the EC detector.
4. The UV detection of the analytes was performed simultaneously at two wavelengths between 190 nm and 210 nm, whereas the excitation and emission wavelengths were set at 220 and 320 nm, respectively, for the native fluorescence detection of the analyzed standards.
5. Laboratory-made multiple channel recording software written in Visual Basic collects and displays different detector signals (see Note 4) after adding in the PC system an A/D-D/A converter for analogue data acquisition.
6. Solutes of concentration 10 μg/mL were injected individually or together using appropriate mixtures. All experiments were carried out at 25°C using a Shimadzu CTO-10ASVP column oven.

2.4. Laboratory-Modified Thin Channel Flow Cell

1. A modification of the Gilson EC detector (Model 141) cell (see Note 5) was constructed by inserting two carbon microfibers (30 μm thick; 0.5 cm exposed length).
2. The microfibers were placed at the Teflon block of the existing cell (see Note 6) into which a disk glassy carbon electrode (GCE) of conventional size (3 mm diameter) was embedded, see Fig. 1.

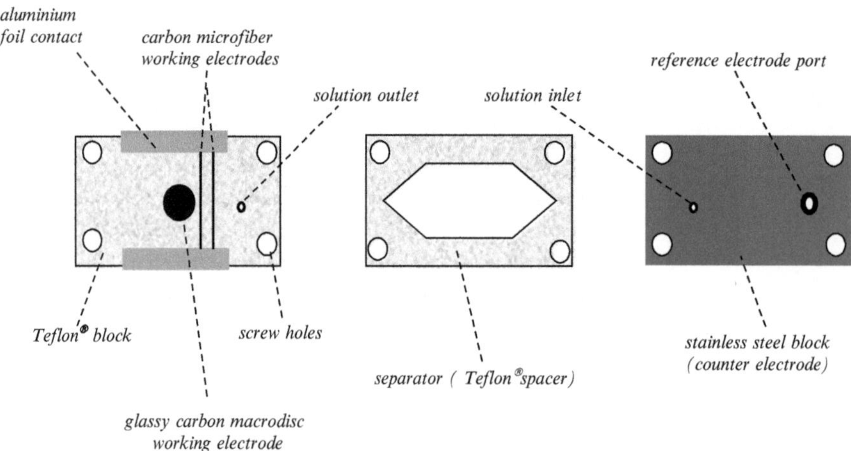

Fig. 1. Schematic representation of modified thin channel flow cell components containing a dual carbon (macro-disk and two microfibers) working electrode. (Reprinted from ref. 22 with permission).

3. The microfibers were glued with tiny drops of nail varnish at points between the active electrode length exposed to the solution (through the spacer opening) and the electrode ends (where electrical contacts were made). They were spaced 1.5 mm from each other and perpendicular to the analyte flow.
4. To achieve electrical contact, two strips of thin aluminum foil covered both fiber ends, sandwiched between the Teflon block and the rim of the spacer.
5. A stack of two 100 μm thick spacers were used to minimize the chance of short-circuiting the fibers with the counter electrode of the opposite block.

3. Methods

3.1. Optimal Detection Conditions

The optimal detection conditions of analyzed standards are given in Subheading 2.3. These were established by a preliminary study for each detection mode as follows:

1. UV detection: Most AAs (with the exception of those containing aromatic substituents, which also absorb at 254 nm) can only be detected at low UV wavelengths thanks to their carboxylic group, where, however, sensitivity and baseline linearity are poor. Thus, the effect of the UV wavelength on the detection of each AA and related compounds was determined between 190 and 230 nm in the mobile phase MP_3 described in Subheading 2.2. It was found that the 200 nm wavelength provided the detection of the most components of the mixture being analyzed with similar sensitivity, but UV detection at 190 nm permits a better detection of weakly retained analytes, such as his, cysta, car, and cys.
2. Fl detection: The phenolic- or catechol- and indole-amino acids (i.e., tyr, m-tyr, me-tyr/dopa, me-dopa/5htp, trp) were shown a native FL signal in contrast to NO_2Tyr, which is not an FL-detectable AA, due to the strong fluorescence quenching characteristic of the nitro-group.
3. EC detection: The optimum potential value to be applied for the EC detection was selected by hydrodynamic voltammograms of all analytes obtained in a mobile phase identical to that used for the study of solutes UV-absorbing behavior, i.e., mobile phase MP_3 described in Subheading 2.2 (see Note 7). It was found that among the 18 solutes tested only 11 can be oxidized at carbon electrodes. The catechol-amino acids, i.e., dopa and me-dopa, as well as 5htp, a hydroxyindole-amino acid, were the most readily oxidized solutes at ~0.5 V versus Ag/AgCl (3 M NaCl), whereas met, i.e., S-methyl-L-homocysteine, was oxidized only at high positive potentials > 1.0 V versus Ag/AgCl

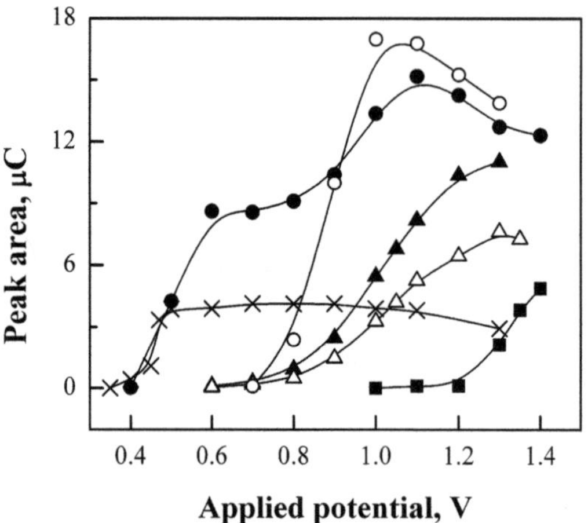

Fig. 2. Hydrodynamic voltammograms of me-dopa (×), 5htp (*filled circle*), trp (*open circle*), cys (*filled triangle*), hcy (*open triangle*), and met (*filled square*). *See* Subheading 3.1 for conditions. (Reprinted from ref. 22 with permission).

(3 M NaCl). The electroactivity of the rest of oxidizable compounds appeared at intermediate potentials between 0.5 and 1.0 V versus Ag/AgCl (3 M NaCl) and it was due to the oxidation of the phenolic group for tyr, m-tyr, me-tyr, and NO_2Tyr, the thiol group for cys and hcy, whereas for trp oxidation occurs at the indole ring. The hydrodynamic voltammograms obtained are shown in Fig. 2 for some of the above electroactive analytes. Note that, the voltammogram obtained for 5htp displays two waves, with the first half-wave potential at ~0.5 V versus Ag/AgCl (3 M NaCl) corresponding to the oxidation of the 5-hydroxy group of the indole ring, whereas the second one at ~0.9 V versus Ag/AgCl (3 M NaCl) to the oxidation of the indole ring itself. Note also that, in Fig. 2 peak areas (i.e., oxidation charges) instead of peak heights (i.e., oxidation currents) are presented versus applied potential, since the peak area of a solute is independent of its retention time, and consequently it is a better indicator than the peak height for its oxidative quantification. Additionally, in all these hydrodynamic voltammetry experiments the UV detector was coupled in series with the EC detector in order to indicate a possible fouling of the electrode surface during the electrochemical oxidation of the compounds tested (see Note 8). Considering the results displayed in Fig. 2, a potential of 1.1 V versus Ag/AgCl (3 M NaCl) was selected as the optimum working potential. Oxidative detection at 1.1 V versus Ag/AgCl (3 M NaCl), however, may result in high noise. In addition, the selectivity of such a high potential detection is also low. Thus, with this research project

we achieved the required specificity by the use of three detectors in series and minimized EC baseline drift by the use of microfiber electrodes (see in Subheading 3.2 for details).

3.2. Advantages of Multimode Detection

Figure 3 shows a typical chromatogram of 18 underivatized amino acids and related compounds (see Note 9) recorded by using three detectors in series and under optimal gradient conditions described

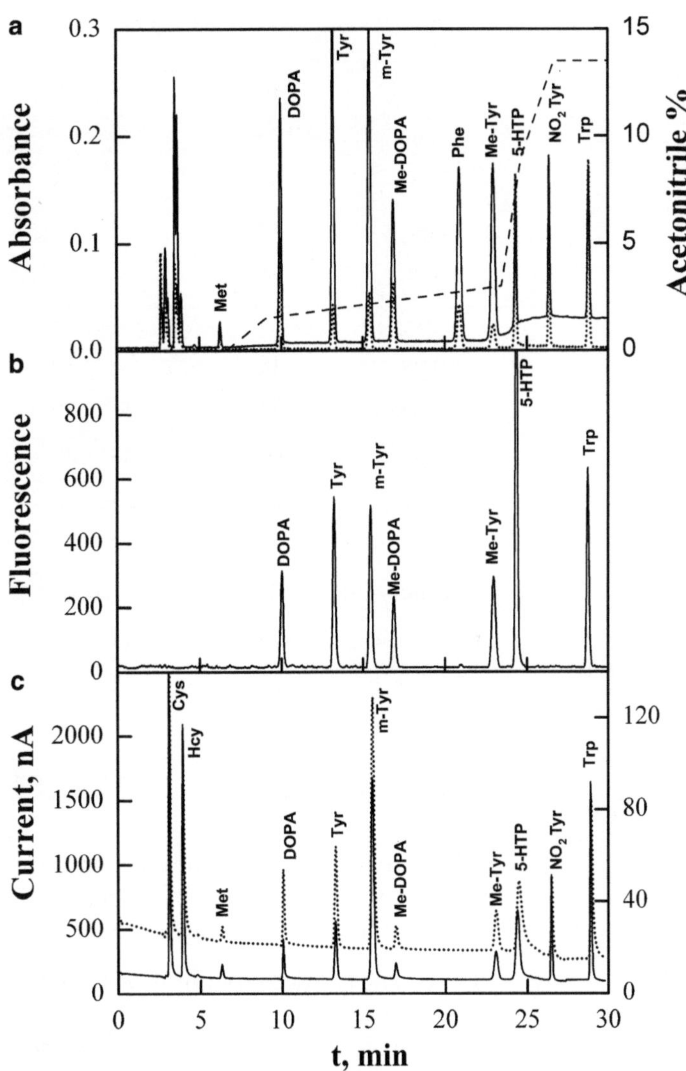

Fig. 3. Chromatogram of a mixture of 18 underivatized-AAs and related compounds with online (**a**) dual UV detection at 190 nm (*solid line*) and 210 nm (*dotted line*), respectively, (**b**) fluorescence detection and (**c**) dual electrochemical detection at glassy carbon macro-disk electrode (*solid line*) and two carbon microfibers (*dotted line*), respectively, under optimal gradient conditions. The *dashed line* in (**a**) shows the gradient profile of % MeCN versus *t* when it reaches the UV detector. The left axis of (**c**) corresponds to oxidative current recorded at the macro-electrode, whereas the right axis of the same figure corresponds to the current recorded at microfibers. *See* Subheading 2.3 for other conditions. (Reprinted from ref. 22 with permission).

Table 1
Detection limits (ng/mL) obtained for underivatized amino acids with different detectors. The sample injection volume was 20 μL. (Reprinted from ref. 22 with permission)

	Detection method			
Solutes	**UV at 190 nm**	**FL**	**EC at macroGCE**	**EC at microfibers**
his	194	–	–	–
cysta	346	–	–	–
car	130	–	–	–
cys	264	–	3	3
cre	49	–	–	–
crn	55	–	–	–
hcy	–	–	5	4
met	548	–	89	59
dopa	53	41	30	9
tyr	40	23	20	7
m-tyr	41	25	6	3
me-dopa	92	58	82	33
phe	74	–	–	–
me-tyr	73	44	42	18
5htp	95	12	17	10
n-tyr	79	–	11	8
trp	87	20	6	4

in Subheading 3.3. From this figure as well as from Table 1, which gives the solute limits of detection (LODs) obtained by different detectors, the following observations arise:

1. The UV detector can also be used to monitor both the chromatographic separation of species and the possible passivation of the working electrodes of the EC detector, since all solutes tested are UV-absorbing.
2. The UV detection of solutes is enhanced by using dual-wavelength UV absorbance. Thus, under the gradient elution used in this study the baseline stability is better with UV detection at 210 nm than that obtained with UV absorbance detection at 190 nm, but the 190 nm wavelength allows a better UV detection of cysta and cys, which cannot be properly detected at 210 nm, see Fig. 4a.

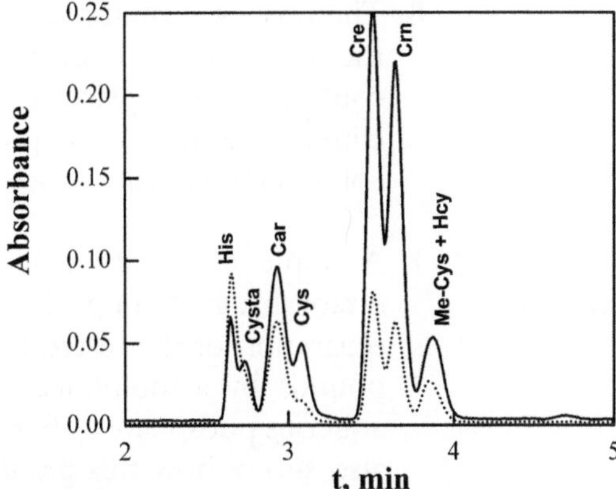

Fig. 4. The first 5 min of the chromatogram of Fig. 3a. (Reprinted from ref. 22 with permission).

3. Both the glassy carbon macro-disk and the two carbon microfibers give good quality responses for all electroactive analytes at 1.1 V versus Ag/AgCl (3 M NaCl), although the oxidation currents of met, me-dopa, and me-tyr are small. However, the advantage of using a microfiber placed vertically in the mobile phase flow is that the microfibers show better S/N characteristics, as predicted also theoretically (27). Thus the limits of EC detection at a macro-electrode for the 11 oxidizable amino acids were evaluated to be between 89 and 3 ng/mL (S/N = 3), and they were improved by a factor of 1.1–3.2 by using microfibers, see Table 1.
4. The universal UV detector is the least sensitive for detection of underivatized AAs, whereas the EC detector with microfibers provides the best detection limits (see Note 10).
5. The combined detection method is in addition particularly suitable for the investigation of unknown mixtures present in complex samples since the UV, fluorescent, and electrochemical peak heights do not covary across different analytes and thus the ratio of native fluorescence and/or UV absorbance to electrochemical activity can be used to characterize the different analytes.
6. The use of multiple-detection could solve problem concerning chromatographic resolution as in case of the co-eluted me-cys and hcy in our chromatographic system, which can successfully be separated by the simultaneous recording of the EC-detected chromatogram where only hcy exhibits a response.
7. Important practical advantages of carbon microelectrodes with respect to conventional glassy carbon electrodes were demonstrated when they are used for HPLC experiments.

8. The proposed multiple detection solves problems concerning the direct detection of underivatized AAs and consequently it could partially replace AA analysis by derivatization methods, eliminating in this way problems introduced by pre- or post-column derivatizing procedures.

3.3. Optimal Separation Conditions

1. A simple optimization algorithm (see Note 11) was applied to establish a gradient profile (Table 2) that leads to the optimum separation of the tested mixture of 18 AAs and related compounds by a multilinear variation of % MeCN, in 0.02 M aqueous phosphate buffer with pH 2.5 (see Note 12). Table 2 also shows how this gradient elution can be implemented if the mobile phases MP_1 and MP_2 (Subheading 2.2) are used in the two flow lines of the HPLC pump system. As is shown in Figs. 3 and 4, the standard mixture was separated with an acceptable degree of resolution in 27 min under the current experimental conditions except me-cys and hcy, which co-elute (see Note 13). The gradient profile of MeCN used versus t is depicted in Fig. 3a (time delay to reach the UV detector is 3.66 min after the gradient created in the pump unit, see Note 14).
2. In case a different chromatographic column is used for the separation of a mixture of underivatized AAs, a reliable determination of the optimum gradient profile may be achieved according the following steps:
   (a) Perform at least three preliminary gradient runs, where the gradient profiles of % MeCN in 0.02 M aqueous phosphate buffer (pH 2.5) versus t are linear or multilinear. The retention times recorded for each solute in each of the above gradients as well as the chromatographic

Table 2
Multilinear gradient profile used for the separation of 18 AAs and related compounds

Optimal gradient profile selected by an optimization algorithm		Gradient program created in the pump unit		
			Flow line A	Flow line B
Time, min	% MeCN	Time, min	Mobile phase (MP_1)	Mobile phase (MP_2)
0	0	0	100	0
1	0	1	100	0
7	1.5	7	90	10
20	3	20	80	20
21	9	21	40	60
24	15	24	0	100

parameters t_0 and t_D (see Note 14) are used as the input data for the fitting algorithm (see Note 15). This algorithm allows for the determination of the dependence of retention upon the composition of mobile phase, i.e., the determination of adjustable parameters of the retention equation (see Note 16).

(b) Test the validity of the previous retention equation by prediction the retention of solutes recorded at least in a linear or multilinear gradient run involving variation of % MeCN in 0.02 M aqueous phosphate buffer (pH 2.5) versus t different than that used in the fitting procedure (see Note 17).

(c) Feed the optimization algorithm (see Note 11) with the adjustable parameters obtained by the fitting procedure, with the number of linear sections of multilinear variation of concentration of MeCN during the chromatographic run, the minimum and the maximum concentration of MeCN as well as with the desired analysis time, $t_{R,max}$, i.e., the retention time of the most distant solute in the sample. Then, the optimization algorithm computes the minor distance of adjacent peaks, δt, for every possible multilinear gradient profile. It is evident that the optimal gradient is that yields the maximum δt value in the shorter analysis time.

(d) Use the determined optimum linear or multilinear gradient profile in the separation of the tested AA sample (see Note 18).

4. Notes

1. A 3 channel online degasser (DGU-20A3) connected to pump unit removes dissolved gases in mobile phases.
2. The high polarity of the most of AAs makes their separation difficult. Reverse-phase liquid chromatography (RP-HPLC) can be applied in the separation of AAs in their underivatized form by using a gradient elution started at 0% of the organic modifier (see Table 1 and Fig. 3a). However, such gradient profile requires the use of a column stable in pure aqueous mobile phases, i.e., Aqua Perfect C18 column.
3. Similar multiple EC detector is also commercially available.
4. The homemade software is available upon request from the authors, although similar software is commercially available.

5. The original Gilson Model 141 two-block flow cell (as that depicted in Fig. 1 but without carbon microfibers) consists of a Teflon® block where a 3 mm diameter glassy carbon (GC) indicator electrode is embedded, while the other block is made of stainless steel and serves as the counter electrode. The two blocks were separated by a 100 μm thick Teflon spacer and an Ag/AgCl (in 3 M NaCl) reference electrode is plugged in a port on the stainless steel block; the solution inlet and outlet ports were on the stainless steel and Teflon parts, respectively. The spacer frame was such that the channel length was 1.2 cm and its width 0.5 cm; these, together with the spacer thickness of 100 μm, result in a total detector volume of 0.006 mL and a dead volume of 0.00225 mL up to the leading electrode edge (when a stack of two spaces is used, these values are doubled).
6. The 30 μm diameter carbon fibers (Goodfellow) were degreased in acetone, dipped in a sulfuric acid + potassium dichromate mixture and thoroughly washed with double distilled water before mounting on the cell.
7. The original Gilson EC detector described in Note 5 was used to obtain the hydrodynamic voltammograms.
8. The reactivation of the passivated working electrode can be achieved by polishing the glassy carbon Teflon-block-embedded electrode with 0.05 μm alumina powder (Buehler).
9. Although creatinine (cre) is not an AA, it was included in the standard AA – dipeptide mixture tested in the present work since it coexists with them in some natural samples.
10. It should be pointed out here that the good performance of microelectrodes in a flow cell demands a well-skilled electrochemist operator.
11. The laboratory-made optimization algorithm with instructions is available from the Web site of the corresponding author of this book chapter http://www.chem.auth.gr/index.php?lang=en&st=55 (folder "HPLC-Algorithms").
12. The mobile phase pH was selected acidic enough to ensure that both the ammonium and carboxyl group are protonated and, consequently, the AAs show higher retention in reversed-phase columns.
13. This problem of co-elution can be solved by using multiple-detection since me-cys and hcy can successfully be separated by EC detected chromatogram (see, Fig. 3c), where only hcy exhibits a response.
14. The time duration of 3.66 min is the sum of t_0 and t_D, where t_0 is the column hold-up time and t_D is the dwell time, i.e., the time needed for a certain change in the mixer of the pump to reach the beginning of the chromatographic column. The

hold-up time was estimated to be $t_0 = 2.56$ min by using water as marker, whereas the dwell time was determined to be $t_D = 1.1$ min by recording the absorbance curve by the UV detector working at 200 nm during a single step gradient, where the concentration of methanol in water jumps from 20 to 0%.

15. The laboratory-made fitting algorithm with instructions is available from the Web site of the corresponding author of this book chapter http://www.chem.auth.gr/index.php?lang=en&st=55 (folder "HPLC-Algorithms").
16. The retention equation used in the fitting algorithm is: $\ln k = c_1 - \frac{c_3\varphi}{1+c_2\varphi}$, where k is the retention factor of a solute with retention time t_R ($k = (t_R - t_0)/t_0$), c_1, c_2 and c_3 are adjustable parameters and φ is the volume fraction of organic modifier (MeCN) in the aqueous buffer.
17. The laboratory-made prediction algorithm with instructions is available from the Web site of the corresponding author of this book chapter http://www.chem.auth.gr/index.php?lang=en&st=55 (folder "HPLC-Algorithms").
18. Experience or trial and error approaches can hardly lead to the optimum gradient profile.

Acknowledgments

This work was supported by European Union/European Social Fund (PhD scholarship, Hrakleitos I).

References

1. Yokoyama Y, Tjuji S, Sato H (2005) Simultaneous determination of creatinine, creatine, and UV-absorbing amino acids using dual-mode gradient low-capacity cation-exchange chromatography. J Chromatogr A 1085:110–116
2. Petritis K, Elfakir C, Dreux M (2002) A comparative study of commercial liquid chromatographic detectors for the analysis of underivatized amino acids. J Chromatogr A 961:9–21
3. Kuban P, Hauser PC (2006) Application of gradient programs for the determination of underivatized amino acids and small peptides in reversed-phase high-performance liquid chromatography with contactless conductivity detection. J Chromatogr A 1128:97–104
4. Smith CI et al (2009) A three-phase liquid chromatographic method for delta C-13 analysis of amino acids from biological protein hydrolysates using liquid chromatography-isotope ratio mass spectrometry. Anal Biochem 390:165–172
5. Zoppa M et al (2006) Method for the quantification of underivatized amino acids on dry blood spots from newborn screening by HPLC-ESI-MS/MS. J Chromatogr B 831:267–273
6. Thiele B et al (2008) Analysis of amino acids without derivatization in barley extracts by LC-MS-MS. Anal BioanalChem 391:2663–2672
7. McCullagh J, Gaye-Siessegger J, Focken U (2008) Determination of underivatized amino acid delta C-13 by liquid chromatography/isotope ratio mass spectrometry for nutritional

studies: the effect of dietary non-essential amino acid profile on the isotopic signature of individual amino acids in fish. Rapid Commun Mass Spectrom 22:1817–1822

8. Armstrong M, Jonscher K, Reisdorph NA (2007) Analysis of 25 underivatized amino acids in human plasma using ion-pairing reversed-phase liquid chromatography/time-of-flight mass spectrometry. Rapid Commun Mass Spectrom 21:2717–2726
9. Scharff-Poulsen AM, Schou C, Egsgaard H (2007) Direct analysis of N-15-label in amino and amide groups of glutamine and asparagine. J Mass Spectrom 42:161–170
10. de Person M, Chairnbault P, Elfakir C (2008) Analysis of native amino acids by liquid chromatography/electrospray ionization mass spectrometry: comparative study between two sources and interfaces. J Mass Spectrom 43:204–215
11. Wood AT, Hall MR (2000) Reversed-phase high-performance liquid chromatography of catecholamines and indoleamines using a simple gradient solvent system and native fluorescence detection. J Chromatogr B 744:221–225
12. Yan D et al (2007) Direct determination of fourteen underivatized amino acids from Whitmania pigra by using liquid chromatography-evaporative light scattering detection. J Chromatogr A 1138:301–304
13. Petritis K et al (2002) Evaporative light scattering detection for in-line monitoring of stopped-flow liquid chromatography-nuclear magnetic resonance analysis of compounds with weak or no chromophore groups. J Sep Sci 25:593–600
14. Cobb Z et al (2001) Evaporative light-scattering detection coupled to microcolumn liquid chromatography for the analysis of underivatized amino acids: Sensitivity, linearity of response and comparisons with UV absorbance detection. J Microcolumn Sep 13:169–175
15. Accinni R et al (2000) Screening of homocysteine from newborn blood spots by high-performance liquid chromatography with coulometric array detection. J Chromatogr A 896:183–189
16. Majidi MR H et al (2006) Simultaneous voltammetric determination of cysteine, tyrosine and tryptophan by using principal component artificial neural networks. Asian J Chem 18:2445–2457
17. Casella IG, Contursi M (2003) Isocratic ion chromatographic determination of underivatized amino acids by electrochemical detection. Anal Chim Acta 478:179–189
18. Sato K et al (2000) Nickel-titanium alloy electrodes for stable amperometric detection of underivatized amino acids in anion-exchange chromatography. Talanta 53:1037–1044
19. Casella IC, Gatta M, Cataldi TRI (2000) Amperometric determination of underivatized amino acids at a nickel-modified gold electrode by anion-exchange chromatography. J Chromatogr A 878:57–67
20. Zen J-M et al (2004) Amino acid analysis using disposable copper nanoparticle plated electrodes. Analyst 129:841-845
21. Pei JH, Li XY (2000) Determination of underivatized amino acids by high-performance liquid chromatography and electrochemical detection at an amino acid oxidase immobilized CuPtCl6 modified electrode. Fresen J Anal Chem 367:707–713
22. Agrafiotou P, Sotiropoulos S, Pappa-Louisi A (2009) Direct RP-HPLC determination of underivatized amino acids with online dual UV absorbance, fluorescence, and multiple electrochemical detection. J Sep Sci 32: 949–954
23. Pappa-Louisi A, Nikitas P, Agrafiotou P (2006) Column equilibration effects in gradient elution in reversed-phase liquid chromatography. J Chromatogr A 1127:97–107
24. Nikitas P, Pappa-Louisi A, Agrafiotou P (2006) Multilinear gradient elution optimization in reversed-phase liquid chromatography using genetic algorithms. J Chromatogr A 1120:299–307
25. Yan D, Li G (2006) Direct determination of 14 underivatized amino acids in leech by reversed phase high performance liquid chromatography. Chin J Anal Chem 34:705–708
26. Wu ZP et al (2004) Determination of phenylalanine isotope ratio enrichment by liquid chromatography/time-of-flight mass spectrometry. Eur J Mass Spectrom 10:619–623
27. Agrafiotou P, Sotiropoulos S (2003) Characterisation of a simple electrochemical detector for high-performance liquid chromatography and flow-injection analysis based on carbon microcylinder electrodes. Anal Chim Acta 497:175–189

Chapter 11

Absolute Quantitation of Proteins by Acid Hydrolysis Combined with Amino Acid Detection by Mass Spectrometry

Olga A. Mirgorodskaya, Roman Körner, Yuri P. Kozmin, and Peter Roepstorff

Abstract

Amino acid analysis is among the most accurate methods for absolute quantification of proteins and peptides. Here, we combine acid hydrolysis with the addition of isotopically labeled standard amino acids and analysis by mass spectrometry for accurate and sensitive protein quantitation. Quantitation of less than 10 fmol of protein standards with errors below 10% has been demonstrated using this method (1).

Key words: Amino acid analysis, Mass spectrometry, Absolute protein quantitation, MALDI, Quantitative mass spectrometry, Isotope-labeled amino acids

1. Introduction

Absolute quantitation of proteins is an important prerequisite to accurately model reaction kinetics in biological systems, and several analytical strategies have been applied to address this demand. Quantitation using fluorescent dyes or radioactive labeling has been widely used in combination with gel-based protein separation (2) but need special equipment and the calibration of the dye-based method for different proteins is not straightforward. Mass spectrometry is a highly sensitive analytical method. However, internal standards are needed for absolute quantitation of peptides and proteins since signal intensities are strongly sequence dependent. To generate internal standards, stable isotopes can be incorporated in vivo by growing cells in media containing isotopically labeled amino acids, such as ^{13}C, ^{15}N, or deuterium-labeled arginine, lysine, and leucine (3). Relative quantitation of samples by mass spectrometry can then be achieved based on the relative

Michail A. Alterman and Peter Hunziker (eds.), *Amino Acid Analysis: Methods and Protocols*, Methods in Molecular Biology, vol. 828, DOI 10.1007/978-1-61779-445-2_11, © Springer Science+Business Media, LLC 2012

signal intensities of labeled and unlabeled peptides. Alternatively, peptides may also be isotopically labeled in vitro, for example by incorporation of ^{18}O at their C-terminus upon enzymatic digestion in ^{18}O-water (4). The relative peptide amounts in a labeled versus an unlabeled sample can then be quantified by mass spectrometry. Absolute quantitation using isotopically labeled synthetic peptides as internal standards has been introduced by Stemmann and colleagues for the absolute quantitation of a phosphopeptide (5). Based on this principle, selected reaction monitoring based mass spectrometry allows for the quantitation of dozens of proteins in mixtures in a single analysis, but at least one isotopically labeled peptide must be synthesized for each protein (6). A further challenge of this method is the accurate quantitation of the isotopically labeled standard peptides. Usually, amino acid analysis is applied for this task and the chemical stability of the peptide standards must be tested under typical storage conditions. Amino acid analysis requires relatively pure peptide or protein samples, but is considered as the most accurate method for protein quantitation (7). After total acid hydrolysis of the protein sample, amino acids are usually derivatized on a precolumn and then quantified by high-performance liquid chromatography (HPLC). The drawback of this method is a modest sensitivity, typically in the pmol range.

Here, we describe a mass spectrometry-based amino acid quantitation method with sensitivity in the femtomole range (1). After hydrolysis of the protein sample, isotopically labeled amino acids are added as internal standards and mass spectrometry is used to quantify the absolute amount of the respective amino acid in the sample. Protein quantitation is then based on the known amino acid composition of the protein. By using more than one standard, the dynamic range of the method can be extended and the components of simple protein mixtures may also be quantified simultaneously. Importantly, no amino acid derivatization steps are required for this method.

2. Materials

1. Isotopically labeled amino acid standards (>98% isotope enrichment) (Cambridge Isotope Laboratories or Aldrich).
2. Acetonitrile (HPLC gradient grade, Merck or Fisher Scientific).
3. Water (LiChrosolv grade, Merck).
4. Trifluoroacetic acid (Uvasol grade, Merck).
5. Hydrochloric acid (30% solution, Suprapur grade, Merck).
6. Thioglycolic acid (pro analysis grade, VWR International).
7. Phenol (pro analysis grade, VWR International).

8. 2,5-Dihydroxybenzoic acid (DHB) (Aldrich).
9. α-Cyano-4-hydroxylcinamic acid (CHCA) (Fluka).
10. Mininert valves (24 mm diameter) and glass vials (Alltech).

3. Methods

For the presented standard protocol, the protein amount should be known roughly within a factor of 10. In cases with completely unknown protein concentration, the experiment may be performed in parallel with a dilution series of added isotopically labeled lysine standard (each step 10× dilution), or alternatively the "extended dynamic range standard" (see Note 3) may be used. Importantly, the amino acid sequence of the protein must be known and the protein preparation should be essentially pure from other contaminating proteins. A modification of the protocol which can be applied to a simple mixture of proteins is described in Note 4. Protein purification may be performed by gel electrophoresis or other means. If gel electrophoresis is used, we recommend performing in-gel digestion and peptide extraction according to the method of Shevchenko et al. (8). However, for gel-separated proteins the extraction efficiency (in our hands roughly 70%) has to be considered for the final protein quantitation. The following "standard protocol" describes quantitation of one protein of known sequence with an isotopically labeled lysine standard; variations of the protocol for specific purposes are presented in Subheading 4.

1. Estimate the amount of lysine in your purified protein based on the estimated protein concentration and the amino acid sequence of your protein.

   Estimated amount of lysine = (number of lysines in the sequence) × (estimated protein amount). Instead of lysine, arginine (see Note 1), and leucine (see Note 2) may be used for the protein quantitation.
2. Dissolve the isotopically labeled D_4-lysine standard in water at a concentration of 1 pmol/μL (stock solution). Add the isotopically labeled lysine standard to your protein sample. The amount of added standard should correspond to the estimated amount of lysine in your sample. In case of a gel-separated protein, the standard should be added to the sample after in-gel digestion, but before peptide extraction (8).
3. Dry your protein or peptide sample in a vacuum centrifuge.
4. Remove the lid from the plastic tube containing the dry sample and place it inside of a 50-mL glass vial.

5. Add 800-µl mixture of 6 M HCl with 0.1% phenol and 0.1% thioglycolic acid to the bottom of the glass vial. Flush the glass vial with argon.
6. Close the vial with a 24-mm mininert valve lid and evacuate the glass vial to approximately 1 mbar using a low-vacuum pump.
7. Place the closed vial in an oven at 107°C for 18 h.
8. Wait until the glass vial has cooled to room temperature. Release the vacuum carefully using the mininert valve and open the glass vial.
9. Remove the plastic vial containing the sample and place it in a vacuum centrifuge for about 15 min to remove remaining traces of acid.
10. Dissolve the sample in 3.5 µL of 0.1% TFA for MALDI-MS.
11. Prepare a 20 mg/mL 2,5-dihydroxybenzoic acid MALDI matrix solution in 70% acetonitrile and 0.1% TFA.
12. For MALDI sample preparation, mix 0.3 µL of DHB matrix solution with 0.3 µL of the sample on the MALDI target plate. For low protein amounts, the sample may also be dissolved in smaller volumes down to 0.5 µL (step 10).
13. As a control to confirm that no matrix signals interfere with the Lys signal, mix 0.3 µL of DHB matrix solution with 0.3 µL of 0.1% TFA on the MALDI target plate.
14. Place the MALDI target in a vacuum chamber to dry the DHB matrix quickly. This results in smaller DHB crystals and a more uniform crystal distribution.
15. Acquire a spectrum of the control sample in positive mode from about 1,000 laser shots at optimal resolution, preferentially in reflectron mode. Adjust the laser energy to the minimum and ensure that the region around 147.1 Da (lysine) and 151.1 Da (D_4-lysine) is free of matrix signals.
16. Acquire a spectrum of the sample in positive mode from about 1,000 laser shots at optimal resolution, preferentially in reflectron mode.
17. Quantify the lysine ratio of the unlabeled (147.1 Da) versus the isotopically labeled lysine (151.1 Da) based on the peak areas.

$$\text{lysine ratio} = (\text{peak area lysine}) / (\text{peak are labeled lysine})$$

18. Calculate the absolute quantity of your protein sample as:

$$\text{protein quantity} = (\text{lysine ratio}) \times (\text{amount of added labeled lysine standard}) / (\text{number of lysines in the protein sequence})$$

4. Notes

1. Instead or in addition to the D_4-lysine standard (see Subheading 3); $^{13}C_6$ and $^{13}C_6{}^{15}N_4$ arginine can be used as internal standards. The arginine can be dissolved in water at a concentration of 1 pmol/μL and then be added to the protein or peptide sample prior to vacuum centrifugation and acid hydrolysis (see above). However, since DHB matrix signals interfere with the arginine peaks (unlabeled: 175.1 Da; 185.1 Da labeled), CHCA should be used as a MALDI matrix.
   (a) Prepare a 20 mg/mL solution of CHCA in 70% acetonitrile and 0.1% TFA.
   (b) Mix 0.3 μL of CHCA matrix solution with 0.3 μL of the sample on the MALDI target plate.
   (c) As a control to confirm that no matrix signals interfere with the arginine signals, mix 0.3 μl of CHCA matrix solution with 0.3 μL of 0.1% TFA on the MALDI target plate.
   (d) Place the MALDI target in a vacuum chamber to dry the CHCA matrix quickly. This results in smaller CHCA crystals and a more uniform crystal distribution.
   (e) Acquire a spectrum of the control sample in positive mode from about 1,000 laser shots at optimal resolution, preferentially in reflectron mode. Adjust the laser energy to the minimum and ensure that the region around the signals for arginine, lysine and their stable isotope labeled counterparts are free of matrix signals.
   (f) Acquire the sample spectrum and quantify the protein as described for the lysine standard above.
2. Instead or in addition to lysine and arginine standards, (see Subheading 3 and Note 1), D_3-leucine can also be used as an internal standard. Both, DHB and CHCA MALDI matrices are compatible with the leucine standard. However, for the quantitation step (Subheading 3 step 18) it has to be considered that leucine and isoleucine are isobaric amino acids, so the added amounts of leucine and isoleucine are quantified. The final protein amount can then be calculated as: protein quantity = (measured leucine ratio) × (amount of added labeled leucine standard)/(number of leucines and isoleucines in the protein sequence).
3. A mixture composed of differently labeled forms of the amino acid standard can be used to increase the dynamic range of protein quantitation. This can be beneficial in cases where it is difficult to estimate the initial protein concentration. Labeled arginine, for example can be added as $^{15}N_2$-, $^{13}C_6$-, and

$^{13}C_6{}^{15}N_4$-arginine at concentrations varying by a factor of 10 (for example: $^{15}N_2$-arginine: 100 fmol, $^{13}C_6$-arginine: 1 pmol, and $^{13}C_6{}^{15}N_4$-arginine: 10 pmol). For the final calculation of the protein concentration, the standard for which the concentration turns out to be closest to the proteins arginine concentration should be used. As a result, the dynamic range of the quantitation standard can be largely increased. For the example mentioned above, protein arginine concentrations ranging from 10 fmol to 100 pmol could be covered with one single standard.

4. Simple mixtures of proteins (of known sequences) can also be quantified using special quantitation standards composed of several labeled amino acids (Ref. 1). The number of labeled amino acids in the standard should correspond to the number of proteins to be quantified simultaneously. Using a mixture of isotopically labeled arginine and leucine, for example, a mixture of two proteins can be quantified. The concentrations of the two proteins can then be calculated from the measured arginine and leucine/isoleucine amounts as the solution of the following equation system:

   (a) $Arg_{measured} = C_{protA} \times N\_Arg_{protA} + C_{protB} \times N\_Arg_{protB}$

   (b) $Leu\_Ile_{-measured} = C_{protA} \times N\_Leu\_Ile_{protA} + C_{protB} \times N\_Leu\_Ile_{protB}$

   $Arg_{measured}$ and $Leu\_Ile_{-measured}$ are the measured absolute arginine and leucine/isoleucine amounts, repectively, C_{protA} and C_{protB} are the protein amounts of the two proteins A and B, respectively (the solutions of the equation system), N_Arg_{protA} and N_Arg_{protB} are the numbers of arginines in the sequences of proteins A and B, respectively, and $N\_Leu\_Ile_{protA}$ and $N\_Leu\_Ile_{protB}$ are the added numbers of leucines and isoleucines in the sequences of proteins A and B, respectively.

References

1. Mirgorodskaya OA, Körner R, Novikov A, Roepstorff P (2004) Absolute quantitation of proteins by a combination of acid hydrolysis and matrix-assisted laser desorption/ionization mass spectrometry. Anal Chem. 76:3569–3575
2. Berggren K et al (2000) Background-free, high sensitivity staining of proteins in one- and two-dimensional sodium dodecyl sulfate-polyacrylamide gels using a luminescent ruthenium complex. Electrophoresis 21:2509–2521
3. Ong SE et al (2002) Stable isotope labeling by amino acids in cell culture, SILAC, as a simple and accurate approach to expression proteomics. Mol Cell Proteomics 1:376–386
4. Mirgorodskaya OA et al (2000) Quantitation of peptides and proteins by matrix-assisted laser desorption/ionization mass spectrometry using ^{18}O-labeled internal standards. Rapid Commun Mass Spectrom 14:1226–1232
5. Stemmann O et al (2001) Dual inhibition of sister chromatid separation at metaphase. Cell 107:715–726
6. Lange V et al (2008) Selected reaction monitoring for quantitative proteomics: a tutorial. Mol Syst Biol 4:1–14
7. Macchi FD et al (2000) Amino acid analysis, using postcolumn ninhydrin detection, in a biotechnology aboratory. Methods Mol Biol 159:9–30
8. Shevchenko A et al (1996) Mass spectrometric sequencing of proteins from silver-stained polyacrylamide gels. Anal Chem 68:850–858

Chapter 12

Amino Acid Analysis by Means of MALDI TOF Mass Spectrometry or MALDI TOF/TOF Tandem Mass Spectrometry

Natalia V. Gogichaeva and Michail A. Alterman

Abstract

Here, we describe two different amino acid analysis protocols based on the application of matrix-assisted laser desorption ionization time-of-flight (MALDI TOF) mass spectrometry (MS). First protocol describes a MALDI TOF MS-based method for a routine simultaneous qualitative and quantitative analysis of free amino acids and protein hydrolysates (Alterman et al. Anal Biochem 335: 184–191, 2004). Linear responses between the amino acid concentration and the peak intensity ratio of corresponding amino acid to internal standard were observed for all amino acids analyzed in the range of concentrations from 20 to 300 μM. Limit of quantitation varied from 0.03 μM for arginine to 3.7 μM for histidine and homocysteine. This method has one inherent limitation: the analysis of isomeric and isobaric amino acids. To solve this problem, a second protocol based on the use of MALDI TOF/TOF MS/MS for qualitative analysis of amino and organic acids was developed. This technique is capable of distinguishing isobaric and isomeric compounds (Gogichayeva et al. J Am Soc Mass Spectrom 18: 279–284, 2007). Both methods do not require amino acid derivatization or chromatographic separation, and the data acquisition time is decreased to several seconds for a single sample.

Key words: Amino acid analysis, Mass spectrometry, Absolute protein quantitation, Quantitative mass spectrometry, MALDI TOF, MALDI TOF/TOF, Metabolomics

1. Introduction

Amino acids play two different and exceptionally important roles in nature: building blocks of proteins/peptides and products/metabolites of cell and energy metabolism. The list of amino acids of interest in bioanalytical and clinical chemistry is not limited to essential or proteinogenic amino acids. There are a number of the so-called unusual amino acids that are constituents of microbial peptides or products of cell metabolism or metabolic disorders. As a result, an interest in a fast, effective, and reliable method for

Michail A. Alterman and Peter Hunziker (eds.), *Amino Acid Analysis: Methods and Protocols*, Methods in Molecular Biology, vol. 828, DOI 10.1007/978-1-61779-445-2_12, © Springer Science+Business Media, LLC 2012

analysis of amino acids and their derivatives is significant in practically every aspect of biochemical/biomedical research, biotechnology, agriculture, clinical medicine, and food technology. Area of application of amino acid analysis (AAA) has recently extended to metabolomics and proteomics fields.

AAA is generally applied either to the analysis of free amino acids (e.g., in body fluids in clinical chemistry, in cell culture media in fermentation, or in food and nutritional supplements) or to bound amino acids (e.g., peptides and proteins in analytical biochemistry). Such complexity in the number of components in amino acid analysis represents a great challenge in analytical chemistry.

AAA is an essential part of analytical and clinical biochemistry for 60 years (1). However, its main technological outline has not changed much over the years. It still consists of precolumn or postcolumn derivatization of amino acids for detection purposes (or converting AA into volatile compounds for GC analysis) followed by separation of amino acid (HPLC, GC, or CE) (2–5). At its best, a single sample analysis time is 15–25 min in case of GC-MS, CE-MS, or LC-MS and 30–180 min in case of HPLC and that without accounting for a modification time (6).

Electrospray ionization (ESI) and matrix-assisted laser desorption/ionization (MALDI) are the two major ionization techniques used in mass spectrometry (MS) of biological molecules. Though an analysis of large biopolymers is successfully performed by both techniques, analysis of low-molecular-weight (LMW) molecules, like amino acids, pharmaceuticals, hormones, etc., is predominantly performed by ESI mass spectrometry.

There are several reasons for this prevalence. First, MALDI matrices (themselves LMW compounds) produce a number of interfering peaks in LMW region (100–600 amu). Second, until the introduction of time of flight TOF/TOF instruments, MALDI TOF mass spectrometers were not capable of performing a real tandem MS and thus delivering structural information of the analyte. Third, until recently, MALDI was not considered to be a technique capable of providing robust quantitative data. New developments in MALDI hardware and increased attention paid to the sample preparation have changed the situation (7, 8). Recent years have seen an explosion of papers describing the application of MALDI TOF for quantitative studies (9–14). A direct comparison of MALDI TOF and ESI LC/MS quantitation schemes did not reveal a clear advantage for either (14–16). MALDI source demonstrated good comparability with ESI source on the same instrument in quantitative analysis of LMW compounds (17).

AAA is one of the areas, where ESI LC/MS has complete dominance over MALDI approach. ESI LC/MS and/or MS/MS have become a typical analytical tool in the analysis of LMW compounds (18–21). Nonetheless, application of MALDI TOF for amino acid analysis has clear advantages over ESI LC/MS/MS, GC/MS, or CE/MS. Foremost, it is an absence of a need for AA modification and/or separation step. Another apparent advantage of MALDI TOF MS over ESI LC/MS is in the area of high-throughput analysis since MALDI TOF data acquisition step is much shorter than ESI LC/MS analysis time (seconds vs. minutes).

Here, we describe two different AAA protocols based on the application of MALDI mass spectrometry. First protocol describes a MALDI TOF MS-based method for a routine simultaneous qualitative and quantitative analysis of free amino acids and protein hydrolysates. This method does not require amino acid derivatization or chromatographic separation, and the data acquisition time is decreased to several seconds for a single sample. No significant ion suppression effects were observed with the developed sample deposition technique and the method was found to be highly reproducible. Linear responses between the amino acid concentration and the peak intensity ratio of corresponding amino acid to internal standard were observed for all amino acids analyzed in the range of concentrations from 20 to 300 μM and correlation coefficients were between 0.983 for arginine to 0.999 for phenylalanine. Limit of quantitation varied from 0.03 μM for arginine to 3.7 μM for histidine and homocysteine. This method was applicable to the mixtures of free amino acids as well as HCl hydrolysates of proteins (22). Second protocol describes method based on the use of MALDI TOF/TOF MS/MS for qualitative analysis of amino and organic acids. This technique is capable of distinguishing isobaric and isomeric compounds (23).

2. Materials

2.1. Chemicals

1. Acetonitrile and trifluoroacetic acid (HPLC gradient-grade quality).
2. Water (HPLC gradient-grade quality).
3. Amino acids (Sigma, St. Louise, MO, USA).
4. α-Cyano-4-hydroxycinnamic acid (CHCA) (Aldrich, Milwaukee, WI, USA) (see Note 1).
5. All other chemicals and reagents should be of the highest purity available.
6. All solutions were prepared at room temperature and stored in refrigerator at 4°C.

2.2. Preparation of Amino Acid Standards

1. Prepare a 10 mM stock solution of AA of interest in deionized water.
2. For Tyr, stock solution concentration should be 3 mM.
3. Methyltyrosine (internal standard needed for quantitative MALDI TOF analysis) stock solution concentration should be 1 mM.

4. Mixtures of amino acid standards at concentrations 300, 200, 100, 80, 60, and 20 μM for MALDI TOF and 200 μM for MALDI TOF/TOF analyses are prepared by dilution in deionized water from a 10 mM stock solution.

2.3. Hydrolysis of Proteins or Peptides

1. Protein/peptide to be analyzed.
2. 6 M (constant boiling) HCl with 1% phenol by volume.
3. Nitrogen (prepurified grade).
4. Methanol.
5. Dry ice.
6. Microliter syringes or adjustable micropipets.
7. Pico-Tag Workstation or Eldex Hydrolysis/Derivatization Module.
8. 6×50 mm sample tubes (Corning # 9820-6, Kimax # 45048-650, or Kimble # 45060-650) or 6×31-mm, 250-μl flat-bottom glass inserts.
9. 2-propanol.
10. A vacuum pump, such as the Precision Scientific Model P-100 Series or similar pump.
11. Teflon forceps.
12. Plastic gloves and insulated gloves.
13. Silicon grease or other high-vacuum grease.

2.4. Preparation of Matrix Solution

1. CHCA is recrystallized before use (see Note 1).
2. For recrystallization, mix the CHCA with a small amount of methanol, heat under reflux until it boils, and slowly add methanol (while still boiling) until the CHCA is just completely dissolved. Filter the solution and let it stand for cooling down to room temperature. To enhance the yield of recrystallization, you can place the solution in the fridge. After crystallization, filter off the CHCA.
3. 8.5 ml of acetonitrile is diluted by 2.47 ml of water and 30 μL of 10% of trifluoroacetic acid is added to this solution. Final concentrations of acetonitrile and trifluoroacetic acid were 85 and 0.03%, respectively (see Note 2).
4. CHCA is added to 1 ml of this solution till saturation.
5. Matrix solution should be stored at –20°C while not in use (see Note 3).

2.5. Data Analysis

All calculations of mol% or residues per molecule and/or quantification of unknown samples were done by Excel spreadsheets available at http://www.abrf.org/ResearchGroups/AminoAcidAnalysis/EPosters/Archive/9c.html.

3. Methods

3.1. Hydrolysis of Proteins or Peptides

1. Clean the sample tubes with an acid (e.g., 6 M HCl). Rinse thoroughly with deionized water, then 100% ethanol, and dry under vacuum. Mark the sample tubes with a diamond pencil before placing them in the reaction vial for identification purposes.
2. Use a syringe or micropipet to place a sample solution into the sample tubes. As many as ten sample tubes may be placed in a single reaction vial. Use forceps to handle the sample tubes.
3. Screw the reaction vial cap onto the vial. Slide the button on the cap to the open (green in) position.
4. Turn on the vacuum pump.
5. Turn the oven to 105°C.
6. Install the reaction vial into the workstation by positioning the top of the vial in the vacuum port and slide the bottom of the vial in the black "shoe."
7. Open the vacuum valve slowly, gradually increasing vacuum to avoid "bubbling" of the samples and its "escape" from the tube.
8. Once the vacuum gauge reaches about 50–55 mTorr, the samples are properly evaporated.
9. Allow the system to stay at this reading for additional 20–30 min for complete drying.
10. Close the vacuum valve and remove the reaction vial from the workstation.
11. Slowly open the vacuum valve (green in) and unscrew the cap from the vial.
12. 200 μL of HCl/phenol should be pipetted into the bottom of the vial. Take care not to introduce HCl directly into any of the tubes.
13. With the reaction vial cap in the open (green in) position, screw the cap onto the vial and install the vial onto the workstation (see step 6).
14. Check that the nitrogen valve is closed and carefully open the vacuum valve until the vacuum gauge reads about 1–2 Torr and HCl begins to bubble. The procedure takes about 20–30 s.
15. Close the vacuum valve.
16. Open the nitrogen valve and purge for 5 s. Close the nitrogen valve.
17. Repeat the vacuum–nitrogen cycle three to four times. Leave the vacuum valve open after the last cycle.

18. Be careful not to evaporate HCl to dryness. If the vacuum gauge falls below 500 mTorr, discontinue purging.
19. While the vial is under vacuum, close the vial cap valve (red in).
20. Close the vacuum valve and remove the vial from the workstation.
21. Place the reaction vial in the oven. Use forceps to avoid contacting the oven interior.
22. After 20–24 h, remove the vial from the oven to cool.
23. Open the vial cap valve (green in). Remove the reaction vial cap and carefully remove the sample tubes with Teflon-coated forceps. Use laboratory wipes to remove HCl from the outside of each tube. Transfer sample tubes to a fresh reaction vial.
24. Repeat steps 3–9 (except step 5).
25. If the sample tubes are not processed immediately, store them in a refrigerator in the reaction vial in the dry state under nitrogen.

3.2. Mass Spectrometry

1. Use Applied Biosystems 4700 Proteomics Analyzer or any latest model of TOF/TOF instrument for MS/MS AAA or any TOF instrument for MS AAA.
2. Use following instrument parameters for 4700 MS mode: reflectron; laser intensity 4,500; positive ion mode; data collected over the mass window 50–300; to cover as much target area as possible, develop a spiral search pattern with shots at each firing position, 75 firing positions per spot, 450 shots total; focus mass 145.
3. Use following instrument parameters for 4700 MS/MS mode: positive MS-MS mode at 2 kV, collision gas, air; metastable suppressor, on (with optimized precursor function); 4,500 total shots (60 shots per subspectra and 75 subspectra); laser intensity for MS and MS-MS data collection, 4,500; precursor mass range, 76–205 (see Note 4).
4. The peak intensity values are determined using Data Explorer version 4.5 or any other available software.
5. For quantitative analysis, export the mass peak list and process it in Microsoft Excel. Then, plot the average ratio of the peak intensities against the amino acid concentration.
6. For calibration curve creation, measure mass spectra of mixtures containing all amino acids plus internal standard (see Note 5).

3.3. Sample Preparation

1. 100 μL of each amino acid stock solution diluted by deionized water. For quantitative MALDI TOF MS amino acid analysis, samples were diluted 1:32, 1:49, 1:99, 1:124, 1:166, and 1:199 to get 300, 200, 100, 80, 60, and 20 μM solutions, respectively.
2. For qualitative MALDI TOF/TOF MS/MS analysis, stock solutions were diluted 1:49 to get 200 μM solutions of amino acids.

3. For quantitative MALDI TOF MS analysis, add 2 μL of internal standard stock solution to 10 μL of amino acid mixture.
4. For hydrolyzed protein/peptide, add an appropriate volume of 0.1% TFA to the sample tubes.

3.4. Spotting

1. In each experiment, 1 μL of CHCA matrix solution is spotted onto each position on the matrix plate and air dried; after that, 1 μL of the mixture of amino acid sample solution and internal standard is loaded on the matrix spot (see Note 6).
2. Prepare four to six spots per each sample.

4. Notes

1. At the time, we have purchased CHCA from Aldrich and recrystallized it before use (Subheading 2.4, step 2). However, it is entirely possible that nowadays one can purchase CHCA of much better quality (in terms of purity) and use it without additional recrystallization. Prepare a matrix solution and see if you have any contaminant peaks interfering with AA peaks.
2. We evaluated different matrices for their suitability for quantitative amino acid analysis. In addition to CHCA, we tested 2, 5-dihydroxybenzoic acid (DHB) and meso-tetra (pentafluorophenyl) porphine (F20TTP) matrices (22). The best results in terms of amino acid ionization and resolution were achieved with CHCA (Fig. 1). Remarkably, the relative amino acids' peak intensities did not change in the range of concentrations from 20 to 300 μM. Comparison of upper and lower panels in Fig. 1 shows that 15-times increase in the amount of amino acids loaded on the target essentially did not change the relative responses of amino acids.
3. To improve the quality of experiments, fresh matrix solution was prepared every week.
4. In case of MALDI tandem MS, formation of immonium ions is dominant fragmentation reaction for all, even substituted and modified, protonated α-amino acids (with the exception of two highly basic amino acids Arg and Orn, Table 1). Characteristic fragmentation patterns that included additional elimination of NH_3 from the immonium ion were seen in case of two isomeric aliphatic amino acids Leu and Ile (Table 1, Fig. 2) that allow clear distinction between those amino acids (23). In contrast, β-, γ-, and δ-amino acids did not form any immonium ion and their fragmentation proceeds through loss of H_2O and/or NH_3 and formation of some other seemingly unrelated structural fragments (Table 1). We found that the presence of specific indicator peaks in MALDI tandem MS allows distinguishing isomeric and isobaric amino acid (Fig. 3).

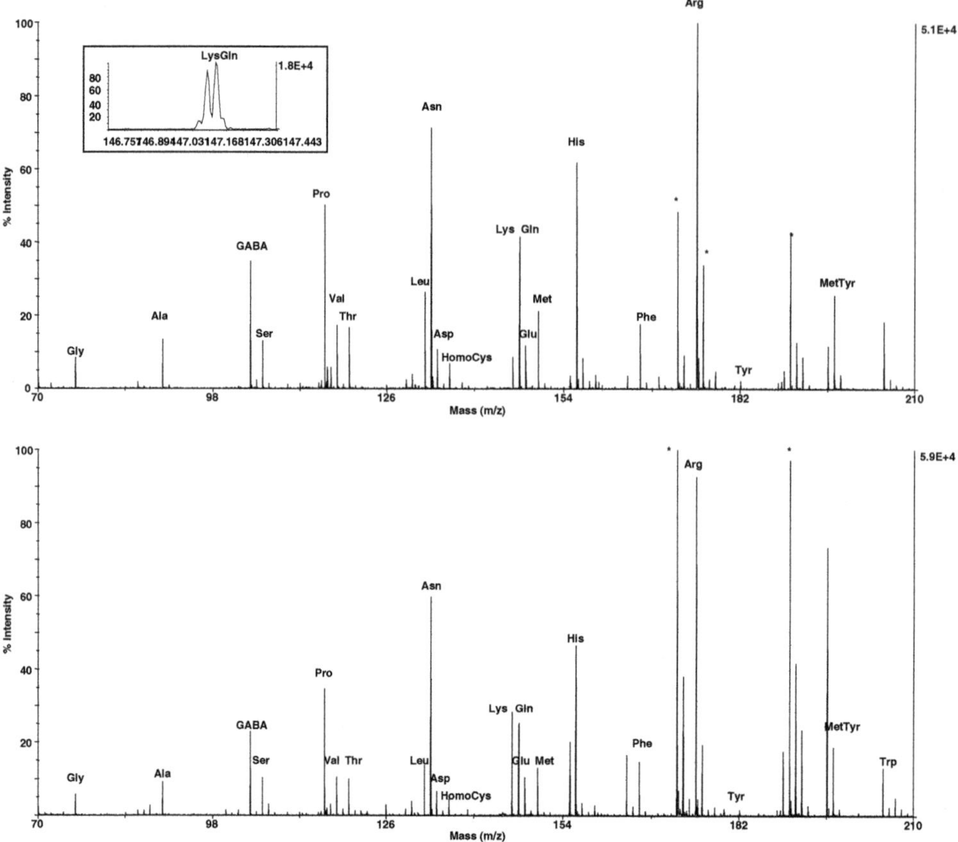

Fig. 1. Positive-ion MALDI TOF MS spectrum of 21 amino acid standard mixture with CHCA as a matrix. *Upper panel*: Final concentration of amino acids 300 μM (250 pmol of each amino acid on target). *Lower panel*: Final concentration of amino acids 20 μM (16.7 pmol of each amino acid on target). The relative peak intensity was not affected by amino acid concentration. *Inset*: Expanded view showing resolution attained, separation of Lys (147.14) and Gln (147.18) peaks. Reprinted from ref. 22 with permission

5. Figure 4 shows some representative calibration curves obtained by application of MALDI MS-based AAA. The internal standard used was methyltyrosine at a constant concentration 166.7 μM. We have obtained linear responses between the amino acid concentration and the peak intensities ratio of corresponding amino acid to internal standard for all proteinogenic amino acids with excellent correlation coefficients >0.983 (22).
6. We have explored a number of different sample deposition techniques. Dried-droplet sample deposition as well as sandwich or overlay methods did not produce consistent quantitative results. After numerous experiments, we have found that a preformed matrix layer technique with 85% AcCN/0.03% TFA provides consistent quantitative results and improves resolution and sensitivity of this method. No significant ion suppression effects were observed with the developed sample deposition technique and the method was found to be reproducible (22).

Table 1
Fragmentation of protonated amino acids by MALDI TOF/TOF tandem mass spectrometry

AA	MH^+	$MH^+ - CO_2H_2$	$MH^+ - H_2O$	$MH^+ - NH_3$	$MH^+ - CO_2H_2 - NH_3$	$MH^+ - CO_2H_2 - H_2O$	Other fragments
Gly	76	**30**					
Ala	90	**44**					
Ser	106	**60**			43	42	
Pro	116	**70**					43
Val	118	**72**					
Thr	120	**74**				56	45
Cys	122	**76**		105	59		112
Leu	132	**86**					44
Ile	132	**86**			69		41, 57
Asn	133	**87**					
Asp	134	**88**					74
Gln	147	**101**		130	**84**		44, 56, 75
Lys	147	101		130	**84**		56, 74
Glu	148	**102**				84	56, 75
Met	150	**104**					61
His	156	**110**					82
Phe	166	**120**					
Arg	175						30, 43, **60**, 70, **116**, 130

(continued)

Table 1 (continued)

AA	MH^+	$MH^+ - CO_2H_2$	$MH^+ - H_2O$	$MH^+ - NH_3$	$MH^+ - CO_2H_2 - NH_3$	$MH^+ - CO_2H_2 - H_2O$	Other fragments
Tyr	182	**136**					
Trp	205	**159**		**188**			**130**
$(CH_3)_2$Gly	104	**58**					42, 44
Hse	120	**74**				56	31, 44
Hyp	132	**86**				68	
Hcy	136	**90**	**118**				
Cit	176	**130**		**159**	**113**		30, **70**, 116
S-(carboxymethyl) Cys	180	**134**		**163**			
Met sulfone	182	**136**					56
Met sulfoxide	166	**120**		149		102	75, 74
Gly-Gly	133	30					**76 (y-type ion)**
Orn	133		115	116			30, 43, **70**
β-Ala	90		72				**30**, 45
Abu	104	**58**					
β-βAIB	104		86		41		**30**, 45
GABA	104		**86**	**87**	41		30, 45, 69
δ-Aminolevulinic acid	132		**114**				
Carnitine	162						**58, 60, 103**

The most abundant ions are marked by *bold font*. Reprinted from ref. 23 with permission

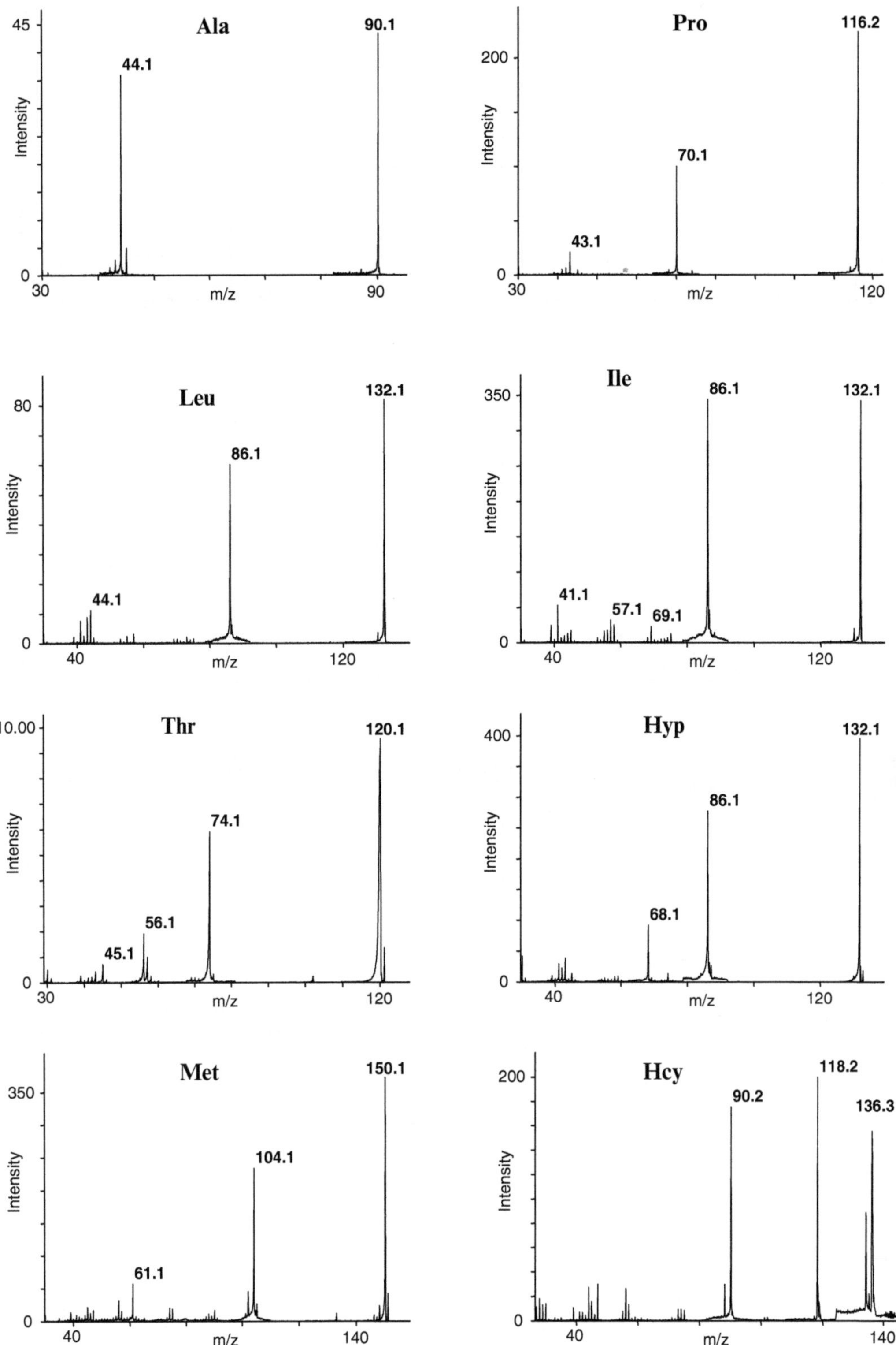

Fig. 2. Representative MALDI TOF/TOF CID spectra of amino acid. Reprinted from ref. 23 with permission.

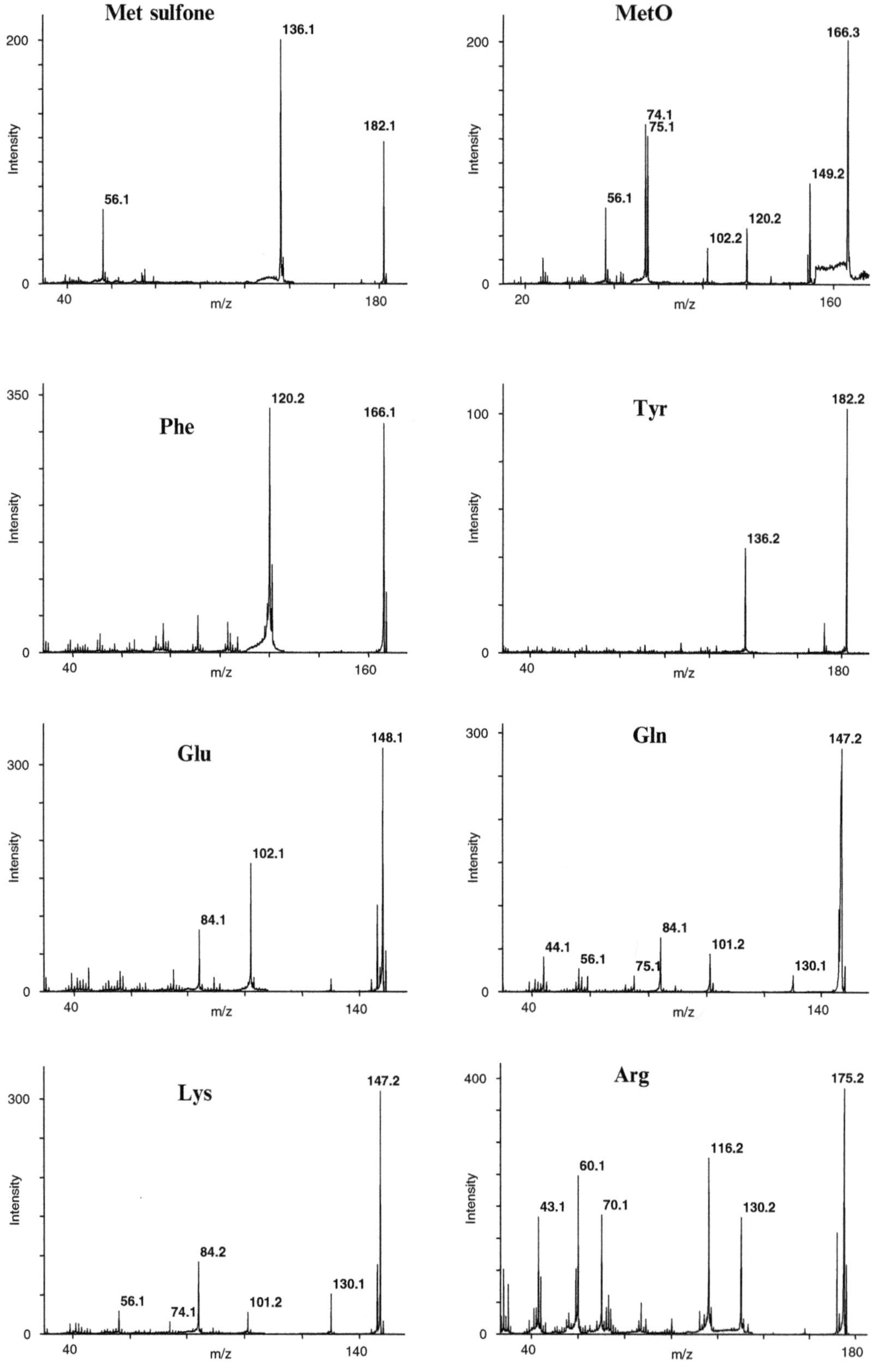

Fig. 2. (continued)

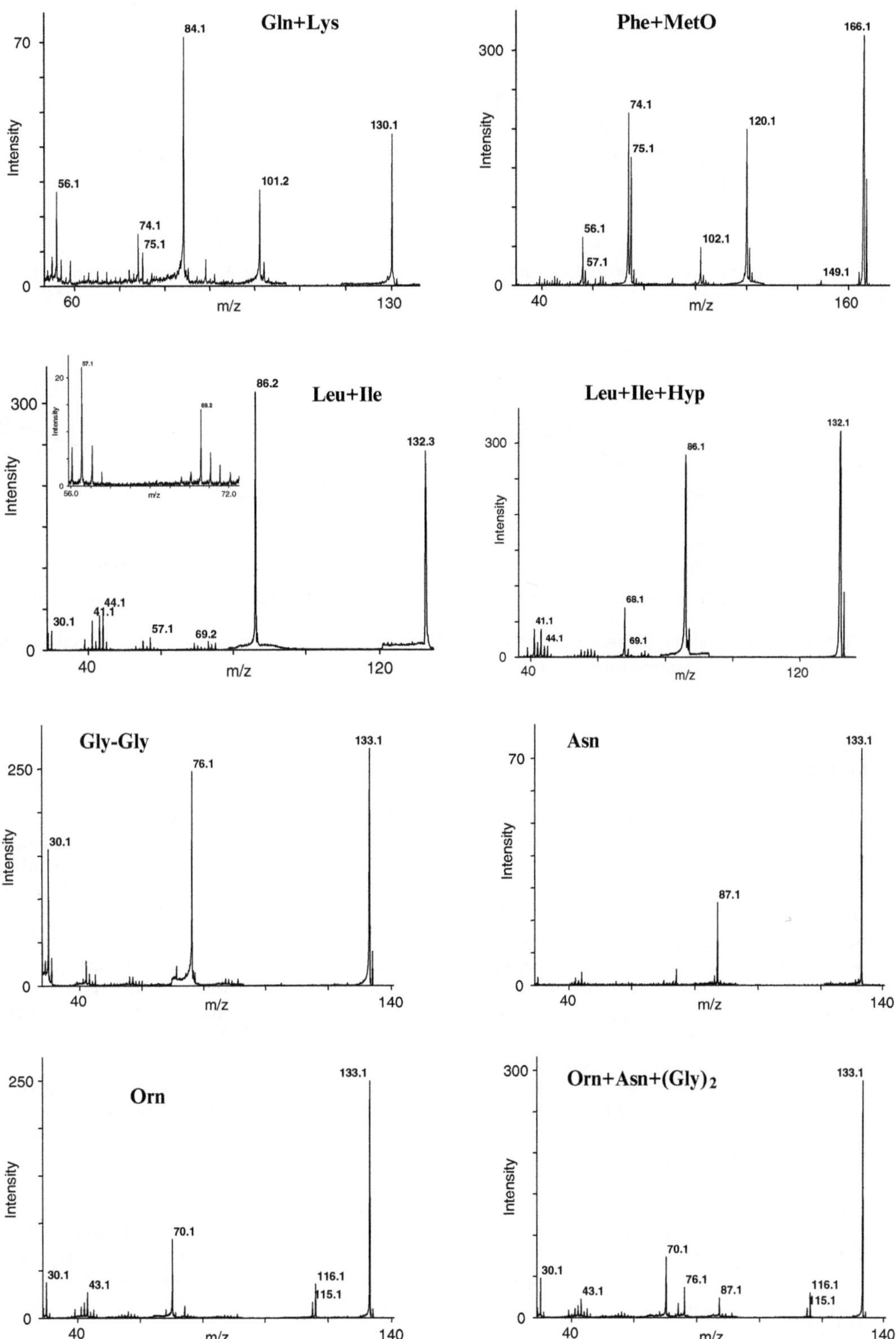

Fig. 3. Representative MALDI TOF/TOF CID spectra of isobaric and isomeric amino acid mixtures. Reprinted from ref. 23 with permission.

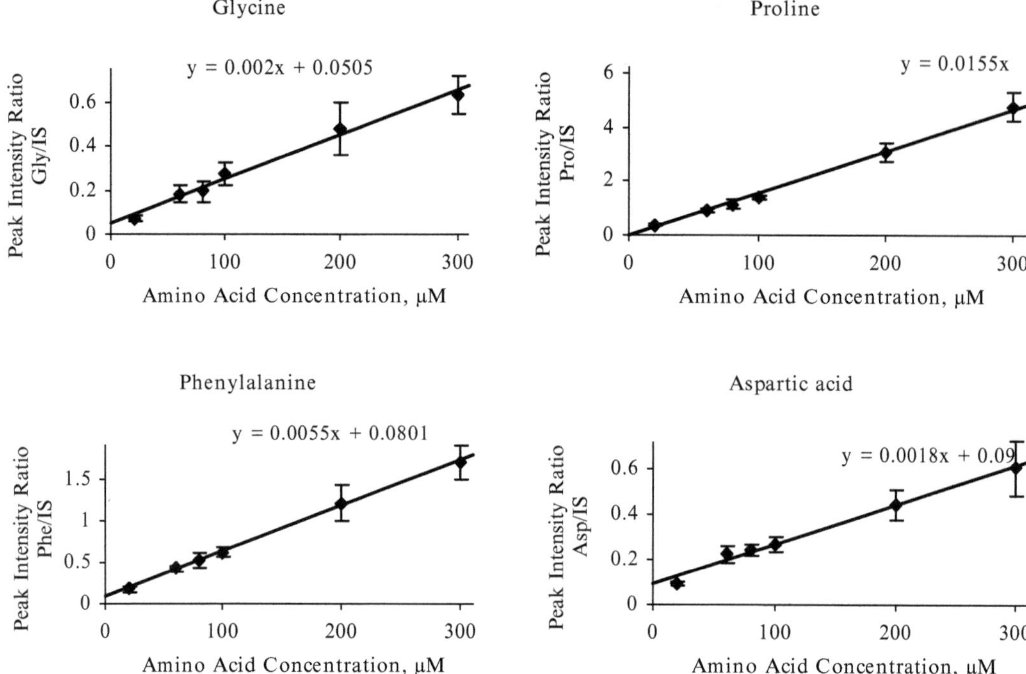

Fig. 4. Representative calibration curves of some amino acids. Each data point represents the mean ± S.D. of data collected in four to six experiments. Observed linear range: from 20 to 300 μM.

References

1. Moore S, Stein WH (1951) Chromatography of amino acids on sulfonated polystyrene resins. J Biol Chem 192:663–681
2. Furst P et al (1990) Appraisal of four pre-column derivatization methods for the high-performance liquid chromatographic determination of free amino acids in biological materials. J Chromatogr 499:557–569
3. Fierabracci V et al (1991) Application of amino acid analysis by high-performance liquid chromatography with phenyl isothiocyanate derivatization to the rapid determination of free amino acids in biological samples. J Chromatogr 570:285–291
4. Duncan M, Poljak A (1998) Amino Acid Analysis of Peptides and Proteins on the Femtomole Scale by Gas Chromatography/Mass Spectrometry. Anal Chem 70:890–896
5. Zahou E, Jornvall H, Bergman T (2000) Amino acid analysis by capillary electrophoresis after phenylthiocarbamylation. Anal Biochem 281:115–122
6. Husek P, Simek P (2001) Advances in Amino Acid Analysis. LCGC 19:986–999
7. Gobey J et al (2005) Characterization and performance of MALDI on a triple quadrupole mass spectrometer for analysis and quantification of small molecules. Anal Chem 77: 5643–5654
8. Bienvenut W et al (2002) Matrix-assisted laser desorption/ionization-tandem mass spectrometry with high resolution and sensitivity for identification and characterization of proteins. Proteomics 2:868–876
9. Hlongwane C et al (2001) Comparative quantitative fatty acid analysis of triacylglycerols using matrix-assisted laser desorption/ionization time-of-flight mass spectrometry and gas chromatography. Rapid Commun Mass Spectrom 15:2027–34
10. Ayorinde F, Garvin K, Saeed K (2000) Determination of the fatty acid composition of saponified vegetable oils using matrix-assisted laser desorption/ionization time- of-flight mass spectrometry. Rapid Commun Mass Spectrom 14:608–615
11. Bucknall M, Fung K, Duncan M (2002) Practical quantitative biomedical applications

of MALDI-TOF mass spectrometry. J Am Soc Mass Spectrom 13:1015–1027

12. Alterman M A et al (2005) Quantitative Analysis of Cytochrome P450 Isozymes by Means of Unique Isozyme-Specific Tryptic Peptides: A Proteomic Approach, Drug Metab Dispos. 33:1399–1407
13. Helmke S et al (2004) Simultaneous quantification of human cardiac alpha- and beta-myosin heavy chain proteins by MALDI-TOF mass spectrometry. Anal Chem 76:1683–1689
14. Kang M, Mathur S, Hartmann R W (2004) Quantitative analysis of 5 alpha-reductase inhibitors in DU145 cells using matrix-assisted laser desorption/ionization time-of-flight mass spectrometry and high-performance liquid chromatography/tandem mass spectrometry. J Mass Spectrom 39:762–769
15. Shi Y et al (2004) A simple solid phase mass tagging approach for quantitative proteomics. J Proteome Res 3:104–111
16. Sleno L, Volmer D (2005) Toxin screening in phytoplankton: detection and quantitation using MALDI triple quadrupole mass spectrometry. Anal Chem 77:1509–1517
17. Medzihradszky K et al (2000) The characteristics of peptide collision-induced dissociation using a high-performance MALDI-TOF/TOF tandem mass spectrometer. Anal Chem 72:552–558
18. Soga T, Heiger D (2000) Amino acid analysis by capillary electrophoresis electrospray ionization mass spectrometry. Anal Chem 72:1236–1241
19. Piraud M et al (2003) ESI-MS/MS analysis of underivatised amino acids: a new tool for the diagnosis of inherited disorders of amino acid metabolism. Fragmentation study of 79 molecules of biological interest in positive and negative ionisation mode. Rapid Commun Mass Spectrom 17:1297–1311
20. Pitt JJ, Eggington M, Kahler SG (2002) Comprehensive Screening of Urine Samples for Inborn Errors of Metabolism by Electrospray Tandem Mass Spectrometry. Clin Chem 48:1970–1980
21. Zytkovicz TH et al (2001) Tandem Mass Spectrometric Analysis for Amino, Organic, and Fatty Acid Disorders in Newborn Dried Blood Spots: A Two-Year Summary from the New England Newborn Screening Program. Clin Chem 47:1945–1955
22. Alterman M, Gogichayeva N, Kornilayev B (2004) Matrix-assisted laser desorption/ionization time-of-flight mass spectrometry-based amino acid analysis. Anal Biochem 335:184–191
23. Gogichayeva N, Williams T, Alterman M (2007) MALDI TOF/TOF tandem MS as a new tool for amino acid analysis, J Am Soc Mass Spectrom 18:279–284

Chapter 13

Heptafluorobutyl Chloroformate-Based Sample Preparation Protocol for Chiral and Nonchiral Amino Acid Analysis by Gas Chromatography

Petr Šimek, Petr Hušek, and Helena Zahradníčková

Abstract

Gas chromatography (GC) is a commonly used technique in amino acid analysis (AAA). However, one of the requirements of the application of GC for AAA is a need for the polar analytes to be converted into their volatile, thermally stable derivatives. In the last two decades, alkyl chloroformates have become attractive derivatization reagents. The reagents react immediately with most amino acid functional groups in aqueous matrices and the process can easily be coupled with liquid–liquid extraction of the resulting less-polar derivatives into immiscible organic phase. Here, we describe a simple protocol for in situ derivatization of amino acids with heptafluorobutyl chloroformate followed by subsequent chiral as well as nonchiral GC/mass spectrometric analysis on a respective nonpolar fused silica and an enantioselective Chirasil-Val capillary column.

Key words: Amino acid, Serum, Plasma, Derivatization, Heptafluorobutyl chloroformate, 2,3-Dimercapto-1-propanesulfonic acid, Gas chromatography, Mass spectrometry, Enantiomer, Chiral analysis, Amino acid analysis

1. Introduction

Separation of amino acids by gas chromatography (GC) has been a popular approach capable to satisfy diverse demands on the determination of free amino acids, including their enantiomeric analysis. However, most of amino acid protic functional groups have to be blocked by derivatization prior to GC, changing thus their polarity and affinity to organic solvents.

The derivatization has been achieved in many ways; two-step acylation–esterification, silylation, and use of alkyl chloroformates (RCF) have been the most popular (1–3). The latter approach is a very attractive as being fast and cost-effective and delivers comfortable

Michail A. Alterman and Peter Hunziker (eds.), *Amino Acid Analysis: Methods and Protocols*, Methods in Molecular Biology, vol. 828, DOI 10.1007/978-1-61779-445-2_13,

sample derivatization directly in aqueous media. Furthermore, the RCF reagents are not only capable to treat amino acids in situ, but the derivatization process can be easily coupled to the common preceding steps, such as precipitation of protein–lipid–glycan polymeric structures, reductive release of thiol-bound amino acids, and liquid–liquid extraction of the amino acid derivatives into immiscible solvents. A wide choice of the RCF reagents is available; RCFs with methyl to hexyl and those with fluorinated alkyls have been applied in numerous applications (4, 5). Recently, we described heptafluorobutyl chloroformate (HFBCF) as a perspective and versatile derivatization reagent for amino acid analysis (AAA) (6). It reacts efficiently in aqueous environment with most protic functional groups and, in comparison to the traditional RCFs with aliphatic alkyls, the carboxyl moiety is directly esterified without the presence of the analogous alcohol.

Enantiomeric AAA has been another important application field performed favorably on chiral GC stationary phases. A Chirasil-Val phase, representing dimethylpolysiloxane-bonded valine tert. butyl amide (in either L or D form) and functionalized cyclodextrins are the most important used chiral GC columns. From the current GC-based methods, only two approaches enable chiral analysis of most proteinogenic amino acids in a single analysis run: (1) a two-step method comprising esterification of carboxylic group and subsequent acylation of the other protic groups with fluorinated anhydrides (7) and (2) in situ derivatization with fluoroalkyl chloroformates (8, 9).

We have combined our recent HFBCF-based sample preparation approach (6) with chiral investigations (8, 9) into a simple protocol useful to both nonchiral GC/MS analysis of more than 50 amino acids in aqueous matrices and, simultaneously, to chiral GC/MS analysis of all proteinogenic amino acid enantiomers, except arginine (not eluted) and proline (not separated) on a Chirasil-Val capillary column.

2. Materials

Use appropriate laboratory wear, glasses, and other personal protective equipment, and follow standard laboratory precautions and local guidelines. Diligently follow all waste disposal regulations.

2.1. Chemicals

1. Prepare all solutions using ultrapure deionized water and analytical-grade reagents.
2. Prepare and store all reagents at 4°C (unless indicated otherwise).
3. Handle the concentrated perchloric acid (PCA) and the HFBCF reagent carefully, and refer to a corresponding material safety data sheet (MSDS).

2.2. Analytical Standards

1. L-Amino acids, D-amino acids, and other analytical standards (puriss or bioultra grade, Sigma–Aldrich, St. Louis, MO, USA, and worldwide offices).
2. ^{13}C or deuterium labeled amino acid standards (Cambridge Isotope Labs, Andover, USA, or Sigma–Aldrich).
3. Prepare a working solution of each amino acid standard by dissolving them at 50 μM in 0.05 M HCl.
4. Dissolve glutamine, asparagine, and tryptophan in water with acetonitrile (4:1, v/v) (see Note 1).

2.3. Glassware

1. The glass culture tubes 6 × 50 mm (Kimble-Kontes, Vineland, NJ, USA) for the sample handling and for the derivatization reactions.
2. The 1.1-ml tapered polypropylene tubes (Continental Lab Products, San Diego, CA, USA) without any closure for the derivatization reactions.
3. Common screw-cap, teflon-lined, 2-ml amber vials for the reagent solutions.
4. Common 2 ml clear screw top vials with 200 μl liners 02-MTV (Chromacol, Welwyn Garden City, UK).

2.4. Volumetric and Mixing Equipment

1. An adjustable 50- and 100-μl transferpettor pipette with a glass capillary (Merck, Darmstadt, Germany) for manipulation with the reactive reagents and their mixtures with organic solvents.
2. The pipette tips with 25 mm capillary (gel-loading type, VWR Int., West Chester, PA, USA).
3. A common vortex mixer.

2.5. Material for the Protein Precipitation Step

1. PCA (70–72% aqueous stock solution, circa 11.8 M).
2. A 1.67 and 1.17 M PCA precipitation solution: Dilute the stock PCA with water in a ratio of 1:6 and 1:9 (v/v), respectively.

2.6. Material for the Step Releasing the Thiol-Bound Amino Acids

1. Sodium 2, 3-dimercapto-1-propanesulfonate monohydrate (DMPS, Sigma–Aldrich).
2. A 1.5% DMPS solution in water (w/w): Weigh 15 mg of the DMPS salt in a 2-ml amber vial and dissolve in 1 ml water. Prepare fresh solution weekly (see Note 2).

2.7. Material for the Derivatization–Extraction Step

1. Inorganic salts, acids, and bases: Sodium carbonate, sodium hydroxide, and sodium chloride.
2. Organic solvents: Pyridine (absolute, over molecular sieve, Sigma–Aldrich), isooctane (2,2,4-trimethylpentane, Suprasolv grade, Merck).

3. Physiological solution: Dissolve 0.9 mg of NaCl in 100 ml water.
4. Carbonate saline buffer: 10 mM sodium carbonate in physiological solution. Mix 0.5 M sodium carbonate with the physiological solution in volume ratio 1:49 (see Note 3).
5. Neutralizing buffer: 1 M NaOH:0.5 M Na_2CO_3=4:1 (v/v). Prepare 1 M sodium hydroxide in water; dissolve 53 g sodium carbonate in 1 L water. Mix the hydroxide with the carbonate solution in a volume ratio of 4:1 (v/v).
6. Catalyst medium: 50 mM Na_2CO_3:pyridine = 3:1. Dilute the 0.5 M sodium carbonate to a 0.05 M solution and mix with pyridine 3:1 (v/v).
7. HFBCF derivatization–extraction medium: The HFBCF reagent (2,2,3,3,4,4,4-heptafluorobutyl chloroformate, Aneclab, České Budějovice, Czech Republic). Dilute HFBCF with isooctane 1:3 (v/v) (see Note 4).

2.8. Material for the Nonchiral and Chiral GC Separation

1. Nonchilar capillary-fused silica column: CP-VF-5 ms, 20 m × 0.25 mm, 0.25 μm film thickness (Agilent, Santa Clara, CA, USA).
2. Chiral-fused silica capillary column: CP-Chirasil-L-Val, 25 m × 0.25 mm i.d., 0.12 μm film thickness; (Agilent, Santa Clara, CA, USA).

2.9. GC/MS Analytical Equipment

1. Gas chromatograph: (Trace Ultra, Thermo Scientific, San Jose, CA, USA). A programmed temperature vaporizing injector (PTV, Thermo Scientific), an injector temperature, 230°C; a 2 mm i.d. baffled inlet liner; 12.5 mm Thermolite septum (both Restek Co., Bellefonte, PA, USA).
2. Mass spectrometer: (DSQ, Thermo Scientific). Electron ionization (EI, 70 eV); ion source temperature, 210°C; ionization energy, 70 eV.

3. Methods

3.1. The Serum/Plasma Sample Preparation

1. Collect venous blood samples by a standardized procedure: Serum into clotting (red topped) test tubes at room temperature for 15–30 min. Plasma into heparin, EDTA, or EDTA–NaF-containing vials, and cool immediately in ice water. Separate serum or plasma by centrifugation at 2,000 × *g* for 15 min at 4°C as soon as possible.
2. Spike serum or plasma sample (75 μL) in 6 × 50 culture tubes with 25 μL of a standard solution containing labeled amino

acids in the carbonate saline buffer if needed, and add 25 μL of 1.5% DMPS solution. Mix gently the content and let to stand for 2–3 min (see Note 5).

3. Precipitate serum/plasma proteins by addition of 25 μL 1.67 M PCA, vortex vigorously for 10–15 s, and after waiting for 2–3 min repeat vortexing and then centrifuge vials at 2,000 × *g* for 10 min (see Note 6).
4. Aspirate a 120-μL aliquot of the supernatant from the culture tube by means of a pipette tip with a 25-mm capillary and transfer the content into another culture tube. Increase the pH by adding 25 μL of the neutralization buffer (see Note 7).
5. Add 50 μL of the HFBCF derivatization–extraction medium and vortex for 10–15 s.
6. Add 25 μL of the catalyst medium, vortex briefly, and repeat vortexing until the milky organic phase is cleared (see Note 8).
7. Add isooctane (80–100 μl), vortex briefly, and after the immiscible phases are separated transfer the upper organic phase (100–120 μl) into an autosampler vial (see Note 9).

3.2. Sample Preparation Protocol for the Amino Acid Standards

1. Replace serum/plasma with a 75 μL physiological solution in a 6 × 50 culture tube; add 25 μL of a standard solution containing labeled amino acids in carbonate saline buffer if needed.
2. Add 25 μL of a 1.5% DMPS solution. Mix gently the content and let to stand for 2–3 min. Add 25 μl of 1.17 M PCA, vortex vigorously, and basify the content by adding 25 μL of the neutralization buffer (see Note 10).
3. Aspirate 120 μl from the above mixture, add 50 μL of the HFBCF derivatization–extraction medium, and follow the steps 6 and 7 in Subheading 3.1 (see Note 11).

3.3. Nonchiral GC/MS Analysis

1. Put a vespel/graphite ferrule on both column ends, install a baffled inlet liner into a GC injector, and install a fused silica column according to a GC/MS Instrument Manual.
2. Evacuate the system, set up a carrier gas helium at a flow rate of 1.3 ml/min and a constant velocity mode, and heat the instrument temperature zones as follows: injector, 230°C; transfer line, 250°C, an electron impact (EI) ion source, 210°C.
3. Set an injection mode to splitless for 0.7 min, purge gas flow closed for 0.8 min, and autosampler injection volume to 0.5–1 μl (see Note 12).
4. Set the oven temperature program as follows: 50°C, hold for 1.5 min, then 20°C/min to 170°C first, and at 30°C/min to 300°C, hold for 1 min. Total analysis time, 13 min (without column cooling and equilibration).

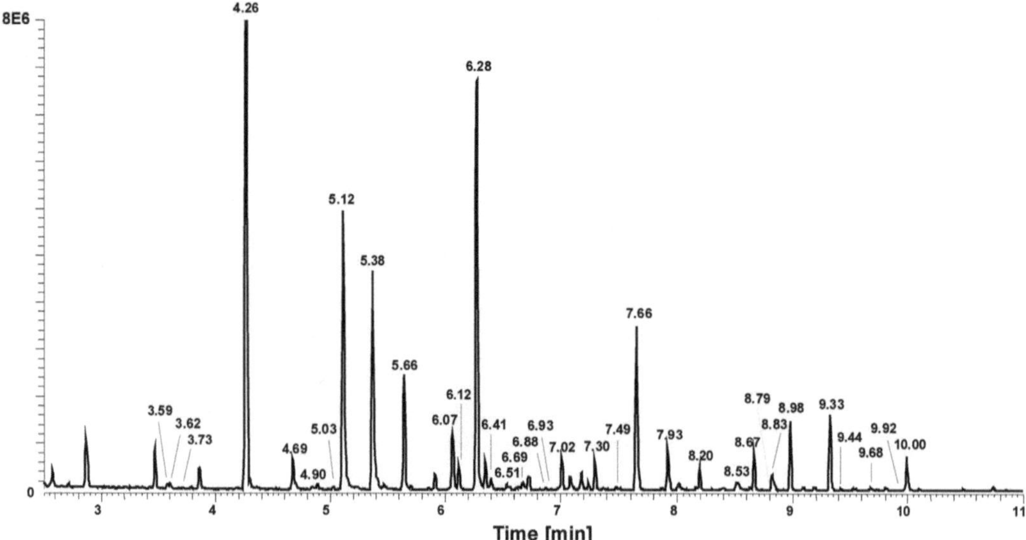

Fig. 1. TIC GC/MS chromatogram obtained by the nonchilar amino acid analysis in human serum. Retention time (RT) in minutes above a particular chromatographic peak denotes an analyte identified in Table 1.

5. Set the operating MS conditions to the full scan mode (typically, 50 – the upper instrument mass range), scan speed, 4–6 scans per second for a quadrupole mass spectrometer (see Note 13).
6. Prepare an analysis sequence of a series of samples (serum/plasma extracts, standards, blanks, labeled internal standards, etc.) and perform the GC/MS analysis. A typical GC/MS analytical run of a serum extract is shown in Fig. 1 (see Note 14).
7. Inspect the GC/MS chromatogram obtained and the occurrence of amino acids according to Table 1. Find the *m/z* fragment ions characteristic for amino acids and other detected components of interest and set up the instrument method to nontargeted (full scan) and/or targeted (selected ion monitoring = SIM) analysis mode as needed. Perform calibration and quantitative determination for a particular analytical problem (see Note 15).

3.4. Chiral GC/MS Analysis

1. Install or change a nonchiral capillary GC column for a CP-Chirasil-Val-L column following a GC/MS Instrument Manual (see Note 16).
2. Set up a helium carrier gas flow at 1–1.2 ml/min, a constant velocity mode, and heat the GC instrument zone temperatures: injector, 230°C; transfer line, 200°C; EI ion source, 210°C.
3. Set injection mode to splitless, and keep the identical injection conditions as described for the nonchiral analysis, except that the injection volume is set to 0.2–0.5 μl (see Note 17).

Table 1
List of serum metabolites observed in the nonchiral GC/MS chromatogram, see Fig. 1, including molecular weight (MW) of the HFBCF-derivatized products and the diagnostic fragment ions observed in their EI spectra. Data on the commonly used labeled amino acids (prospective internal standards) are enclosed

					Diagnostic ions (% of abundance)			
	RT (min)	Analyte amino acid	3-Letter code	MW	*m/z* (100)	*m/z*	*m/z*	(M-227)+
1	3.59	2-Keto-3-methylvaleric acid	KMV	312	57	85(31)	69(6)	85(31)
2	3.62	Ketoisocaproic acid	KIC	312	57	85(69)	69(13)	85(69)
3	3.73	d3-Methylmalonic acid (IS)	d3-MMA	485	286	258(17)	441(5)	258(17)
4	3.75	Methylmalonic acid	MMA	482	283	255(19)	438(5)	255(19)
5	4.26	Lactic acid	2HPA	498	227	271(92)	226(20)	271(92)
6	4.69	2-Hydroxybutyric acid	2HB	512	285	227(54)	133(44)	285(100)
7	4.90	2-Hydroxyisovaleric acid	2HIV	526	55	299(86)	113(58)	299(86)
8	5.02	d3-Sarcosine (IS)	d3-Sar	500	273	272(10)	113(10)	273(100)
9	5.04	Sarcosine	Sar	497	270	113(36)	226(19)	270(100)
10	5.11	d3-Alanine (IS)	d3-Ala	500	273	271(17)	274(12)	273(100)
11	5.12	Alanine	Ala	497	270	113(10)	227(7)	270(100)
12	5.36	d2-Glycine (IS)	d2-Gly	485	258	113(18)	214(10)	258(100)
13	5.38	Glycine	Gly	483	256	113(16)	212(10)	256(100)
14	5.48	Aminobutyric acid	ABU	511	284	113(25)	84(22)	284(100)
15	5.65	Valine	Val	525	298	98(48)	283(39)	298(100)

(continued)

Table 1 (continued)

	RT (min)	Analyte amino acid	3-Letter code	MW	Diagnostic ions (% of abundance)			
					m/z (100)	*m/z*	*m/z*	(M-227)+
16	6.07	allo-Isoleucine	allo-Ile	539	312	283(78)	256(57)	312(100)
17	6.07	Leucine	Leu	539	312	256(93)	270(29)	312(100)
18	6.12	Isoleucine	Ile	539	312	283(99)	256(73)	312(100)
19	6.28	Proline	Pro	523	296	297(8)	113(5)	29(100)6
20	6.41	Threonine1	Thr1	527	283	100(28)	113(20)	300(18)
21	6.39	Pipecolic acid	Pip	537	310	82(8)	113(8)	310(100)
22	6.51	Serine1	Ser1	513	283	286(56)	86(53)	286(56)
23	6.52	γ-Aminobutyric acid	GABA	511	112	256(68)	270(49)	284(17)
24	6.68	Aspartic acid	Asp	723	254	296(37)	496(12)	496(12)
25	6.69	Asparagine	Asn	522	295	113(29)	95(17)	295(100)
26	6.88	Threonine2	Thr2	753	282	227(70)	113(67)	526(8)
27	6.93	Thiaproline	TPN	541	314	287(84)	113(18)	314(100)
28	6.99	d3-Serine2 (IS)	d3-Ser2	742	271	183(17)	315(4)	515(7)
29	7.01	13C3-Serine2 (IS)	13C3-Ser2	742	270	183(10)	514(10)	
30	7.01	Serine2	Ser2	739	268	183(16)	312(8)	512(2)
31	7.30	Glutamic acid	Glu	737	282	310(40)	82(35)	510(24)
32	7.48	d3-Methionine (IS)	d3-Met	560	64	283(26)	560(5)	333(6)
33	7.49	Methionine	Met	557	61	283(43)	557(26)	330(12)
34	7.64	d2-Cysteine (IS)	d2-Cys	757	113	302(16)	530(16)	530(16)

35	7.66	Cysteine	Cys	755	113	300(51)	528(33)	528(33)
36	7.92	d5-Phenylalanine (IS)	d5-Phe	578	335	96(71)	334(21)	351(19)
37	7.93	Phenylalanine	Phe	573	91	330(79)	131(19)	346(7)
38	8.19	d4-Homocysteine (IS)	d4-Hcy	773	285	514(8)	85(69)	545(2)
39	8.20	Homocysteine	Hcy	769	282	82(34)	283(32)	542(4)
40	8.53	Glutamine	Gln	554	84	282(95)	327(66)	327(66)
41	8.66	Ornithine	Orn	766	296	256(21)	566(4)	
42	8.79	1-Methylhistidine1	1MeHis1	803	95	334(71)	94(11)	
43	8.83	Histidine	His	789	307	562(47)	546(45)	562(47)
44	8.86	1-Methylhistidine2	1MeHis2	577	95	96(7)	378(5)	
45	8.98	13C6-Lysine (IS)	13C6-Lys	786	315	257(30)	314(28)	
46	8.98	Lysine	Lys	780	310	256(44)	380(8)	
47	9.33	Tyrosine	Tyr	815	333	289(62)	588(5)	588(5)
48	9.44	Cysteinylglycine	CysGly	812	113	300(78)	528(36)	585(4)
49	9.66	d4-Cystathinone (IS)	d4-Cth	1,042	332	284(52)	514(8)	
50	9.68	Cystathionine	Cth	1,038	328	282(69)	510(8)	
51	9.85	d4-Cystine (IS)	d4-(Cys)2	1,060	498	270(38)	530(10)	
52	9.86	Cystine	(Cys)2	1,056	496	268(31)	528(31)	829(1)
53	9.92	γ-Glutamylcysteine	γ-GluCys	1,066	113	285(56)	270(46)	
54	9.99	Tryptophan	Trp	612	130	385(4)	612(3)	385(4)
55	10.15	d8-Homocystine (IS)	d-8(Hcy)2	1,078	514	318(56)	692(11)	
56	10.17	Homocystine	(Hcy)2	1,070	510	282(50)	314(35)	

4. Set the temperature program to 92°C, held for 1 min, programmed at 2.5°C/min to 200°C, held for 6 min, and then back at 20°C/min to 92°C. Total analysis time, 55 min.
5. Set the MS to a full scan mode (nontargeted analysis, typically 60–600 mass units), scan speed, 4 scans per second, or use a SIM mode (see Note 18).
6. Prepare an analysis sequence for the AAA of a series of samples (serum/plasma extracts, amino acid standards, blanks, labeled internal standards, etc.) and perform the GC/MS analysis. A typical chromatogram obtained by the chiral GC/MS analysis of a serum extract with added and absent D-amino acids is presented in the top and bottom parts of Fig. 2, respectively (see Note 19).

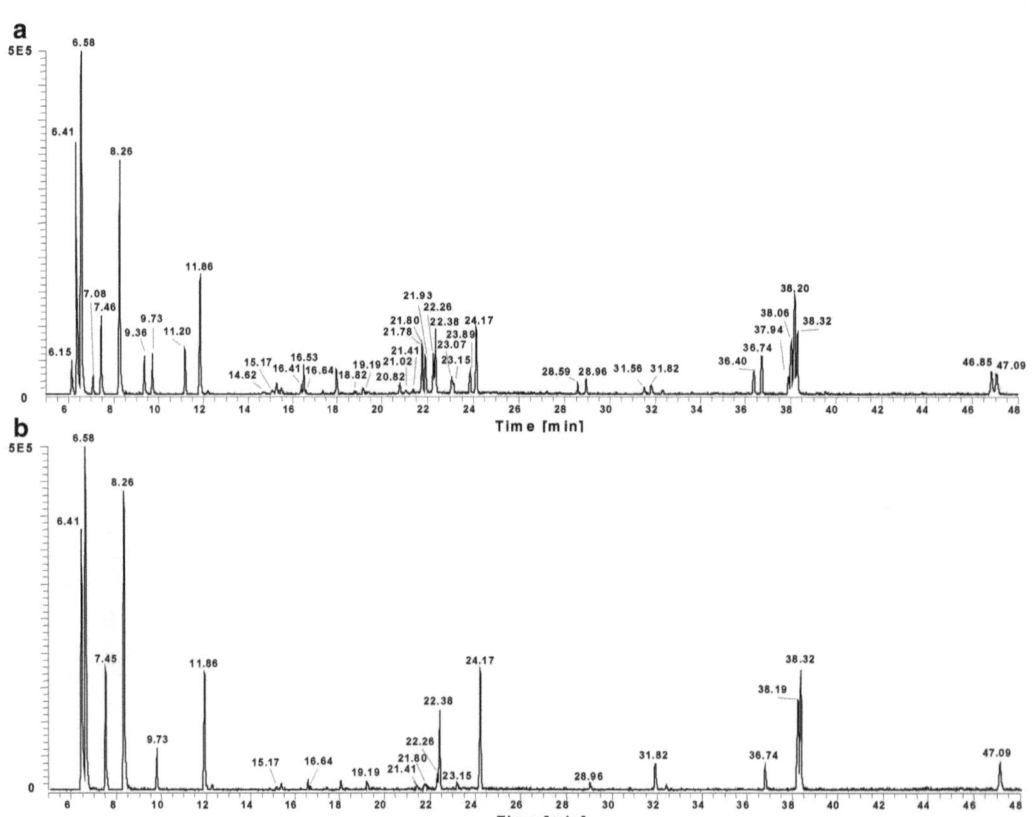

Fig. 2. TIC GC/MS chromatogram of the chiral amino acid analysis obtained from the same serum extract measured by the nonchiral GC/MS method; refer to Fig. 1. Retention time in minutes above a particular chromatographic peak denotes an analyte identified in Table 2. *Top*. Chiral GC/MS analysis of the serum spiked with the following D-amino acid enantiomers (min): D-Ala (6.15), D-Val (7.08), D-Ile (9.36), D-Leu (11.2), D-Asp (16.41), D-Met (20.84), D-Phe (22.38), D-Asn (23.07), D-Cys (23.89), D-His(31.56), D-Orn (36.4), D-Lys (37.97), D-Tyr (38.06), D-Trp (46.85) all at 60 μM; D-Thr (14.62), D-Ser (18.82), D-Gln (21.02), D-Glu (22.26) and D,L-HCy (28.59) all at 120 μmol/ml. (A pure D-Hcy enantiomer was not available.) *Bottom*. Chiral GC/MS analysis of the same serum without the labeled amino acids.

7. Inspect the obtained GC/MS chromatogram and the occurrence of D, L amino acid enantiomers in the data set by means of Table 2. Find the diagnostic *m/z* fragment ions of amino acids and optimize finally the protocol for a particular application (see Note 20).

Table 2
List of the resolved D,L-enantiomers and some other amino acids recorded by the chiral GC/MS in serum; refer also to Fig. 2. The resolution of each L-enantiomers and earlier eluted D-antipode was calculated according to the formula: $R = RT_L - RT_D/0.5(W_D + W_L)$ (RT – retention time, *W* – peak width)

RT (min)	Analyte amino acid	3-Letter code	Diagnostic (*m/z*)	Resolution (*R*)
2.94	Sarcosine	Sar	270	
6.15	D-Alanine	D-Ala	270	
6.41	D,L-Proline	Pro	296	
6.58	L-Alanine	L-Ala	270	2.1
7.08	D-Valine	D-Val	298	
7.38	D-α-Aminobutyric acid	D-Abu	284	
7.45	L-Valine	L-Val	298	2.22
7.86	L-α-Aminobutyric acid	L-Abu	284	2.18
8.26	Glycine	Gly	256	
8.34	DL-β-Aminobutyric acid	Baib	256	
8.71	D-alloIsoleucine	D-alloIle	312	
9.17	L-alloIsoleucine	L-alloIle	312	2.09
9.36	D-Isoleucine	D-Ile	312	
9.73	L-Isoleucine	L-Ile	312	2.53
9.93	D-Thiaproline	D-TPN	314	
11.2	D-Leucine	D-Leu	312	
11.86	L-Leucine	L-Leu	312	4.25
13.32	γ-Aminobutyric acid	GABA	112	
14.62	D-Threonine2	D-Thr	282	
15.17	L-Threonine2	L-Thr	282	1.69
16.41	D-Aspartic acid	D-Asp	254	
16.64	L-Aspartic acid	L-Asp	254	1.33
18.82	D-Serine2	D-Ser	268	

(continued)

Table 2 (continued)

RT (min)	Analyte amino acid	3-Letter code	Diagnostic (*m/z*)	Resolution (*R*)
19.19	L-Serine2	D-Ser	268	2.47
20.84	D-Methionine	D-Met	61	
21.02	D-Glutamine	D-Gln	84	
21.41	L-Methionine	L-Met	61	4.31
21.78	D-Glutamic acid	D-Glu	282	
21.80	L-Glutamine	L-Gln	84	5.3
21.93	D-Phenylalanine	D-Phe	91	
22.26	L-Glutamic acid	L-Glu	282	2.8
22.38	L-Phenylalanine	L-Phe	91	2.61
23.07	D-Asparagine	D-Asn	295	0.72
23.15	L-Asparagine	L-Asn	295	0.72
23.89	D-Cysteine	D-Cys	113	
24.17	L-Cysteine	L-Cys	113	1.63
26.24	L-1-Methylhistidine1	L-MeHis1	95	
26.81	L-1-Methylhistidine2	L-MeHis2	95	
28.59	D-Homocysteine	D-Hcy	282	
28.96	L-Homocysteine	L-Hcy	282	2.11
31.56	D-Histidine	D-His	307	
31.82	L-Histidine	L-His	307	2.27
36.40	D-Ornithine	D-Orn	296	
36.74	L-Ornithine	L-Orn	296	1.94
37.94	D-Lysine	D-Lys	310	
38.06	D-Tyrosine	D-Tyr	333	
38.20	L-Tyrosine	L-Tyr	333	1.63
38.32	L-Lysine	L-Lys	310	1.52
46.85	D-Tryptophan	D-Trp	130	
47.09	L-Tryptophan	L-Trp	130	2.2
47.30	D-Cystathionine	D-CTH	328	
47.64	L-Cystathionine	L-CTH	328	1.23

4. Notes

1. The ^{13}C or deuterium-labeled amino acids are useful quantitation by the internal calibration method. List of the most commonly used labeled and natural amino acids is in Table 1. Working solutions of an amino acid standard mixture and that of containing labeled analogues are prepared by a serial dilution and mixing of stock amino acid solutions with water or carbonate saline buffer. The composition and the concentration levels in working standard solutions depend largely on a particular analytical problem. Glutamine, asparagine, and tryptophan are not stable in acidic solutions and should be stored in neutral or slightly basic environment.
2. To prolong stability of the DMPS solution, blowing of a freshly prepared solution with argon or nitrogen and storage in a sealed amber vial are recommended.
3. The carbonate saline buffer is a basic (pH about 8.5–9) and does not interfere the sample preparation process. It is used to maintain mild basic pH environment in reaction–extraction media.
4. The 2,2,3,3,4,4,4-heptafluorobutyl chloroformate reagent is a liquid with a boiling point = 105–107°C and density = 1.6 g/cm^3. The reagent (analytical purity) is commercially available from Aneclab, the Czech Republic, http://www.aneclab.com. WARNING! Manipulation with HFBCF must be performed in a well-ventilated area (fume hood). An injection syringe should be carefully rinsed with isopropanol after the reagent transfer to prevent corrosion of the plunger. HFBCF is stable for at least 18 months when stored at 4°C in a tightly closed teflon-lined vial.
5. DMPS (trade name Dimaval) is a nontoxic, chelating agent used in chelation therapy for elimination of mercury, lead, and arsenic from human body. The 1.5% DMPS solution (at a final concentration of 0.3–0.4% (v/v) in serum/plasma) releases efficiently protein-bound thiol amino acids. The present ionic sulfonic group in the structure keeps DMPS and its oxidation products in aqueous phase, away from the amino acid derivatives which enter into the nonpolar isooctane phase.
6. 1.7 M PCA precipitates gently serum/plasma proteins. Lipoprotein structures are not impaired, and potentially interfering analytes, such as fatty acids, steroids, or glycerides are not extracted into the acquired supernatant.
7. The acidic supernatant has to be neutralized to pH = 8–9 after the precipitation step. Its pH can be checked easily by putting a drop on a pH paper strip.

8. The derivatization reaction is essentially catalyzed by pyridine in a mild basic environment (pH = 9–10).
9. The isooctane addition simplifies partition of the upper organic phase.
10. Serum/plasma aliquot is replaced by the same volume of an amino acid standard mixture. Furthermore, the PCA concentration (1.67 M) used in the precipitation–neutralization step should be lowered by about 25% to 1.17 M if the physiological fluid is not present.
11. Aspiration of the organic layer should conform to the serum/plasma extract volume. Treating the amino acid standard mixture (no protein precipitation) results in a larger volume of the organic phase. This can be compensated by aspiration of an adequate volume, i.e., 120 μl, or by means of an appropriate labeled internal standard added into the physiological fluid prior to sample preparation.
12. By experience, inertness of the injector liner and capillary column is the most critical part affecting GC/MS analytical performance. Splitless injection with a baffled inlet liner is recommended for maximum sensitivity.
13. Larger amino acids, such as cystathionine, cystine, and homocystine, acquire molecular weights above 1,000 mass units by derivatization; refer to Table 1. Nevertheless, the bulky heptafluorobutyl groups make initially polar amino acids highly volatile and inert to adsorption sites of an inlet liner and stationary phase. The HFBCF-based derivatives are eluted at much lower temperatures and amino acids difficult for GC analysis, such as cystine, homocystine, or 1-methylhistidine, can be analyzed much easier than with other GC-oriented derivatization methods.
14. The sample preparation protocol provides exceptionally clean extracts of serum/plasma, as documented in Fig. 1. The physiological amino acids are easily detected in the chromatogram with the assistance of Table 1. Note that the protocol has also limits in the analysis of amino acids with functional groups that remain untouched by the HFBCF-mediated derivatization. It is, for example, a guanidino or ureido group in arginine and citrulline, respectively. This disables their GC analysis by the described protocol.
15. Details of the calibration and validation procedure for the determination of amino acids in human serum by the described protocol are reported in ref. 6.
16. A CP-Chirasil-L-Val column (an Agilent-Varian-Chrompack product line) is a 50% cross-linked phase stable to 200°C. As a

rule, the D-enantiomers are eluted before their L-counterparts. A Chirasil-D-Val column is also commercially available and can be used complementarily. In this case, an L-enantiomer is eluted before its D-form. Although the CP-Chirasil-Val columns are relatively robust, some precautions are recommended to prolong their life. Temperatures higher than 200°C and temperature programs up and down faster than 30°C/min should be avoided. Some batch-to-batch differences were observed in separation efficiency and stability of a particular column, probably owing to a sophisticated phase coating process. It is, therefore, recommended to test a new, relatively expensive chiral column carefully prior to routine work.

17. Due to its nature, Chirasil-Val column is prone to overloading and higher column bleeding than fully cross-linked nonpolar phases. Splitless injection is, therefore, generally not recommended. However, final sample extract is dissolved in nonpolar isooctane here and this feature is highly advantageous. More than 1,000 splitless injections (typically, 0.5 μl) of human serum extracts and peptide hydrolysates on a Chirasil-Val column without any remarkable change in the column performance have shown that the method is suitable for trace enantiomeric amino acid analysis.
18. Enantiomeric amino acid analysis has also been performed by GC with a cost-effective flame ionization detection (FID). Separation characteristics of the HFBCF-based amino acid derivatives on a Chirasil-L-Val operated with hydrogen as a carrier gas have been reported in refs. 8, 9.
19. The records obtained from a subsequent nonchiral and chiral GC/MS of a typical human serum extract are reported in the respective Figs. 1 and 2 (bottom part) and demonstrate that the sample preparation protocol is suitable for both nonchiral and chiral amino acid analysis. Table 2 summarizes the chiral GC/MS separation data on 22 resolved enantiomeric pairs, 3 nonchiral (Sar, Gly, GABA), 2 not resolved (D,L-Pro, D,L-Baib), and 2 amino acids (D-TPN, L-1Me-His), for that a corresponding antipod was not available. Interestingly, organic acids, such as the dominating lactic acid (4.26 min in Fig. 1), are not recorded on the chiral chromatogram in Fig. 2, and thus do not interfere with the chiral amino acid analysis.
20. From 20 proteinogenic amino acids, 18 D,L-enantiomeric pairs are very well-separated in less than 50 min, except D,L-Pro (not separated) and arginine (not eluted). D,L-Asn (23.07, 23.15 min) is the most difficult pair to resolve ($R = 0.72$, Table 2). For maximum resolution, carrier gas flow and temperature program should be optimized primarily on this amino acid.

Acknowledgments

This work was supported by the Fund for Research Support, EEA/ Norway grant No. A/CZ0046/1/0018, the Czech Science Foundation, project No.213/09/2014, and the Internal Grant Agency of Ministry of Health of the Czech Republic, No. NS9755-4/2008.

References

1. Gehrke CW, Kuo CT, Zumwalt R W (1987) Amino acid analysis by gas chromatography. Vol. I-III CRC Press, Boca Raton
2. Hušek P, Šimek P (2001) Advances in amino acid analysis. LC-GC North America 18: 986–999
3. Molnar-Perl I (2005) Quantitation of amino acids and amines by chromatography. J Chromatogr–Library, Vol. 70, Elsevier, Amsterdam
4. Hušek P, Šimek P (2001) Alkyl chloroformates in sample derivatization strategies for GC analysis. Current Pharm Anal 2:21–41
5. Hušek P et al (2008) Fluoroalkyl chloroformates in treating amino acids for gas chromatographic analysis. J Chromatogr A 1186: 391–400
6. Šimek P, Hušek P, Zahradníčková H (2008) Gas chromatographic-mass spectrometric analysis of biomarkers related to folate and cobalamin status in human serum after dimercaptopropanesulfonate reduction and heptafluorobutyl chloroformate derivatization. Anal Chem 80:5776–5782
7. Patzold R, Bruckner H (2005) Chiral separations of amino acids. In: Molnar-Perl I (ed.) Quantitation of amino acids and amines by chromatography. J Chromatogr–Library vol. 70, Elsevier, Amsterdam
8. Zahradníčková H, Hartvich P, Šimek P, Hušek P (2008) Gas chromatographic analysis of amino acid enantiomers in Carbetocin peptide hydrolysates after fast derivatization with pentafluoropropyl chloroformate. Amino Acids 35: 445–450
9. Zahradníčková H, Hušek P and Šimek P (2009) GC separation of amino acid enantiomers via derivatization with heptafluorobutyl chloroformate and Chirasil-L-Val column. J Sep Sci 25:3919–3924

Chapter 14

The EZ:Faast Family of Amino Acid Analysis Kits: Application of the GC-FID Kit for Rapid Determination of Plasma Tryptophan and Other Amino Acids

Abdulla A.-B. Badawy

Abstract

Plasma tryptophan (Trp) and other amino acids (AA) can be determined rapidly by gas (GC) or liquid (LC) chromatography using the Phenomenex EZ:Faast™ family of kits. Three kits are available: (1) GC-FID or -NPD, (2) GC-MS, (3) LC-MS. The two GC kits can determine 32 AA, whereas the LC-MS can determine five additional AA. All three kits, however, share the same experimental procedure up to and including the preparation of derivatised AA. The method is based on solid-phase extraction (SPE), thus saving time on prior removal of plasma or other proteins and interfering substances, and can be applied to other body fluids and experimental media and to supernatants of extracts of solid material. Briefly, SPE is performed using a proprietary cation-exchange mechanism. The acid solution of the internal standard ensures that the free amino acids are in an anionic form suitable for cationic binding. The alkaline nature of the elution medium ensures that the AA cations are released prior to derivatisation. The latter involves production of chloroformate derivatives of both the amino and carboxylic acid groups. With experience, six plasma samples can be so processed within 12 min. The shortest analytical run is <7 min per sample using the GC-FID/NPD kit. Despite its many steps, the procedure becomes second nature and an enjoyable task. I have now used the GC-FID kit with manual injection to process >1,600 plasma and other samples. Limit of detection of AA is 1 μM or less. The procedure has been validated and optimised for Trp and its main five brain uptake competitors.

Key words: Amino acids, EZ:Faast™ Kits, Gas chromatography, Liquid chromatography, Plasma, Solid-phase extraction, Tryptophan, Amino acid analysis

1. Introduction

Amino acid (AA) analysis in biological samples is an important laboratory aid in many clinical and research situations, including diagnosis of inherited disorders of AA metabolism, monitoring of bone and muscle disorders and of patients requiring long-term

Michail A. Alterman and Peter Hunziker (eds.), *Amino Acid Analysis: Methods and Protocols*, Methods in Molecular Biology, vol. 828, DOI 10.1007/978-1-61779-445-2_14, © Springer Science+Business Media, LLC 2012

nutritional support, evaluation of new baby milks and parenteral and enteral feeds, detection of renal drug toxicity, and monitoring the AA precursors of the cerebral monoamines serotonin and the catecholamines in relation to psychiatric and other behavioural disorders. Many problems are encountered in AA analysis (1). One such problem is the relatively long time required for both sample preparation and analysis. The former requires protein precipitation, removal of interfering substances, and derivatisation, all of which take time, as much as 40 min. Analysis time varies with the commonly used existing techniques between 2 and 4 h with dedicated cation-exchange AA analysers and up to 1 h with high-performance liquid chromatography (HPLC) (1).

Gas chromatography has greater advantages over the above techniques, notably its speed, greater power of resolution (an important asset when analysing complex physiological fluids), and the wider choice of detector systems [such as flame-ionisation (FID), nitrogen-phosphorus (NPD), electron-capture (ECD), flame photometric (FPD), and mass-spectrometric (MS)]. The speed of GC is illustrated by a study using GC-MS (2) in which processing, derivatisation, and analysis of AA extracted from blood spots can be completed within 30 min. An even-faster GC procedure is that using the Phenomenex EZ:faast family of AA analysis kits (3). The greatest advantage of these kits is their speed of sample processing and analysis. The simple solid-phase extraction (SPE) and rapid derivatisation of AA combine to shorten preparation time considerably, and analysis time varies between 7 and 15 min. Three kits are available, utilising (1) gas chromatography (GC) with flame ionisation (FID) or nitrogen-phosphorus (NPD) detection, (2) GC with mass spectrometric (MS) detection, (3) liquid chromatography (LC-MS). The two GC kits can measure 32 AA, whereas the LC kit measures 37 AA, including Arg, Cys, Cit, 1- and 3-methylhistidines, and GABA (γ-aminobutyric acid), which are not measurable by the GC kits (GABA co-elutes with serine in the GC-FID kit; see Note 9). Sample preparation with the GC-FID/NPD kit is the shortest (see below), whereas an extra ~5 min are required with the two MS kits for evaporating the derivatised AA solvent, redissolving the residue and transferring the new solution into an injection vial. The two GC kits each takes 7 min for a sample analysis run, whereas the LC kit takes 15 min. We have successfully applied the GC-FID kit primarily for rapid determination of plasma Trp and other AA which compete with it for entry into the brain in a study (4) in which ~1,000 plasma samples were processed and, to date, >1,600 plasma, serum and other samples have been analysed in a number of studies, some of which have already been published (5–8). The GC-FID kit has already been used in two other studies, one for measurement of GABA in Tall Fescue herbage (9) and the other for 33 AA and dipeptides in spent cell culture media (10). The kits are not limited to body fluids, but can be used to determine free AA in other material, e.g. beer, fermentation broth, orange juice, cheese, dried tomatoes, and corn meal, for which the manufacturer has application notes, and also with other experimental supernatants (see Notes 10 and 11). I believe that this simple, elegant, rapid, and enjoyable procedure merits wider application in both research and diagnostic settings.

2. Materials and Equipment

The kit(s) contain three AA standards and six reagents. All components are guaranteed for 12 months from the date of purchase when stored at recommended temperatures and used as described in the kit manuals. The kits also include a column and other material, such as sorbent tips and syringes.

2.1. AA Standards

There are three standard AA solutions containing a 200 μM concentration of each AA. Two vials (2 ml each) of every standard solution are provided. All three standards should be stored in a freezer and should undergo the minimum thawing and use time before refreezing.

Standard 1: For the GC-FID/NPD and GC-MS kits, Standard 1 contains a mixture of 23 AA. These are AAA (α-aminoadipic acid), ABA (α-aminobutyric acid), aIle (allo-isoleucine), Ala, Asp, βAIB (β-aminoisobutyric acid), C-C, Glu, Gly, His, Hyp (4-hydroxyproline), Ile, Leu, Lys, Met, Orn, Phe, Pro, Sar, Ser, Thr, Tyr, and Val. For the LC-MS kit Standard 1 contains 27 AA. These are 22 of the 23 in the above Standard (all except aIle) plus five extra AA, namely, Arg, Cit, GABA, and 1- and 3-methylhistidines (see Note 9).

Standard 2: This is common to all three kits and includes 3 AA which are unstable in acid solution, namely, Asn, Gln, and Trp.

Standard 3: This is also common to all three kits and includes six urine-specific AA, namely, APA (α-aminopimelic acid), CTH (cystathionine), GPR (Gly-Pro dipeptide), HYL (hydroxylysine), PHP (proline-hydroxyproline dipeptide), and TPR (thioproline). This Standard mixture is used only for urine analysis or if one or more of its six components requires assessment in other media.

2.2. Reagents

Reagents 1, 3B and 4 should be stored at 4°C. Reagents 2, 3A, 5 and 6 should be stored at room temperature. For convenience, the bottom of the reagent box has been designed as a tray that can be easily lifted from the work station and placed in the refrigerator when the kit is not in use for an extended period of time. The reagents are stable for 12 months if stored as described.

Reagent 1: The norvaline internal standard (NVIS) (50 ml).
This is a 200 μM solution (0.0022%) of norvaline in dilute HCl. It should be added to all standards, blanks (if necessary), and samples.

Reagent 2: The washing solution (90 ml).
This is a mixture of 1-propanol and H_2O.

Reagent 3A: Elution medium component I (60 ml).
This is an aqueous NaOH solution.

Reagent 3B: Elution medium component II (40 ml).
This is a mixture of 1-propanol and 3-picoline.

Preparation of the Elution Medium (Combined Reagents 3A and 3B)

The elution medium (200 μL per sample) should be prepared in a fume cabinet freshly on the day of the experiment, in economical amounts, depending on the total number of samples to be analysed, by mixing three parts of Reagent 3A with two parts of Reagent 3B (120 μl of Reagent 3A and 80 μl of Reagent 3B for a total volume of 200 μl) in an appropriate size capped vial. The prepared mixture is stored at room temperature and any unused amount should be discarded by the end of the experiment. Table 1 shows the amounts to be pipetted for various numbers of samples. In all instances, there should be at least an extra 200 μl to compensate for any pipetting losses.

Reagent 4: Organic derivatisation solution I (4 vials 6 ml each).
This is a mixture of $CHCl_3$, 2,2,4-trimethylpentane and propylchloroformate.

Reagent 5: Organic solution II (50 ml).
This is a mixture of 2,2,4-trimethylpentane and $CHCl_3$.

Reagent 6: Acid solution (50 ml).
This is a dilute HCl solution.

Table 1
Composition of the elution medium

No. of samples	Reagent 3A	Reagent 3B	Total volume
2	360 μl	240 μl	600 μl
5	0.72 ml	0.48 ml	1.20 ml
8	1.08 ml	0.72 ml	1.80 ml
10	1.32 ml	0.88 ml	2.20 ml
12	1.56 ml	1.04 ml	2.60 ml
16	2.04 ml	1.36 ml	3.40 ml
20	2.52 ml	1.68 ml	4.20 ml
25	3.12 ml	2.08 ml	5.20 ml

2.3. Other Kit Materials

The following are materials supplied with the GC-FID/NPD kit. Some of the additional materials for the GC/MS and LC/MS kits are described in Subheading 3.2 below and others can be seen in the User Guide (3)

1. Sorbent tips in racks (4×96).
2. Sample preparation glass vials (4×100).
3. One vial rack.
4. One adjustable Drummond Dialamatic micro-dispenser (20–100 μL) (The Drummond Scientific Co., Broomall, PA 19008, USA).
5. Syringes (0.6 ml and 1.5 ml) (10 of each).
6. One ZB-AAA GC column (10 m×0.25 mm) with a certified chromatogram of its performance under the manufacturer's specified experimental conditions using all three standards for each kit.
7. One EZ:faast demonstration video and reference CD.
8. One user manual.
9. 5 Focus liners™.

2.4. Other Materials Required, But Not Supplied in the Kit

These are:

1. 100 μL–1 mL pipette.
2. 30–300 μL pipette.
3. Pipette tips.
4. Vortex.
5. Injection vials of appropriate volumes with caps.
6. Pasteur pipettes for sample transfer.
7. Container for proper waste disposal.
8. Septa.

2.5. Equipment

The equipment used in this Laboratory was the Perkin-Elmer Clarus 500 gas chromatograph with a flame ionisation detector (FID) and manual injection [1 μl of the derivatised AA using a 1-μL SGE syringe (SGE Europe Ltd., UK: http://www.uk@sge.com)]. The Phenomenex column supplied was a Zebron ZB-AAA capillary GC column (10 m×0.25 mm). The column oven temperature programme was as follows: 32°C per min from 110 to 320°C. The FID detector temperature was 320°C and 1 μl was injected at an injection temperature of 250°C and a split level of 1:2. The carrier gas was H_2 at a pressure of 8 psi (a flow rate of 1 mL/min). The flow rate, pressure setting and split ratios will need adjusting, depending on GC instrument specifications, the carrier gas chosen and other practical considerations. Data handling and processing were performed by the Perkin-Elmer Total Chrome software.

3. Methods

3.1. Procedure Common to All Kits

1. For each sample to be analysed, line up 1 sorbent tip and 1 glass sample preparation vial in the vial rack, ensuring that the tip can reach the bottom of the vial (see Notes 1 and 2).
2. Pipette 100 µL of the sample (serum, plasma, or other) (see Notes 7, 10, 11 and 12) and 100 µL of Reagent 1 [the norvaline internal standard (NVIS)] into each sample preparation vial.
3. Place the vial rack into the work station and place the latter inside a fume cabinet, wherein all subsequent steps should be performed (see Note 2).
4. Attach a sorbent tip to a 1.5-mL syringe and loosen the syringe piston, then immerse the tip and let the solution in the sample preparation vial pass through the sorbent tip by slowly pulling back the syringe piston in small steps (over a ~1 min period, see Note 3).
5. Pipette 200 µL of Reagent 2 (wash solution) into the same sample preparation vial and pass this solution slowly through the sorbent tip and into the syringe barrel (see Note 3). Drain the liquid from the sorbent bed by pulling air through the sorbent tip, then detach the latter and leave it in the sample preparation vial, discarding the liquid accumulated in the syringe.
6. Pipette 200 µL of the elution medium (prepared freshly each day by mixing three parts of Reagent 3A with two parts of Reagent 3B; see Table 1) in the same sample preparation vial.
7. Pull back the piston of a 0.6-mL syringe halfway up the barrel and attach the sorbent tip used in steps 3–5.
8. Wet the sorbent with the eluting medium; watch as the liquid rises through the sorbent particles and stop when the liquid reaches the filter plug in the sorbent tip.
9. Eject the liquid and sorbent particles out of the tip and into the sample preparation vial. Repeat steps 7 and 8 until all the sorbent particles in the tip are expelled into the sample preparation vial. Only the filter disc should remain in the empty tip. Keep the syringe, as it can be reused with many other samples.
10. Using the adjustable Drummond Dialamatic Micro-dispenser provided transfer 50 µL of Reagent 4 (organic derivatisation solution 1) into the sample preparation vial, ensuring that the tip of the Micro-dispenser does not touch the inner wall of the vial.
11. Emulsify the liquid in the vial by repeated vortexing for ~5–8 s, holding the vial straight and firmly between fingers. Allow the reaction to proceed for at least 1 min. The emulsion will gradually separate into two layers.

12. Re-emulsify the liquids in the vial by vortexing again for 5 s and allow the reaction to proceed for another min.
13. Transfer with the Micro-dispenser 100 μL of Reagent 5 (organic solution II) (50 μL twice, for convenience) and mix for 5 s. Allow the reaction to proceed for 1 more min.
14. Pipette 100 μL of Reagent 6 (not with the Micro-dispenser) and vortex for ~5 s then allow the emulsion to separate into two layers. The upper layer contains the derivatised AA to be analysed by GC or LC (see Note 4).
15. For standards, follow the whole procedure using 100 μL of each standard and 100 μL of the NVIS. Each AA in all these standards is present at a 200 μM concentration (see Notes 4, 5, 6 and 8).

3.2. Procedures Specific for Each Kit

1. *GC-FID/NPD Kit*: This is the simplest of all three kits. Transfer part of the upper organic layer in step 14 above using a Pasteur pipette into an autosampler (or other) vial for direct automatic or manual GC injection.
2. *GC-MS Kit*: Transfer part of the upper organic layer in step 14 above (about 50–100 μL) using a Pasteur pipette into an autosampler vial (included) and avoid transferring any of the aqueous layer along with the organic layer. Evaporate the solvent slowly under a gentle stream of nitrogen, stopping when the sample is almost dry. Do not leave samples under the nitrogen stream for more than 3 min. Re-dissolve the AA derivatives in 100 μL (or less) of Reagent 6. Transfer the reconstituted sample into an insert (included) using a Pasteur pipette, and place the insert in the same autosampler vial. The reconstituted sample is ready for GC/MS analysis.
3. *LC-MS Kit*: Transfer part of the upper organic layer in step 14 above (about 50 μL) into an autosampler vial (included) using a Pasteur pipette, avoiding the transfer of any of the aqueous layer along, and evaporate to near dryness in a gentle stream of nitrogen (for no more than 3 min). Re-dissolve in 70–100 μl of the HPLC mobile phase (10 mM ammonium formate in water: 10 mM ammonium formate in methanol 1:2, v/v). Transfer the reconstituted sample into an insert (included) and place the insert into the same autosampler vial. The sample is ready for LC/MS analysis.

3.3. Sample Chromatograms Using the GC-FID Kit

The chromatograms in Fig. 1 depict typical separations of free AA under the conditions described in Subheading 2.5 above in five different samples each of which containing the norvaline internal standard (NVIS): (a) a six-component mixture of Val, Leu, Ile, Phe, Tyr, and Trp (200 μM each) prepared in our laboratory for the purpose of our study (4), (b) a mixture of the Phenomenex

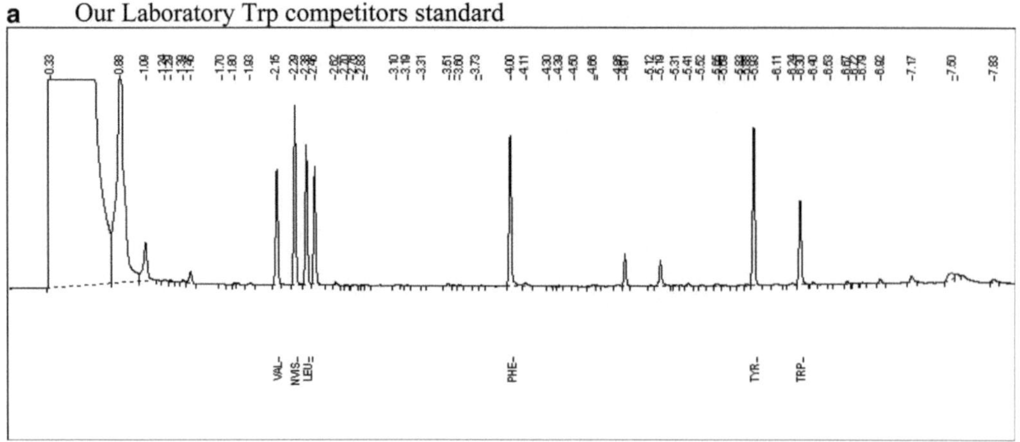

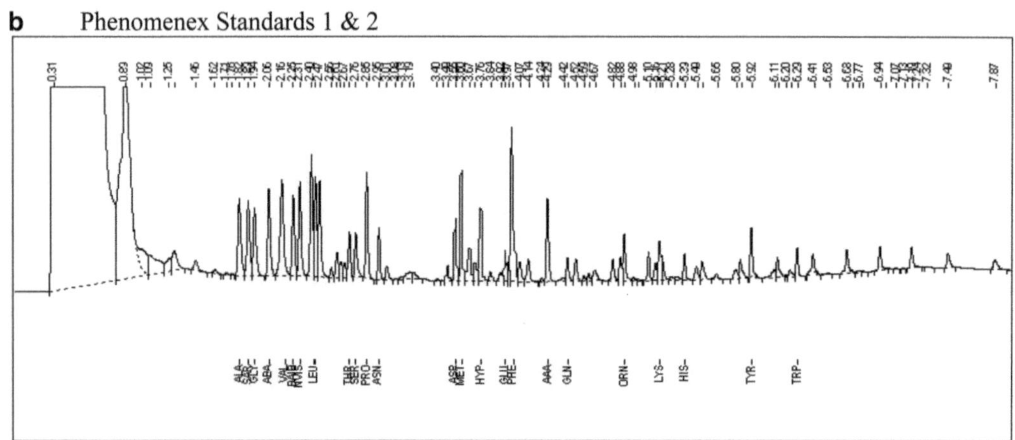

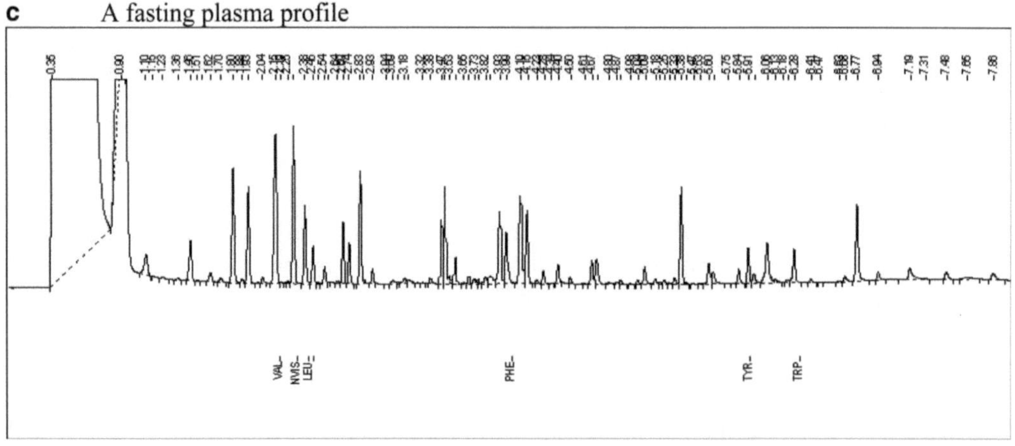

Fig. 1. Sample chromatograms of amino acids analysed by the Phenomenex EZ:faast GC-FID kit; (**a**) A six-amino-acid mixture prepared in the laboratory; (**b**) the Phenomenex Standards 1 and 2; (**c**) A fasting human plasma profile; (**d**) plasma profile of a subject undergoing the acute tryptophan depletion test; (**e**) A plasma profile of a subject undergoing the acute tryptophan loading test. All chromatograms include the norvaline internal standard (NVIS) which elutes immediately after valine. Reproduced here from ref. 4 with permission.

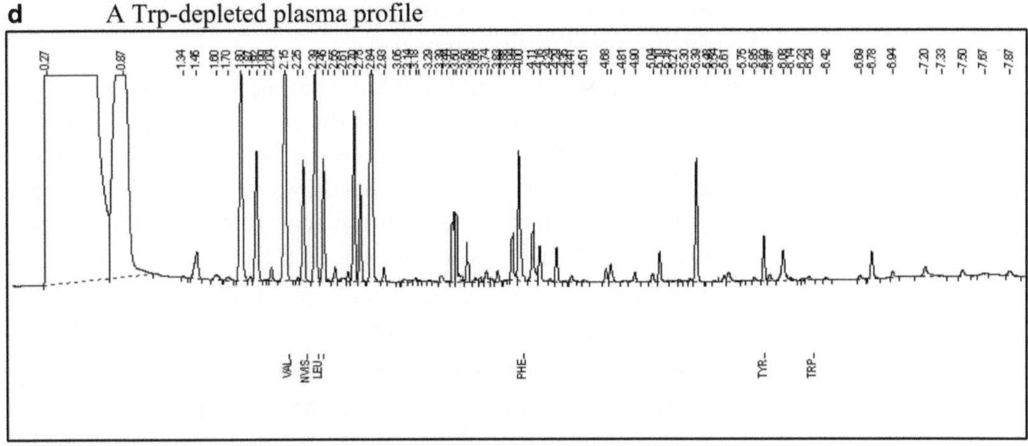

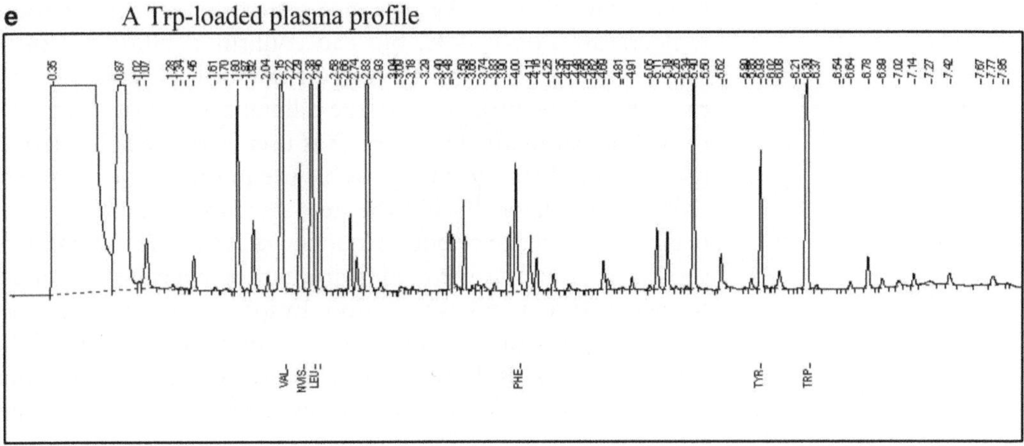

Fig. 1. (continued)

Standards 1 and 2 (26 AA) and three human plasma samples marked only for the above 6 AA (Trp and its five competitors), (c) a fasting plasma, (d) plasma from a subject undergoing the acute Trp depletion test, (e) plasma from a subject undergoing acute Trp loading.

4. Notes

1. Despite its many steps, I found the procedure simple, elegant, and enjoyable. Initially, it may take the analyst > 10 min to process two samples. With experience, I am able to process six samples simultaneously within ~12–15 min.
2. Because of the toxicity of some components, the procedure should be performed in a fume cabinet. Particular care should

be taken to avoid inhaling fumes when preparing the elution medium and during steps 6–14.

3. As stated in steps 4 and 5 in Subheading 3.1 above, withdrawal of the sample (step 4) and of the wash solution (step 5) by pulling back the syringe piston must be performed slowly over a 1-min period each. Rapid withdrawal will result in inefficient SPE of AA (step 4) and inefficient removal of interfering substances by the wash solution (step 5).
4. The derivatised AA solutions obtained in step 14 or beyond should be placed in a cold autosampler for immediate chromatography or at 4°C until their manual injection.
5. Stability of derivatised AA: The manufacturer states that the derivatised AA are stable for 1 day at room temperature and for several days at 4°C. We have not tested the stability at room temperature nor at 4°C, but can confirm stability at −24 to −32°C for up to 48 h (the longest duration tested). I would recommend the processing of the calibrant reference (standard) in each experiment. However, it is useful to store the derivatised standard solution for possible use in the next experiment a day or two later, should the new standard be faulty for any reason (e.g. spillage or poor extraction of AA). The old standard should be stored at the coldest freezer temperature and effectively sealed to avoid evaporation. Evaporation, however, does occur, despite cold storage, resulting in concentration of the AA standard. This does not present any problems, as the software can adjust AA concentrations relevant to the internal standard value for each test sample.
6. Sometimes, the laboratory may run out of the kit standards. These can be ordered from the manufacturer independently of the whole kit. Investigators could always make and use their own standards if their work involves only a small number of amino acids. If so, the "in-house" standard should always be compared initially with the Phenomenex standard(s) containing the relevant amino acids.
7. The buffy layer seen sometimes after the thawing of frozen plasma can occasionally block the sorbent tip. To avoid this, the thawed plasma sample should be centrifuged at 2,000–3,000 × *g* for 5 min and a clear portion is pipetted. This is not necessary with freshly isolated plasma.
8. Minor changes in GC conditions could affect the retention times of AA. It is therefore important in every experiment to pay particular attention to the standard chromatogram to ensure that all AA are correctly identified and calibrated by the software used before running the first test sample.
9. Taurine cannot be measured by any of the EZ:faast kits, because its strong sulphonic acid group (pK_a 1.9) prevents its retention

onto the sorbent material. GABA co-elutes with serine (4) and its measurement in plasma samples is therefore not possible as its concentration (0.15 μM, by HPLC (11)) is far too negligible in comparison with that of Ser. However, in experimental settings where Ser is absent, GABA can be determined by the GC kits. The Arg derivative is not volatile, hence the use of LC, rather than GC, to measure Arg.

10. Reproducible, high efficiency solid phase extraction requires that the sample be in liquid form prior to loading onto an SPE device and that the pH of the sample-internal standard mixture should be in the 1.5–6.0 range. The ideal SPE sample should meet the following conditions: (a) liquid of low viscosity (to pass through the cartridge), (b) low solids or particulate contaminants (to prevent clogging), (c) solvent composition that is suitable for retention (each mechanism has different matrix solvent composition requirements for proper retention).
11. For processing solid biological samples, the following steps should be generally adopted: (a) homogenization of the sample in the appropriate aqueous or organic solvent, depending on analyte stability, (b) decantation, filtration, centrifugation, and/or sonication, (c) collection of the supernatant, (d) any additional organic separation, followed by collection of the appropriate phase, its evaporation or dilution with water.
12. For further information, investigators should refer to the Phenomenex SPE user guide.

Acknowledgements

This work was funded by the Wellcome Trust (069301). I thank Chris Morgan and Jennifer Turner for technical assistance and Ben Atkins and Lindsey Starr (Phenomenex, UK) for helpful discussions and useful technical information.

References

1. Walker V, Mills GA (1995) Quantitative methods for amino acid analysis in biological fluids. Ann Clin Biochem 32:28–57
2. Deng C, Li N, Zhang X (2004) Rapid determination of amino acids in neonatal blood samples based on derivatisation with isobutyl chloroformate followed by solid phase microextraction and gas chromatography/mass spectrometry. Rapid Commun Mass Spectrom 18:2558–2564
3. Phenomenex EZ:faast [easy fast] amino acid sample testing kit (2003) User guide, Phenomenex, 411 Madrid Avenue, Torrance, CA 90501–1430, USA (http://www.phenomenex.com/)
4. Badawy AA-B, Morgan CJ, Turner JA (2008) Application of the Phenomenex EZ:faast™ amino acid analysis kit for rapid gas-chromatographic determination of concentrations of plasma tryptophan and its brain uptake competitors. Amino Acids 34:587–596
5. Badawy AA-B et al (2007) The acute tryptophan depletion and loading tests: Specificity issues. Int Congr Series 1304C, 159–166

6. Dougherty DM et al (2008) Comparison of 50- and 100 g *L*-tryptophan depletion and loading formulations for altering 5-HT synthesis: pharmacokinetics, side-effects, and mood states. Psychopharmacol 198:431–445
7. Badawy AA-B et al (2009) Activation of liver tryptophan pyrrolase mediates the decrease in tryptophan availability to the brain after acute alcohol consumption by normal subjects. Alcohol Alcohol 44:267–271
8. Badawy AA-B, Dougherty DM, Richard DM (2010) Specificity of the acute tryptophan and tyrosine plus phenylalanine depletion and loading tests Part II. Normalisation of the tryptophan and the tyrosine plus phenylalanine to competing amino acid ratios in a new control formulation. Int J Tryptophan Res 3:35–47
9. Kagan IA et al (2008) A validated method for gas chromatographic analysis of gamma-aminobutyric acid in tall fescue herbage. J Agric Food Chem 56:5538–5543
10. Mohabbat T, Drew B (2008) Simultaneous determination of 33 amino acids and dipeptides in spent cell culture media by gas chromatography-flame ionisation detection following liquid and solid phase extraction. J Chromatogr B Analyt Technol Biomed Life Sci 862:86–92
11. Cowley DS et al (1996) Effect of diazepam on plasma γ-aminobutyric acid in sons of alcoholic fathers. Alcohol Clin Exp Res 20:343–347

Chapter 15

Amino Acid Analysis in Physiological Samples by GC–MS with Propyl Chloroformate Derivatization and iTRAQ–LC–MS/MS

Katja Dettmer, Axel P. Stevens, Stephan R. Fagerer, Hannelore Kaspar, and Peter J. Oefner

Abstract

Two mass spectrometry-based methods for the quantitative analysis of free amino acids are described. The first method uses propyl chloroformate/propanol derivatization and gas chromatography–quadrupole mass spectrometry (GC–qMS) analysis in single-ion monitoring mode. Derivatization is carried out directly in aqueous samples, thereby allowing automation of the entire procedure, including addition of reagents, extraction, and injection into the GC–MS. The method delivers the quantification of 26 amino acids. The isobaric tagging for relative and absolute quantification (iTRAQ) method employs the labeling of amino acids with isobaric iTRAQ tags. The tags contain two different cleavable reporter ions, one for the sample and one for the standard, which are detected by fragmentation in a tandem mass spectrometer. Reversed-phase liquid chromatography of the labeled amino acids is performed prior to mass spectrometric analysis to separate isobaric amino acids. The commercial iTRAQ kit allows for the analysis of 42 physiological amino acids with a respective isotope-labeled standard for each of these 42 amino acids.

Key words: Amino acids, Gas chromatography–mass spectrometry, Propyl chloroformate, Liquid chromatography–tandem mass spectrometry, Derivatization, Stable isotope dilution, iTRAQ, Amino acid analysis

1. Introduction

Several analytical methods are available to quantitatively analyze free amino acids in physiological samples (1,2). The widely accepted reference method is cation-exchange chromatography using a lithium buffer system followed by postcolumn derivatization with ninhydrin and UV detection (3–5). Ninhydrin forms a purple product (Ruhemann's purple, $\lambda_{max} = 570$ nm) with primary amines, which is

Michail A. Alterman and Peter Hunziker (eds.), *Amino Acid Analysis: Methods and Protocols*, Methods in Molecular Biology, vol. 828, DOI 10.1007/978-1-61779-445-2_15, © Springer Science+Business Media, LLC 2012

detected by UV. The reaction with secondary amines gives a yellow product (λ_{max} = 440 nm). Since the protocol yields two nonspecific dyes, amino acids must be baseline separated for quantification. However, this results in analysis times of up to 2.5 h per sample. Interferences by other ninhydrin-reactive compounds in complex biological samples and the need for protein precipitation are further drawbacks of the method.

Here, we present two alternative methods for the quantitative analysis of free amino acids in biological fluids. Both methods involve amino acid derivatization, relying on either gas or liquid chromatographic separation and mass spectrometric detection (6).

Alkyl chloroformates, together with an alcohol, are frequently used reagents for amino acid derivatization with subsequent gas chromatography (GC) analysis (7). Carboxylic groups are converted to esters and amino groups to carbamates. The reaction can be catalyzed by pyridine or picoline. The general reaction is shown in Fig. 1. One should note that the ester formed depends on the alcohol used in the reaction medium. The derivatization of the carboxylic group proceeds via a mixed carboxylic–carbonic anhydride, which undergoes an exchange reaction with the alcohol present in the reaction medium, forming the respective ester (8). In addition to the alcohol added to the reaction mixture, traces of an alcohol can also originate from hydrolysis of the chloroformate reagent. To a lower extent, decarboxylation of the anhydride also occurs, forming an ester containing the alkyl group from the alkyl chloroformate (8). Alkyl chloroformate derivatization is fast and the amino acids can be derivatized directly in aqueous solution. The derivatives are extracted with an organic solvent, e.g., chloroform, and an aliquot is injected directly into the GC. Here, we use propyl chloroformate/propanol derivatization, which is completely automated and requires low sample volumes in the range of 20–50 µl (9). The method is based in part on the EZ:faast Amino Acid Analysis kit from Phenomenex (Phenomenex Inc., Torrence, CA, USA).

The second method for the quantitative analysis of amino acids employs the isobaric Tagging for Relative and Absolute Quantification

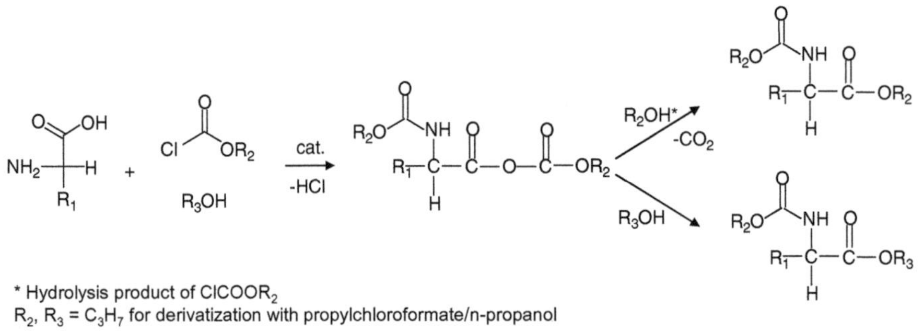

* Hydrolysis product of $ClCOOR_2$
R_2, R_3 = C_3H_7 for derivatization with propylchloroformate/n-propanol

Fig. 1. Reaction scheme for the derivatization of amino acids with alkyl chloroformates.

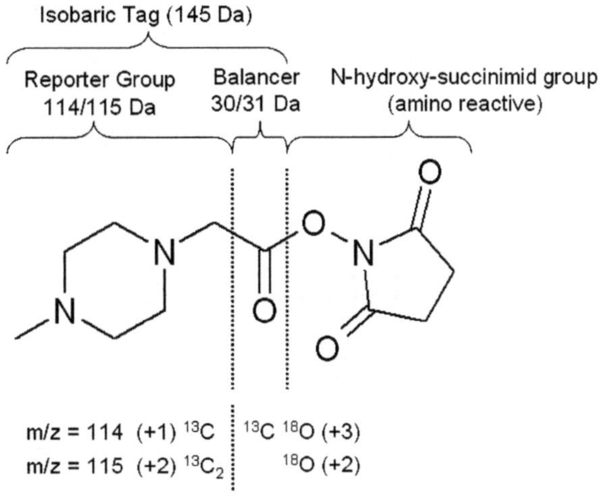

Isobaric Tag (145 Da)

Reporter Group 114/115 Da | Balancer 30/31 Da | N-hydroxy-succinimid group (amino reactive)

m/z = 114 (+1) ^{13}C | ^{13}C ^{18}O (+3)
m/z = 115 (+2) $^{13}C_2$ | ^{18}O (+2)

Fig. 2. Structure and isotope patterns of iTRAQ® labeling reagents.

(iTRAQ) technology offered by AB SCIEX (Foster City, CA, USA). Originally developed for isotope labeling of peptides (10), an iTRAQ kit is now available for the analysis of 42 free amino acids by reversed-phase liquid chromatography (RP-LC)–tandem mass spectrometry (MS/MS). Amino groups are labeled covalently with tags of varying isotope patterns. The tag consists of a reporter group (with the masses *m/z* 114 and 115), a neutral balance linker (*m/z* 31 and 30), and an amino-reactive group (*N*-hydroxy-succinimide ester). The neutral balance group and the reporter ion mass add up to *m/z* 145 for both markers so that the reagents and consequently the labeled amino acids are isobaric. The structure of the reagent and isotope patterns of ^{13}C and ^{18}O are shown in Fig. 2. The 115-labeling reagent is used to derivatize the amino acids in the sample, whereas the 114-labeling reagent is employed by AB SCIEX to provide a 114-labeled standard mix containing 42 analytes for absolute quantification. The 114-labeled internal standard is added to the sample derivatized with the 115-labeling reagent and RP-LC on a C18 column is performed followed by electrospray ionization in positive mode (ESI) and MS/MS in multiple-reaction monitoring (MRM) mode. Since the two derivatives of an amino acid have the same mass, they coelute from the LC column and experience the same matrix effects during ESI. Applying the MRM mode of a triple quadrupole mass spectrometer, the first quadrupole filter (Q1) selects the quasi-molecular ion of the respective amino acid, which enters the collision cell (Q2). Upon collision-induced dissociation, the reporter ions (*m/z* 114 and 115) are cleaved and filtered by the third quadrupole (see Fig. 3). Each amino acid is quantified based on the ratio between the reporter ions.

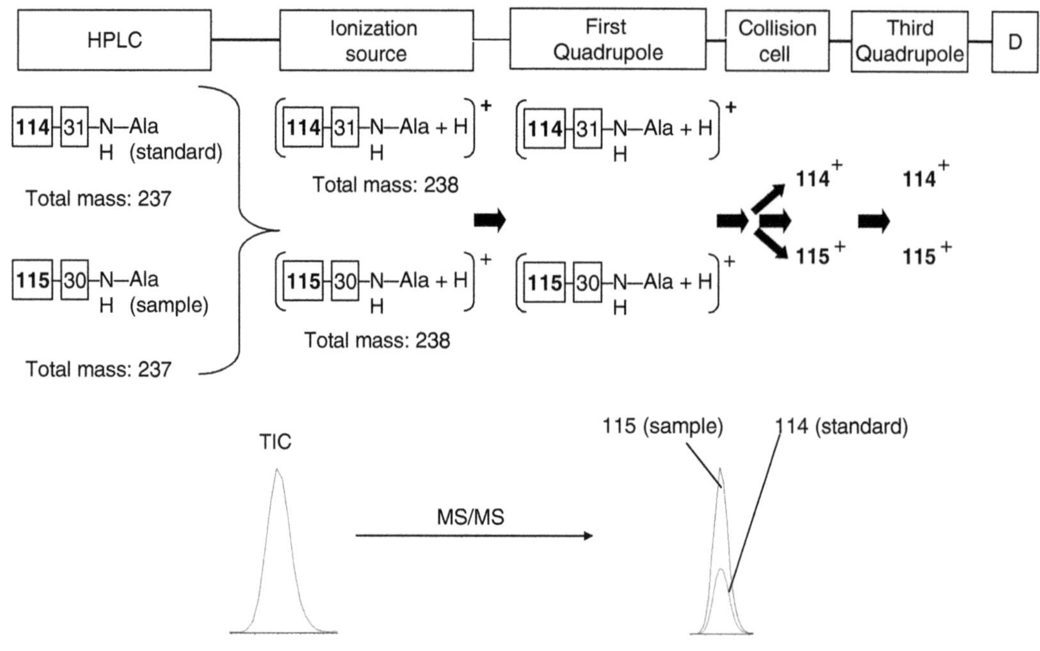

Fig. 3. Scheme for the LC–ESI–MS/MS analysis of iTRAQ derivatives.

2. Materials

2.1. Samples

Serum, plasma, cell culture media, urine, as well as cell and tissue extracts can be analyzed by these methods.

2.2. Reagents and Standards for Method A (GC–MS Analysis)

1. Part of reagents and standards used are from the Phenomenex EZ:faast Amino Acid Analysis kit (Phenomenex Inc.).
2. Solvents: *n*-propanol (LC–MS grade), chloroform, isooctane, acetonitrile (HPLC or LC–MS grade), and water (Milli-Q) (see Note 1).
3. *N*-methyl-*N*-(trimethylsilyl)-trifluoroacetamide (MSTFA, Macherey-Nagel, Dueren, Germany).
4. Derivatizing reagent 1 – DR1 (alcohol solution): Phenomenex reagent 3B containing 77% *n*-propanol and 23% 3-picoline. Store in the refrigerator.
5. Derivatizing reagent 2 – DR2 (propyl chloroformate solution): Phenomenex reagent 4 containing 17.4% propyl chloroformate, 11% isooctane, and 71.6% chloroform (see Note 2). Store in the refrigerator.
6. Stabilization reagent: Mix 1 ml *n*-propanol, 1 ml phenol solution (25 mg in 4 ml water), 328 μl thiodiethylene glycol, and top up with water to a final volume of 10 ml. Store at –20°C.

7. Internal standard:
   (a) [U-^{13}C, U-^{15}N] cell free amino acid mix (Euriso-top, Saint-Aubin Cedex, France) containing alanine, arginine, aspartic acid, asparagine, cystine, glycine, glutamic acid, glutamine, serine, histidine, isoleucine, leucine, lysine, methionine, phenylalanine, proline, threonine, tyrosine, tryptophan, and valine (see Note 3). Prepare a stock solution by dissolving 30 mg of the labeled amino acid mix in 10 ml water.
   (b) [2,5,5-$^2H_3$] α-aminoadipic acid (C/D/N Isotopes Inc., Quebec, Canada). Prepare a stock solution in water with a final concentration of 10 mM.
   (c) [2,3,4,5,6-$^2H_5$] hippuric acid (Sigma–Aldrich, Taufkirchen, Germany). Prepare a stock solution in acetonitrile/water (30:70, *v*/*v*) with a final concentration of 10 mM.
   (d) Prepare the internal standard solution by mixing 500 μl labeled amino acid mix, 100 μl [2,3,4,5,6-$^2H_5$] hippuric acid stock solution, 100 μl [2,5, 5-$^2H_3$] α-aminoadipic acid stock solution, and 300 μl of stabilization reagent.
   (e) Store all standard solutions at −80 to −20°C.
8. Calibration standards:
   (a) SD1: Phenomenex amino acid standard solution 1 containing alanine, sarcosine, glycine, α-aminobutyric acid, valine, β-aminoisobutyric acid, leucine, allo-isoleucine, isoleucine, threonine, serine, proline, aspartic acid, methionine, hydroxyproline, glutamic acid, phenylalanine, α-aminopimelic acid, ornithine, lysine, histidine, tyrosine, and cystine at a concentration of 200 μM.
   (b) SD2: Phenomenex amino acid standard solution 2 containing asparagine, glutamine, and tryptophan at a concentration of 200 μM.
   (c) SD3: Phenomenex amino acid standard solution 3 containing alpha-aminoadipic acid, cystathionine, glycyl-proline, hydroxylysine, proline–hydroxyproline, and thiaproline at a concentration of 200 μM.
   (d) SD4: Hippuric acid (Sigma–Aldrich). Prepare a stock solution in acetonitrile/water (30:70, *v*/*v*) with a final concentration of 2 mM.
   (e) SD5: Amino acid standard solution (Sigma–Aldrich) containing alanine, arginine, aspartic acid, cystine, glycine, glutamic acid, histidine, isoleucine, leucine, lysine, methionine, phenylalanine, proline, threonine, tyrosine, serine, and valine at a concentration of 1 mM each in 0.1 M HCl.
   (f) Phenomenex norvaline solution (200 μM).

(g) Store all standard solutions at −80 to −20°C.

(h) Master mix 1: Mix 300 μl each of SD1, SD2, and SD3, followed by the addition of 30 μl of SD4 and 70 μl of stabilization reagent. The final concentration of all analytes in master mix 1 is 60 μM. Prepare a 1:10 and a 1:100 dilution in water (see Note 4).

(i) Master mix 2: Mix 500 μl SD 5, 250 μl SD 4, and 250 μl of stabilization reagent. The final concentration of all analytes in master mix 2 is 500 μM.

2.3. Supplies for Method A (GC–MS Method)

1. Autosampler vials (12 x 32 mm) with magnetic crimp caps (Gerstel, Muehlheim, Germany).
2. Chemically inert SILTEC liner (Gerstel).

2.4. Equipment for Method A (GC–MS Method)

1. Agilent Technologies Model 6890 GC equipped with an MSD model 5975 Inert XL, PTV injector, and an MPS-2 Prepstation sample robot (Gerstel) for automated sample derivatization and handling. The robot is equipped with an agitator for sample incubation and two syringes of different volumes.
2. A 10-μl syringe is used for internal standard addition and sample injection while reagents are added using a 250-μl syringe. Between adding steps, the syringes are washed five times each with *n*-propanol and isooctane. Samples are kept in a cooled tray at 5°C.
3. ZB-AAA column (15 m x 0.25 mm ID x 0.1 μm film thickness) from the Phenomenex EZ:faast Amino Acid Analysis kit.

2.5. Reagents and Standards for Method B (LC–MS/MS Method)

1. NIST amino acid standard (see Note 5).
2. Standard 1 and 2 from the Phenomenex EZ:faast Amino Acid Analysis kit (see above and Note 6).
3. AB SCIEX Amino Acid Analysis kit:
   (a) AB SCIEX Amino Acid Analysis column (C18 RP, 5 μm, 4.6 x 150 mm).
   (b) AA 45/32 Phys iTRAQ 115 labeling reagent.
   (c) AA 45/32 Phys Labeling Buffer (0.45 M borate buffer, pH 8.5, containing 20 pmol/μl norvaline).
   (d) AA 45/32 Phys Sulfosalicylic Acid (10% sulfosalicylic acid, containing 400 pmol/μl norleucine).
   (e) AA 45/32 Phys Hydroxylamine (1.2% hydroxylamine solution).
   (f) AA 45/32 Phys Mobile Phase Modifier A (100% formic acid).
   (g) AA 45/32 Phys Mobile Phase Modifier B (100% heptafluorobutyric acid).
   (h) Isopropanol.

(i) AA 45/32 Phys 114-labeled standard (see Note 7).

(j) AA 45/32 Phys 115-labeled standard (see Note 8).

(k) AA 45/32 Phys unlabeled standard (see Note 9).

(l) AA 45/32 Phys Standard Diluent (0.5% formic acid solution).

(m) Allo-Isoleucine.

(n) AA 45/32 Phys Control Plasma.

2.6. Supplies for Method B (LC–MS/MS Method)

1. Three 0.5-ml Eppendorf cups and one autosampler vial per sample or three 96-well plates (see Note 10).
2. Pipettes in the range of 5–40 µl.

2.7. Equipment for Method B (LC–MS/MS Method)

1. Agilent 1200 SL LC-system, equipped with a binary pump, thermostated autosampler, and column oven.
2. AB SCIEX 4000 QTrap mass spectrometer.

3. Methods

3.1. Sample Preparation for GC–MS Method

1. Label autosampler vials for blanks, quality control samples, and samples to be analyzed.
2. Add 20–50 µl of sample to the autosampler vials (20 µl for serum and cell culture media or 50 µl for urine).
3. Add 10 µl of stabilization reagent.
4. Add 10 µl of internal standard solution (this step can also be performed automatically by the sample robot).
5. Cap the cups, vortex, and place them in the cooled tray of the autosampler. Make sure to use magnetic crimp caps to allow sample handling by the robot.
6. Automated sample derivatization:

   (a) Dilute the sample with water to a total volume of 225 µl.

   (b) Add 10 µl of norvaline solution.

   (c) Add 80 µl DR1 (alcohol solution).

   (d) Move the sample to the agitator.

   (e) Mix the solution at 750 rpm for 0.2 min at 35°C.

   (f) Add 50 µl DR2 (propyl chloroformate solution).

   (g) Mix the solution at 750 rpm for 0.2 min at 35°C, equilibrate for 1 min, and mix again for 0.2 min.

   (h) Add 250 µl of isooctane.

   (i) Mix the solution at 750 rpm for 0.2 min at 35°C.

   (j) Inject 2.5 µl from the upper organic phase directly into the PTV.

3.2. GC–MS Instrument Operating Conditions

1. Instrument operating conditions are listed in Table 1.
2. Initially, a high concentrated standard (master mix 2) is analyzed in scan mode (50–420 m/z) to determine the retention times of the amino acid derivatives. A representative chromatogram is shown in Fig. 4.
3. The mass spectrometer is operated in single-ion monitoring (SIM) mode for quantification. Two characteristic mass fragments are used for most amino acids, except for the labeled amino acids. The ion traces are listed in Table 2 (see Note 13).
4. Change the PTV liner every 30–70 samples depending on the sample matrix.
5. Condition each new liner prior to sample analysis using the following sequence:
   (a) Run solvent blank (2.5 μl isooctane).
   (b) Inject 2.5 μl MSTFA twice.
   (c) Derivatize and analyze two high-concentrated standards (master mix 2, 225 μl).
   (d) Run two solvent blanks (2.5 μl isooctane).

Table 1
GC–MS operating conditions

Oven	Initial temperature	70°C
	Initial time	1 min
	Ramp	30°C/min to 300°C, hold for 3 min
	Run time	11.7 min
Inlet	Mode	Split, split ratio 1:15
	PTV injector	50°C for 0.1 min, 12°C/s to 320°C hold for 5 min
	Liner	SILTEC liner from Gerstel
Injector	Injection volume	2.5 μl
Column	Capillary column	Phenomenex ZB-AAA, 15 m × 0.25 mm ID, 0.1 μm film thickness
	Carrier gas	Helium
	Flow rate	1.1 ml/min
	Mode	Constant flow
	Outlet pressure	Vacuum
Transfer line	Temperature	310°C
qMS	Tune	Auto tune (see Note 11)
	Solvent delay	3.5 min (see Note 12)
	Acquisition	SIM mode (see Table 15.2)
	MS source	230°C
	MS quadrupole	150°C
	Electron energy	70 eV

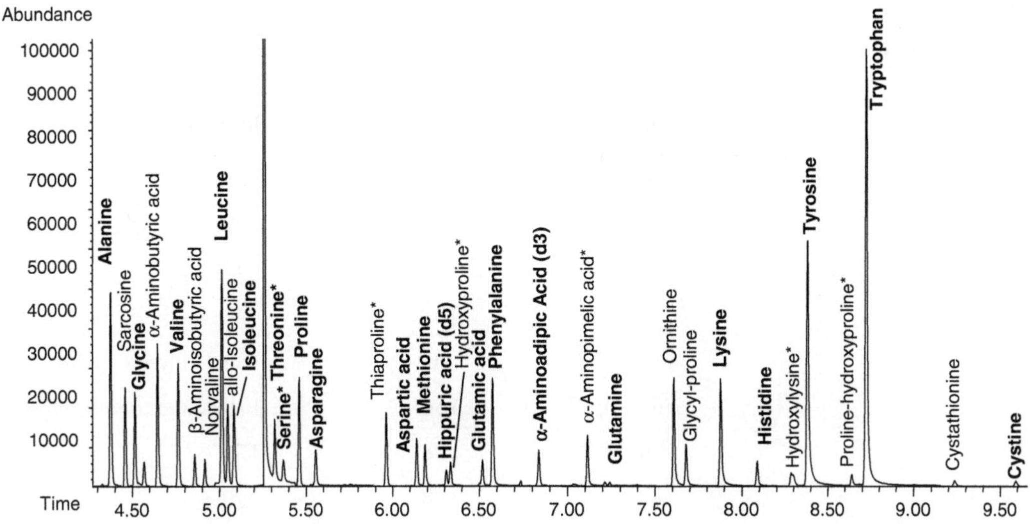

Fig. 4. Full-scan GC–MS chromatogram for the analysis of master mix 1 after derivatization with propyl chloroformate. Amino acids printed in *bold* have a corresponding stable isotope-labeled internal standard. Amino acids marked with a *star* are not considered for quantification (see Note 13).

Table 2
Ion traces selected for the SIM analysis of method A

Amino acid	Quantifier	Qualifier	Internal standard
Alanine	130	88	133
Sarcosine	130	217	Labeled alanine (133)
Glycine	102		105
α-Aminobutyric acid	144	102	Labeled valine (163)
Valine	158	116	163
ß-Aminoisobutyric acid	116		Labeled valine (163)
Norvaline (standard)	158	72	
Leucine	172	130	178
Allo-Isoleucine	172	130	Labeled isoleucine (178)
Isoleucine	172		178
Proline	156		161
Asparagine	155	69	160
Aspartic acid	216	130	220
Methionine	203	277	206
Hippuric acid	134	105	139

(continued)

Table 2 (continued)

Amino acid	Quantifier	Qualifier	Internal standard
Glutamic acid	230	170	235
Phenylalanine	190	206	199
α-Aminoadipic acid	244		247
Glutamine	84	187	89
Ornithine	156	70	Labeled lysine (176)
Glycyl-proline	70	156	Labeled glutamine (89)
Lysine	170	128	176
Histidine	282	168	290
Tyrosine	107	206	114
Tryptophan	130		140
Cystathionine	203	272	Labeled cystine (252)
Cystine	248	216	252

Amino acids printed in *bold* are quantified using the respective stable isotope-labeled amino acid as internal standard. For the remaining amino acids, a closely eluting stable isotope-labeled amino acid is used as internal standard

(e) Run a method blank (10 μl internal standard plus 215 μl water) (see Note 14).

(f) Analyze samples (see Note 15).

3.3. Calibration

1. For calibration, increasing volumes of the diluted and nondiluted standards are pipetted automatically by the autosampler into empty vials and then derivatized as described above (see Table 3 and Note 16).
2. Master mix 2 is used to extend the calibration to higher concentrations.

3.4. Method A: Data Analysis

1. Peak areas for quantifier and qualifier ions are integrated.
2. The ratio of quantifier to qualifier ion is checked to test for potential interferences.
3. Amino acids are normalized by the area of the labeled amino acid or by the area of the closest eluting internal standard compound.
4. Calibration curves are generated.
5. Amino acids in the samples are quantified.

3.5. Sample Preparation (Standard Procedure) for Method B (LC–MS/MS Method)

1. If not stated otherwise, all preparation steps are carried out at room temperature.
2. Mixing steps are performed with a vortexer.

Table 3
Calibration table for method A

	Master mix	Master mix concentration (μM)	Volume (μl)	Absolute amount (nmol)
Cal1	1:100 dilution of master mix 1	0.6	50	0.03
Cal2	1:100 dilution of master mix 1	0.6	150	0.09
Cal3	1:10 dilution of master mix 1	6	25	0.15
Cal4	1:10 dilution of master mix 1	6	100	0.60
Cal5	1:10 dilution of master mix 1	6	175	1.05
Cal6	Master mix 1	60	25.00	1.50
Cal7	Master mix 1	60	50.00	3.00
Cal8	Master mix 1	60	100.00	6.00
Cal9	Master mix 1	60	150.00	9.00
Cal10	Master mix 1	60	175	10.50
Cal11	Master mix 2	500	50	25.00
Cal12	Master mix 2	500	100	50.00
Cal13	Master mix 2	500	175	87.50

3. Immediate analysis of the samples is recommended. If necessary, derivatized samples can be stored up to 1 month at −20°C. Degradation of the derivatives may occur, but the internal standard corrects for this problem.
4. Mix an aliquot of 40 μl of sample with 10 μl of 10% sulfosalicylic acid for 10 s to precipitate proteins in the sample.
5. Centrifuge the cup (2 min, 10,000 × *g*, 4°C) to spin down the precipitated proteins.
6. Transfer 10 μl of the supernatant into the second Eppendorf cup and mix it with 40 μl of labeling buffer to adjust pH to basic conditions.
7. Take 10 μl from the mixture to the third Eppendorf cup and mix it with 5 μl of iTRAQ reagent 115.
8. Let the reaction incubate at room temperature for 60 min.
9. Add 5 μl of 1.2% hydroxylamine solution and mix the solution to quench the labeling reaction.
10. Evaporate the solvent.
11. Reconstitute the sample in 32 μl of iTRAQ reagent 114-labeled amino acids and transfer it to a vial that can be used in the autosampler.

3.6. Sample Preparation for Method B (Samples with No or Little Protein Content)

1. For samples containing no or little protein (e.g., cell culture media, urine, cell, or tissue extracts), a shorter sample preparation protocol can be used. The benefits are the following: less time consumption and smaller sample volumes needed. The complete labeling procedure can be performed in one reaction vessel. No transfer is needed.
2. 5 µl of the sample are mixed with 10 µl of the labeling buffer.
3. 5 µl of the iTRAQ reagent 115 are added.
4. Mix the sample and incubate for 60 min at room temperature.
5. 5 µl of 1.2% hydroxylamine solution are added. Mix the solution to quench the reaction.
6. Evaporate the solvent.
7. Reconstitute the dried sample in 32 µl of iTRAQ reagent 114-labeled amino acids.

3.7. Instrument Operating Conditions for Method B

1. Mobile phase A: 998.9 ml water, 1.00 ml mobile phase modifier A, 100 µl mobile phase modifier B (see Note 17).
2. Mobile phase B: 449.5 ml acetonitrile, 0.50 ml mobile phase modifier A, 50 µl mobile phase modifier B (see Note 17).
3. Inject 2 µl of the derivatized sample onto the LC–MS/MS. A resulting chromatogram is shown in Fig. 5 (see Note 18). For setting up the sequence table, take Notes 14 and 15 into account.

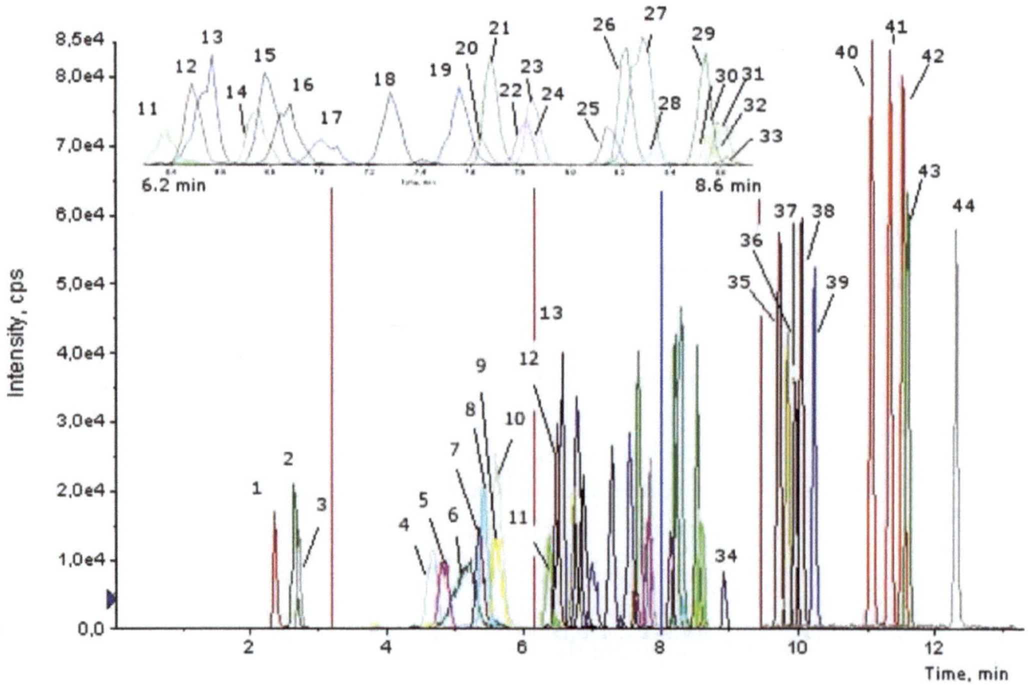

Fig. 5. iTRAQ chromatogram featuring five periods. The last period is not shown, because it is only used for column regeneration. The red line indicates at which time a period change was performed (ABI protocol), the blue line indicates where the third period was divided into two periods (our improvement, see Note 19). The corresponding amino acids are listed in Table 5.

Table 4
HPLC–ESI–MS/MS operating conditions

HPLC		
Column temp.	50°C	
Flow	0.800 ml/min	
Gradient	Time (min)	Percent mobile phase A
	0.0	98
	10.0	72
	10.1	0
	16.0	0
	16.1	98
	25.0	98
MS/MS-settings[a]		
Ion spray voltage (IS)	5 kV	
Aux. gas temp. (TEM)	450°C	
Curtain gas (CUR)	20	
Nebulizer gas (GS1)	40	
Aux. gas (GS2)	40	
Collision gas (CAD)	Medium	
Entrance potential (EP)	10 V	
Declustering potential (DP)	50 V	
Collision energy (CE)	30 V	
Collision cell exit potential (CXP)	12 V	

[a] Tune settings may vary slightly between instruments

4. Instrument operating conditions are listed in Table 4.
5. The MS/MS is operated in MRM using one transition each for the analyte and the corresponding internal standard from the precursor ion to the charged reporter ions (*m/z* 115 for the analyte and *m/z* 114 for the internal standard). The transition monitored for each amino acid is shown in Table 5.
6. In order to increase the number of data points per peak, the LC–MS/MS run is divided into several periods (see Fig. 5 and Note 19).

3.8. Data Analysis: Method B

1. Data files are processed using analyst 1.5.1.
2. Peak areas are integrated.
3. Peak-area ratios of analyte and internal standard are used to quantify the concentration of amino acids (see Note 20).
4. The LLOQ for all amino acids is in the range of 1–5 μM.
5. Typical correlation coefficients of the standard curves are >0.99.
6. Recoveries are between 90 and 110%, and relative standard deviations upon sample derivatization range from 3 to 16%.

Table 5
Amino acids analyzed with method B (iTRAQ LC–MS/MS method)

#	Name	Abbreviation	Mass (amu)	Labeled mass (amu)	Retention time (min)
1	*O*-phospho-L-serine	PSer	185.07	330.12	2.36
2	*O*-phosphoethanolamine	PEtN	141.08	286.13	2.66
3	Taurine	Tau	125.08	270.13	2.72
End of period 1; duration 3.20 min (total time 3.20 min)					
4	L-asparagine	Asn	132.12	277.17	4.66
5	L-serine	Ser	105.11	250.16	4.83
6	Hydroxy-L-proline	Hyp	131.12	276.17	5.19
7	Glycine	Gly	75.10	220.15	5.37
8	L-glutamine	Gln	146.13	291.18	5.43
9	L-aspartic acid	Asp	133.10	278.15	5.63
10	Ethanolamine	EtN	61.12	206.17	5.64
End of period 2; duration 3.00 min (total time 6.20 min)					
11	L-citrulline	Cit	175.16	320.21	6.35
12	Sarcosine	Sar	89.11	234.16	6.66
13	β-Alanine	β-Ala	89.11	234.16	6.82
14	L-Alanine	Ala	89.11	234.16	7.33
15	L-Threonine	Thr	119.12	264.17	6.52
16	L-Glutamic acid	Glu	147.12	292.17	6.78
17	L-Histidine	His	155.13	300.18	6.94
18	1-Methyl-L-histidine	1MHis	169.15	314.20	7.08
19	3-Methyl-L-histidine	3MHis	169.15	314.20	7.60
20	Homocitrulline	Hcit	189.18	334.23	7.66
21	γ-Amino-*n*-butyric acid	GABA	103.13	248.18	7.71
22	D/L-β-Aminoisobutyric acid	β-Aib	103.13	248.18	8.25
23	L-α-Amino-*n*-butyric acid	Abu	103.13	248.18	8.56
24	L-α-Aminoadipic acid	Aad	161.13	306.18	7.88
25	L-Anserine	Ans	240.19	385.24	7.91
26	L-Carnosine	Car	226.17	371.22	7.85
27	L-Proline	Pro	115.13	260.18	8.32
28	L-Arginine	Arg	174.18	319.23	8.20
29	δ-Hydroxylysine[a]	Hyl	162.10	451.32	8.37
30	L-Ornithine[a]	Orn	132.16	421.31	8.58
31	Cystathionine[a,b]	Cth	222.26	256.15	8.60
32	L-Cystine[a,b]	C-C	240.02	265.13	8.63
33	Argininosuccinic acid	Asa	290.28	435.33	7.48
34	L-Lysine[a]	Lys	146.11	435.33	8.96
End of period 3; duration 3.25 min (total time 9.45 min)					
35	L-Valine	Val	117.14	262.19	9.75
36	L-Norvaline	Nva	117.14	262.19	10.06
37	L-Methionine	Met	149.11	294.16	9.87
38	L-Tyrosine	Tyr	181.14	326.19	9.97
39	L-Homocystine[a,b]	Hcy	268.36	279.14	10.24
40	L-Isoleucine	Ile	131.16	276.21	11.09
41	L-Leucine	Leu	131.16	276.21	11.35
42	L-Norleucine	Nle	131.16	276.21	11.54
43	L-Phenylalanine	Phe	165.14	310.19	11.61
44	L-Tryptophan	Trp	204.15	349.20	12.32
End of period 4; duration 3.75 min (total time 13.20 min)					

For each amino acid, two fragments are detected: 114 for the internal standard, 115 for the analyte

[a] Double labeled

[b] Double charged

4. Notes

1. Prepare all solutions using ultrapure water (purified by means of a PURELAB Plus system) or solvents of HPLC or LC–MS grade.
2. Pay attention to the expiration date of propyl chloroformate solution. Propyl chloroformate hydrolyzes over time.
3. The U-^{13}C, U-^{15}N-labeled amino acid mix is derived from algae. The concentration of the individual amino acids may vary from batch to batch and has to be calculated from the data sheet supplied by the vendor. The internal standard can contain minute amounts of unlabeled amino acids, which influence the LOD of the method. Samples containing only the internal standard must be analyzed to determine these blank values.
4. Prepare the master mix fresh for each calibration.
5. Buy NIST-certified amino acid standard containing 17 amino acids at a concentration of 2.5 mM (certified concentration varies in the range of 1–3 mM; most of them are around 2.4–2.5 mM); Product No. SRM 2389 replaced by SRM 2389a (concentrations are now between 2.44 and 2.55 mM, except cysteine 1.23 mM).
6. The amino acid standards are part of the Phenomenex EZ:faast kit, but they can be purchased separately (Product No. AL0-7500; includes standards 1–3).
7. Be sure to use the correct volume of diluent reagent for reconstitution to obtain the correct concentration. The volume is given in the certificate of analysis and is noted on the vial.
8. This standard should be used to correct for cross talk due to the isotopic pattern. If you carry out calibration curves, this correction is not necessary.
9. This unlabeled standard is for testing the protocol and the chromatography. Carry out the protocol and analyze the standard to check the chromatography, like period settings. For this testing, you can also use any other amino acid standard. We check the chromatography with a calibration point.
10. To reduce operating time and effort, 96-well plates can be used. The protocol is the same, but multichannel pipettes can be used and the well plate can be used in a compatible autosampler. The drying process takes a bit longer.
11. The mass spectrometer should be tuned for optimal sensitivity using perfluorotributylamine (PFTBA) as a reference.
12. Data is not collected in the first few minutes to exclude solvent and reagent peaks.

13. The list of amino acids quantified was reduced in contrast to a previously published method (9). Hippuric acid was included in the analysis. Threonine and serine were excluded because the quantification of these analytes proved to be not reproducible as already described (9). Moreover, thiaproline, hydroxyproline, hydroxylysine, α-aminopimelic acid, and proline–hydroxyproline were not quantified because these analytes were not detected in the biological samples, specifically serum, analyzed so far in our laboratory.
14. Make sure to include blanks in the measurement to account for contamination, which can derive from solvents, extraction solutions, impure internal standards, or unconditioned columns. The blanks need to contain everything, except the physiological fluid, and undergo the complete process.
15. Analyze samples in random order to avoid a systematic error. Distribute biological reference samples across the whole sample set to monitor for analytical variances across the measurement.
16. Always run calibration samples from low to high concentrations.
17. The given volumes are used by the manufacturer. We are using 1 L water and 0.5 L acetonitrile and add the phase modifiers A and B in the given volumes. This very small dilution does not influence chromatography.
18. Before measuring samples, perform a test run with a standard to check if the acquisition windows are set correct. Shifts in retention time can cause the analyte to elute outside the preset acquisition window resulting in failure to detect.
19. To reduce the MRMs measured at the same time and to increase the number of data points across a peak, the mass spectrometric measurement of chromatographic run is divided in four periods. Nevertheless, there are still 34 MRMs measured in period three, which results in a dwell time of 30 ms per MRM. That results in one data point every 1.2 s and gives less accurate results because the number of points over the peak is too low. Therefore, this period can be divided into two periods as indicated in Fig. 5 with a blue line. However, very stable retention times are needed because the peak maxima of the two analytes before and after period change are only separated by 20 s.
20. The standard protocol uses isotope dilution for quantification. It was found to deliver inaccurate results with an overquantification of approximately 120%. Therefore, we used individual calibration curves for 32 AAs. The NIST and Phenomenex standards are diluted to different concentrations in the range of 0.5–2,500 and 0.5–200 μM, respectively. All calibration points are measured in triplicates.

Acknowledgments

This work was supported by the Bavarian Genome Network (BayGene) and the intramural ReForM C program.

References

1. Kaspar H et al (2009) Advances in amino acid analysis. Anal Bioanal Chem 393:445–52
2. Waldhier M C et al (2009) Capillary electrophoresis and column chromatography in biomedical chiral amino acid analysis. Anal Bioanal Chem 394:695–706
3. Moore S, Spackman D H, and Stein W H (1958) Automatic recording apparatus for use in the chromatography of amino acids. Fed Proc 17:1107–1115
4. Moller S E (1993) Quantification of physiological amino acids by gradient ion-exchange high-performance liquid chromatography. J Chromatogr 613:223–230
5. Le Boucher J et al (1997) Amino acid determination in biological fluids by automated ion-exchange chromatography: performance of Hitachi L-8500A. Clin Chem 43:1421–1428
6. Kaspar H et al (2009) Urinary amino acid analysis: a comparison of iTRAQ-LC-MS/MS, GC-MS, and amino acid analyzer. J Chromatogr B Analyt Technol Biomed Life Sci 877:1838–1846
7. Husek P (1998) Chloroformates in gas chromatography as general purpose derivatizing agents. J Chromatogr B: Biomedical Sciences and Applications 717:57–91
8. Wang J et al (1994) Analysis of amino acids by gas chromatography-flame ionization detection and gas chromatography-mass spectrometry: simultaneous derivatization of functional groups by an aqueous-phase chloroformate-mediated reaction. J Chromatogr A 663:71–78
9. Kaspar H et al (2008) Automated GC-MS analysis of free amino acids in biological fluids. J Chromatogr B Analyt Technol Biomed Life Sci 870:222–232
10. Ross PL et al (2004) Multiplexed protein quantitation in Saccharomyces cerevisiae using amine-reactive isobaric tagging reagents. Mol Cell Proteomics 3:1154–1169

Chapter 16

Automated Analysis of Primary Amino Acids in Plasma by High-Performance Liquid Chromatography

Durk Fekkes

Abstract

The concentration of primary amino acids (AAs) in plasma can accurately be determined using high-performance liquid chromatography (HPLC). Before the analysis can be performed, several steps have to be regarded. First, the time and method of blood withdrawal, type of blood tube, use of medication, and differences in dietary intake are important factors that should be standardized. Second, the handling of and the way the blood is transported to the laboratory, the time between blood withdrawal and centrifugation, the method of centrifugation, and the temperature and time of plasma storage have to be noticed. Third, the methods used for deproteinization and derivatization may account for varying results between laboratories.

In this chapter, we describe an HPLC method that measures primary amino acids in plasma using automated precolumn derivatization with *ortho*-phthalaldehyde, and that pays attention to the above-mentioned criteria. This method is relatively fast, simple, sensitive, and reliable. Since with this method we can determine over 40 physiological amino acids with a very good resolution, trace amounts of amino acids can also be determined. In addition, interassay resolution times have very low variation and the use of two internal standards guarantees reliable quantification.

Key words: Amino acids, HPLC, Precolumn derivatization, *ortho*-Phthalaldehyde, Plasma, Acid precipitation, Centrifugation, Amino acid analysis

1. Introduction

Many aspects of the method described in this chapter can be found in two earlier publications. The first was in 1995 (1) and describes the validation of the determination of amino acids (AAs) by high-performance liquid chromatography (HPLC) using automated precolumn derivatization with *ortho*-phthalaldehyde (OPA). The second was a review on the state of the art of high-performance liquid chromatographic analysis of amino acids in physiological

Michail A. Alterman and Peter Hunziker (eds.), *Amino Acid Analysis: Methods and Protocols*, Methods in Molecular Biology, vol. 828, DOI 10.1007/978-1-61779-445-2_16,

samples (2). In this review, the different ways of collection, handling, storage, and deproteinization of physiological samples, as well as various methods used for precolumn derivatization and reversed-phase HPLC have been described. In this chapter, we describe in a detailed manner our experience with and how we perform the HPLC analysis of primary amino acids in blood plasma, including sample collection and preparation.

1.1. Blood Sampling and Handling

Since amino acids in blood are under the influence of a diurnal rhythm, blood is routinely collected between 8 and 10 a.m. (3). Blood taken between 10 and 12 a.m. may result in falsely decreased levels of especially the large neutral amino acids (3). It is recommended but not necessary that the subjects have fasted overnight. For the measurement of amino acids in plasma, blood is drawn by venipuncture into tubes containing potassium EDTA as the anticoagulant. After centrifugation, the plasma is stored at –80°C until analysis.

1.2. Sample Preparation

Plasma is deproteinized with 5-sulfosalicylic acid (SSA) containing an internal standard and buffered with LiOH (1, 2). This sample is put in a cooled autosampler.

1.3. Derivatization

The amino acids are derivatized with OPA in the autosampler before chromatography (1). OPA is preferred when also trace amounts of amino acids have to be quantitated. In addition, OPA itself does not fluoresce and no interfering peaks are produced. The relative instability of the derivatives formed is not a serious problem since we use an automated precolumn derivatization method (1).

1.4. HPLC Analysis

A reversed-phase HPLC method is used with a ternary gradient in order to separate also the methylhistidines (Me-His) – accomplished by temporarily decreasing the pH of the elution solvent from 6.6 to 5.8 – and to ensure that small peaks are present throughout the whole chromatography (1). The latter was accomplished by increasing the amount of organic solvent during the run. For reproducible retention times between runs, the column temperature is kept at 25°C and all solvents are degassed before they are pumped through the column. Regeneration between runs is not necessary because the third solvent contains 57% of organic solvents, the amount of which is high enough to ensure that all substances are eluted from the column. Equilibration of the column is already started at the end of the chromatography and is prolonged during derivatization. Quantification is done by measuring peak heights relative to the internal standard. This method is in favor of the area method because not all peaks are completely resolved and the retention times are very stable (see Figs. 2–4).

2. Materials

2.1. Chemicals

1. All chemicals used are of analytical grade and the solvents are of chromatographic grade.
2. The water is purified with a Direct-Q Water purification System (Millipore) and mostly shows a resistivity of 18.2 MΩ.cm (HPLC water).
3. The amino acids, including a physiological amino acid standard solution (product no.A-9906), nitrilotriacetic acid, and propionic acid (Sigma–Aldrich, St. Louis, MO, USA).

2.2. Blood Tubes

Plastic or glass tubes of 4–10 ml containing potassium EDTA as the anticoagulant are used. Routinely, we use 4-ml vacutainer tubes containing 0.15% potassium EDTA.

2.3. Deproteinization

1. Deproteinization solution: 24% (*w*/*v*) SSA, containing 2 mM of the internal standards norvaline and homoserine (Hse):
   (a) Weigh 30 g of SSA and add 90 ml of HPLC water in a measuring cylinder.
   (b) Mix with a magnetic stirrer and fill up till 100 ml with HPLC water (30% SSA).
   (c) Pipette the following solutions in a 50-ml flask: 4 ml of 20 mM homoserine + 4 ml of 20 mM norvaline + 32 ml of 30% SSA (deproteinization solution).
2. Neutralization solution: 0.3 M LiOH.

2.4. Derivatization and Amino Acid Standards

1. Derivatization (OPA) reagent:
   - Pipette 18 ml of a 0.2 M sodium borate buffer (pH 10.4) into an amber glass jar (see Note 1): Weigh 2.47 g of boric acid, add 160 ml of HPLC water, mix and adjust the solution pH to 10.4 with 4 M NaOH, and fill up till 200 ml with HPLC water (0.2 M sodium borate buffer).
   - Bubble this solution for 30 min with nitrogen to remove the oxygen.
   - Weigh 50 mg of OPA in a polystyrene tube of 11 ml.
   - Pipette 1 ml methanol (HPLC quality) in this tube and mix on a vortex mixer.
   - Add the latter solution to the borate buffer and mix with a magnetic stirrer.
   - Bubble the 2-mercaptoethanol for 2 min with nitrogen.
   - Add 50 μl of 2-mercaptoethanol to the OPA solution while stirring.
   - Add 10 mg of nitrilotriacetic acid and bubble this solution with nitrogen (see Note 2).

- Add 50 μl of Brij-35 and stir for another minute (see Note 3).
- Keep this reagent in the dark at 4°C (see Note 4). For each assay, 1 ml of this OPA reagent was transferred to an amber glass autosampler vial with a silicone rubber, PTFE-coated screw cap.

2. Physiological AA calibration solution:
   - Pipette the following amino acid solutions in a polystyrene tube of 11 ml: 625 μl of 2 mM asparagine + 200 μl of 20 mM glutamine (Gln) + 200 μl of 20 mM norvaline (internal standard 1) + 500 μl of 2.5 mM phosphoserine + 200 μl of 20 mM homoserine (internal standard 2) + 3,275 μl of loading buffer (0.2 M lithium citrate buffer pH 2.2). The total volume of this *amino acid mixture A* is 5 ml and the composition is: 250 μM asparagine, 800 μM glutamine, 800 μM norvaline, 250 μM phosphoserine, and 800 μM homoserine.
   - Pipette 1 ml of the *amino acid mixture A* and 1 ml of physiological amino acid standard solution (product no. A-9906) into a polystyrene tube of 11 ml and add 2 ml of solvent A (physiological calibration solution) (see Note 5).
   - Pipette portions of 200 μl of the physiological calibration solution into 1.5-ml glass autosampler vials and store at −30°C until analysis. In general, a vial is stored no longer than 2 months after being used for HPLC analysis.
3. Tryptophan (Trp) standard solutions:
   - Pipette the following amino acid solutions in a polystyrene tube of 11 ml: 500 μl of 2.5 mM phosphoserine (PSer) + 100 μl of 25 mM sulfocysteine + 125 μl of 20 mM serine + 125 μl of 20 mM glutamine + 625 μl of 2 mM phosphoethanolamine (PEA) + 200 μl of 20 mM homoserine + 250 μl of 10 mM citrulline + 125 μl of 20 mM alanine + 200 μl of 20 mM norvaline (internal standard) + 250 μl of 10 mM cystathionine and allo-cystathionine (Allo-Cystat) + 1,250 μl of 2 mM allo-isoleucine (Allo-Ile) + 1,250 μl of loading buffer. The total volume of this *amino acid mixture B* is 5 ml and the composition is: 250 μM PSer and PEA, 500 μM sulfocysteine, serine, glutamine, citrulline, alanine, cystathionine + allo-cystathionine and allo-isoleucine, and 800 μM norvaline and homoserine.
   - Pipette 200 μl of the *amino acid mixture B* and 200 μl of 1 mM tryptophan into a polystyrene tube of 5 ml (500 μM tryptophan standard).
   - Pipette 300 μl of 1 mM tryptophan + 300 μl of solvent A into a polystyrene tube of 5 ml (500 μM tryptophan). Pipette 200 μl of the *amino acid mixture B* and 200 μl of 500 μM tryptophan into a polystyrene tube of 5 ml (250 μM tryptophan standard).

- Pipette 300 μl of 500 μM tryptophan + 300 μl of solvent A into a polystyrene tube of 5 ml (250 μM tryptophan). Pipette 200 μl of the *amino acid mixture B* and 200 μl of 250 μM tryptophan into a polystyrene tube of 5 ml (125 μM tryptophan standard).
- Pipette 300 μl of 250 μM tryptophan + 300 μl of solvent A into a polystyrene tube of 5 ml (125 μM tryptophan). Pipette 200 μl of the *amino acid mixture B* and 200 μl of 125 μM tryptophan into a polystyrene tube of 5 ml (62.5 μM tryptophan standard).
- Add to all four *tryptophan standards* 400 μl of solvent A (see Note 5).
- Pipette portions of 100 μl of the four *tryptophan standards* into four clear 0.2-ml glass microinserts fitted in 1.5-ml autosampler vials and store at −30°C until analysis. Every month, a new vial with a tryptophan standard is used (see Note 6).

2.5. Solvents for HPLC Analysis

1. *Solvent A*: Sodium phosphate (250 mM), propionic acid (250 mM), acetonitrile, tetrahydrofuran, and water, 20/20/7/0.2/51 (*v*/*v*), with an ultimate pH of 6.6–6.7:
   (a) Weigh 35.5 g of Na_2HPO_4 and add 1 L of HPLC water (250 mM).
   (b) Mix 18.5 ml of propionic acid with 982.5 ml of HPLC water (250 mM).
   (c) Pour both solutions into a 5-L brown glass bottle.
   (d) Add approx. 20 ml of 1 M NaOH solution to adjust the solution pH to 6.4.
   (e) Add 350 ml of acetonitrile.
   (f) Add 10 ml of tetrahydrofuran.
   (g) Add approx. 2,620 ml HPLC water to a final volume of 5 L.
   (h) Measure the ultimate pH (should be around pH 6.6, if not adjust the pH).
   (i) Store this solvent at 4°C and filter 2.5 L of the solvent through a 0.45-μm regenerated hydrophilic cellulose membrane before use (see Note 7).
2. *Solvent B*: Sodium phosphate (250 mM), propionic acid (250 mM), acetonitrile, tetrahydrofuran, and water, 20/20/7/0.2/51 (*v*/*v*), with an ultimate pH of 5.8–5.9:
   (a) Weigh 17.75 g of Na_2HPO_4 and add 500 ml of HPLC water (250 mM).
   (b) Mix 9.25 ml of propionic acid with 491 ml of HPLC water (250 mM).

(c) Pour both solutions into a 2.5-L brown glass bottle.
(d) Add dropwise concentrated propionic acid to adjust the solution pH to 5.65.
(e) Add 175 ml of acetonitrile.
(f) Add 5 ml of tetrahydrofuran.
(g) Add approx. 1,320 ml HPLC water to a final volume of 2.5 L.
(h) Measure the ultimate pH (should be around pH 5.8, if not adjust the pH).
(i) Filter the solvent through a 0.45-μm regenerated hydrophilic cellulose membrane.

3. *Solvent C*: Acetonitrile, methanol, dimethylsulfoxide, and water, 28/24/5/43 (*v*/*v*):
   (a) Add the following solutions into a 2.5-L brown glass bottle: 700 ml of acetonitrile + 600 ml of methanol + 125 ml of dimethylsulfoxide + 1,075 ml of HLPC water.
   (b) Filter the solvent through a 0.45-μm regenerated hydrophobic cellulose membrane.

2.6. HPLC Equipment

1. The HPLC system (Fig. 1 depicts the complete system) consists of an HP 1050 Series quaternary pump (Hewlett-Packard, Waldbronn, Germany), an HP 1100 Series online degassing

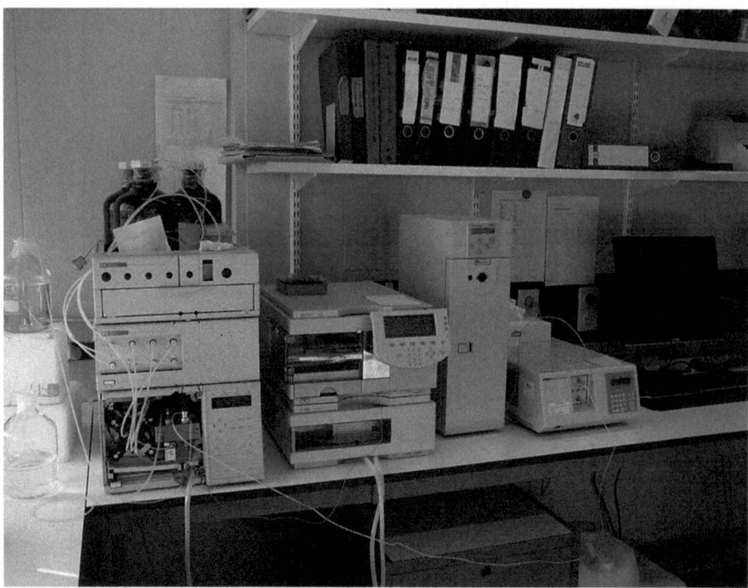

Fig. 1. Equipment used for the HPLC analysis. *Left apparatus from top to bottom*: Solvent cabinet, degasser, and quaternary pump. *Second apparatus*: Autosampler with control module and cooling unit. *Third apparatus*: Column oven with Peltier cooling device. *Fourth apparatus*: Fluorescence detector. *Most right apparatus*: Computer for data acquisition.

device, and a Mistral column oven with a built-in Peltier cooling device (Spark Holland, Emmen, the Netherlands).

2. An automated precolumn derivatization is done on HP 1100 thermostated (4°C) autosampler.
3. Fluorescence is monitored at an excitation wavelength of 337 nm and an emission wavelength of 452 nm with a Jasco Model FP-920 fluorescence detector (B&L Systems, Maarssen, the Netherlands) equipped with a 150-W xenon lamp and a 16-μl flow cell.
4. Data are collected online and processed by a Geminix Integration Pack data acquisition system (Kontron Instruments, Milan, Italy).
5. The separation column is a Spherisorb ODS II cartridge column, 5 μm, 125 mm × 3 mm (Waters), which is protected by a Hypersil ODS guard column, 5 μm, 20 mm × 2.1 mm (Hewlett-Packard) (see Note 8).

3. Methods

3.1. Blood Sampling and Handling

1. Collect blood between 8 and 10 a.m. by venipuncture into tubes containing 0.15% potassium EDTA as the anticoagulant (see Note 9) after an overnight fast or after a light breakfast, e.g., a cup of tea and two slices of bread with cheese and marmalade.
2. Transport the blood to the laboratory within 2–3 h by placing the tube in a closed container or simply in an envelope. We do not recommend the storage of the blood on ice before further preparation because this may change the transfer of amino acids from blood cells to plasma and vice versa. However, when it is sure that the time between blood withdrawal and centrifugation exceeds a period of 2–3 h, the blood should be kept at 4°C (4).
3. Centrifuge the blood at 2,650 × g_{max} for 20 min at 20°C in a temperature-controlled centrifuge using a swing-out rotor (see Note 10).
4. Transfer the supernatant (platelet poor plasma) *carefully* (do not disturb the platelet layer!) into a polypropylene tube.
5. Pipette 500 μl of the plasma from the polypropylene tube into an Eppendorf tube for direct amino acid analysis (see Subheading 3.2) or freeze the tube (see 3.1 step 6).
6. Store both tubes at −80°C.
7. Perform the amino acid analysis within 2 years (see Note 11).

3.2. Sample Preparation

1. Add 100 µl of 24% (*w*/*v*) SSA containing 2 mM of the internal standards norvaline and homoserine to the 500-µl plasma sample (deproteinization).
2. Mix immediately on a vortex mixer and place the tube for 15 min at 4°C.
3. Spin down the precipitate at 18,400 × g_{max} for 15 min at 4°C in a cooled microcentrifuge. We use the Hettich microcentrifuge (EBA 12R).
4. Pipette 400 µl of the supernatant into a 1.5-ml glass autosampler vial.
5. Add 100 µl of 0.3 M LiOH in order to adjust the pH to 2.5 and vortex mix (see Note 12).

3.3. Derivatization

1. Set the temperature of the HP 1100 autosampler at 4°C.
2. Put the sample vials, the amber glass autosampler vial with OPA reagent, and an autosampler vial with HPLC water in the cooled autosampler.
3. Program the HP 1100 autosampler as follows:
   (a) Set stop time at 55 min.
   (b) Set draw position offset at 2 mm.
   (c) Draw 4 µl of the OPA derivatization reagent with a draw speed of 20 µl/min.
   (d) Wash the needle with HPLC water.
   (e) Draw 1 µl of sample or standard with a draw speed of 20 µl/min.
   (f) Wash the needle with HPLC water.
   (g) Mix 15 times along the capillary of the autosampler with a speed of 200 µl/min.
   (h) Inject the mixture (5 µl) onto the column with an eject speed of 200 µl/min.

3.4. HPLC Analysis

1. Program the HP 1050 Series quaternary pump as follows:
   (a) Set stop time at 55 min.
   (b) Set flow rate at 1.0 ml/min.
   (c) Import the gradient program (see Note 13).
   (d) Flush the column after the last run with solvent C for 15 min.
2. Before starting the first run, all solvents (40% solvent A, 30% solvent B, and 30% solvent C) are pumped through the degasser for 30 min at a flow rate of 5.0 ml/min with the column disconnected and purge valve open. This equilibration step is done because the degasser removes some organic solvent from all three solvents during standing, which results in variable retention times of the first two runs.

3. Set the temperature of the Mistral column oven at 25°C.
4. The fluorometer settings are:
   (a) Excitation wavelength: 337 nm (bandwidth 18 nm).
   (b) Emission wavelength: 470 nm (bandwidth 18 nm).
   (c) Gain: 100.
   (d) Response: 3 s.
5. Program the HP 1100 autosampler as follows (with in between washing of the needle with HPLC water):
   (a) Inject the physiological calibration solution two times.
   (b) Inject one of the tryptophan standard solutions.
   (c) Inject a control sample (a plasma sample that has been analyzed several times).
   (d) Inject seven plasma samples.
   (e) Inject the physiological calibration solution.
   (f) Inject seven plasma samples.
   (g) Inject the physiological calibration solution.
   (h) Inject one of the tryptophan standard solutions.
   (i) Inject a control sample, etc.
   (j) Set stop time at 55 min (see Note 14).
6. Program the Geminix Integration Pack data acquisition system as follows:
   (a) Set stop time at 48 min.
   (b) Report peak no, peak height (mV), area (mV × min), half peak width (min), base peak width (min), retention time (min), name, and concentration.
   (c) Quantify the concentration by measuring peak height relative to the internal standard norvaline (see Note 15).
7. Set the flow rate at 1.0 ml/min and pump successively solvents C and A through the column, both during 10 min, and note the back pressure (typical back pressures range from 150–155 Bar with solvents A and B to 175–181 Bar with solvent C).
8. Start the first run when the back pressure is stable.

4. Conclusions

Many details of the HPLC method – coefficients of variation, which amino acids can be analyzed, sensitivity, and linearity – have been described earlier (1) and here we describe our experience with this method during the last 15 years and also indicate additional tricks and nuances. It is highly recommended to perform

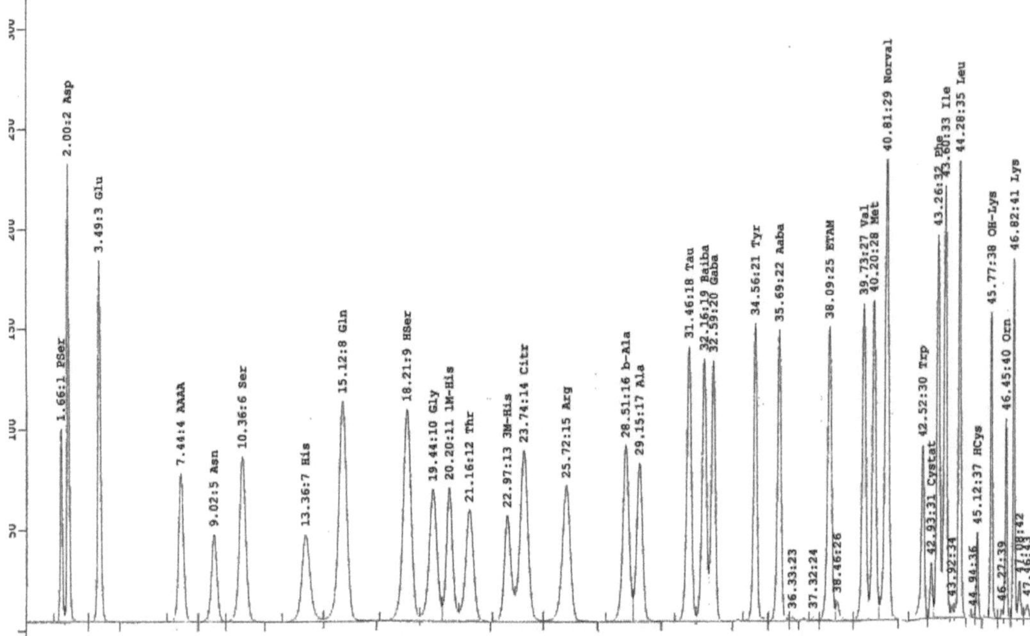

Fig. 2. Chromatogram of the physiological calibration solution (60–120 pmol per amino acid, *see* Subheading 2.3, item 2) after 50 injections. Analytical column: Spherisorb ODS II cartridge, 5 μm, 125 × 3 mm. Guard column: Phenomenex Security Guard cartridge C18, 5 μm, 4.0 × 2.1 mm. Flow rate: 1.0 ml/min. For solvents, *see* Subheading 2.4, item 1. For gradient program, see Table 1.

several test runs after a new analytical column has been connected (see Note 16). A typical amino acid chromatogram of the physiological calibration solution (see Subheading 2.3, item 2) after 50 injections is shown in Fig. 2. It can be seen that most amino acids show very good resolution. However, some amino acids are not resolved (see Note 17). An example of a chromatogram of human plasma is shown in Fig. 3. It can be seen that all amino acids show excellent resolution and that low amounts of 3-methylhistidine (3-Me-His) (the concentration in this sample was 4 μM) can be measured routinely. The concentration of 1-methylhistidine (1-Me-His) in this sample was below 1 μM and therefore, the Geminix data acquisition system was not able to detect this small peak eluting between Gly and Thr. If necessary, the exact concentration of 1-Me-His can be measured by increasing the fluorometer gain from 100 to 1,000 and reanalyzing the sample.

The quantification of the amino acid PEA may be problematic because the retention time of PEA depends on the lifetime of the analytical column (see Note 18). In Figs. 4 and 5, chromatograms of a Trp standard solution (see Subheading 2.3, item 2) after 50 and 110 injections, respectively, are shown. It can be seen that after 110 injections PEA elutes between Hse and Citr (see Fig. 5) while after 50 injections this amino acid coelutes with Hse (see Fig. 4).

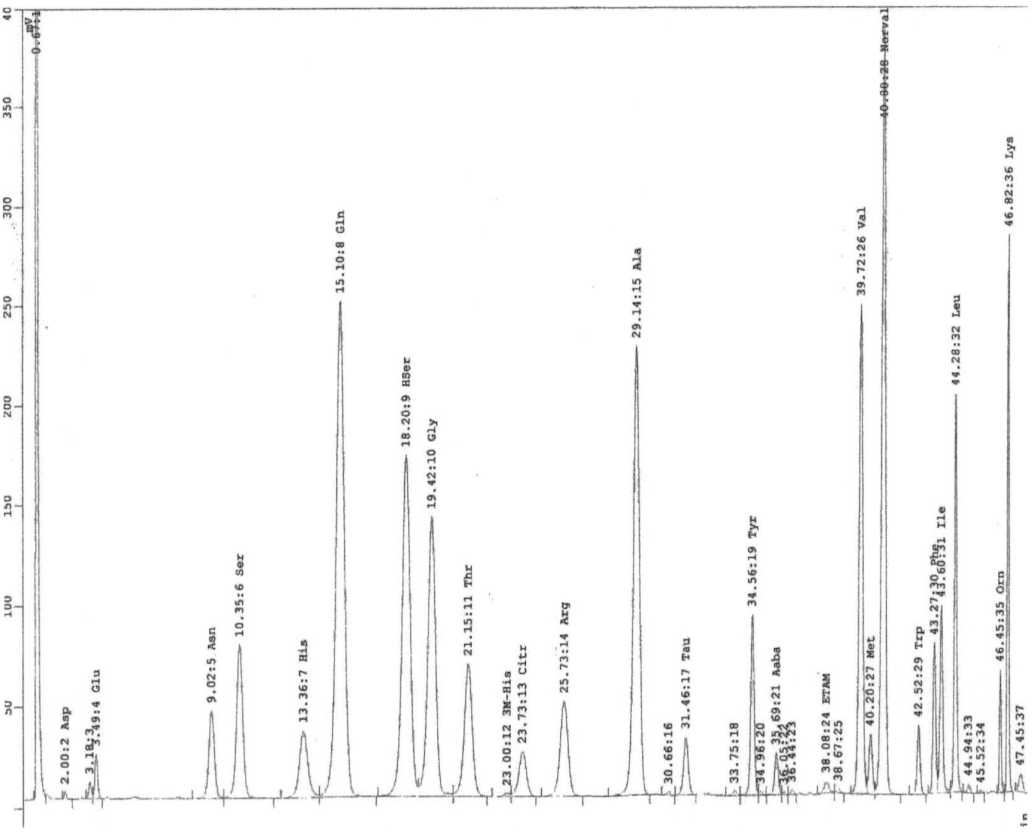

Fig. 3. Chromatogram of plasma amino acids from a healthy volunteer. For details, see legend to Fig. 2.

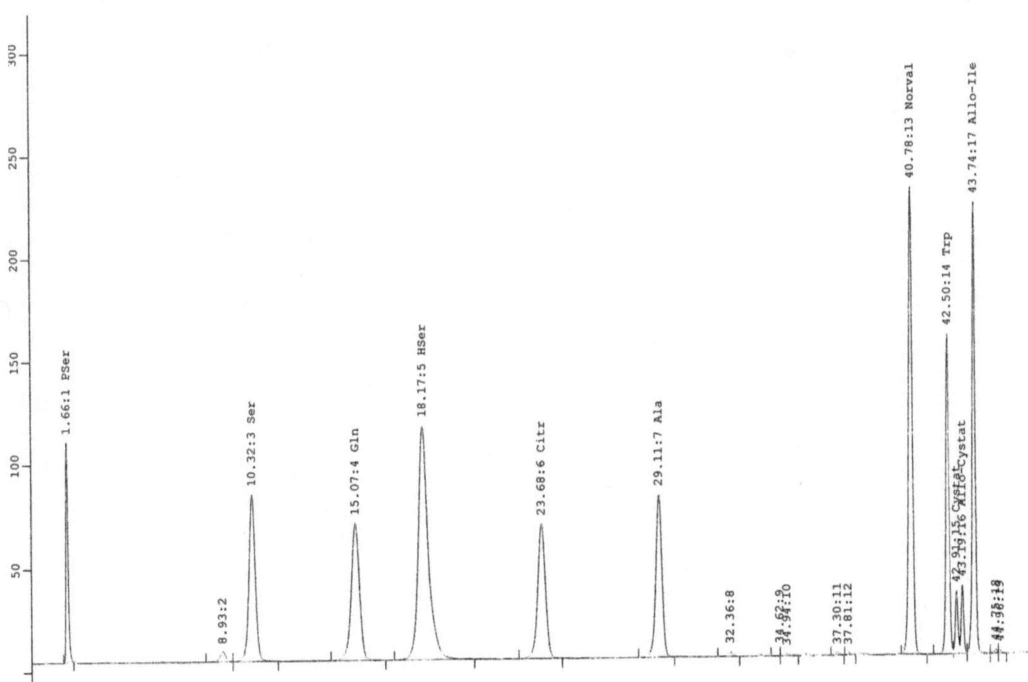

Fig. 4. Chromatogram of 250 μM tryptophan standard solution (see Subheading 2.3, item 2) after 50 injections. For details, see legend to Fig. 2.

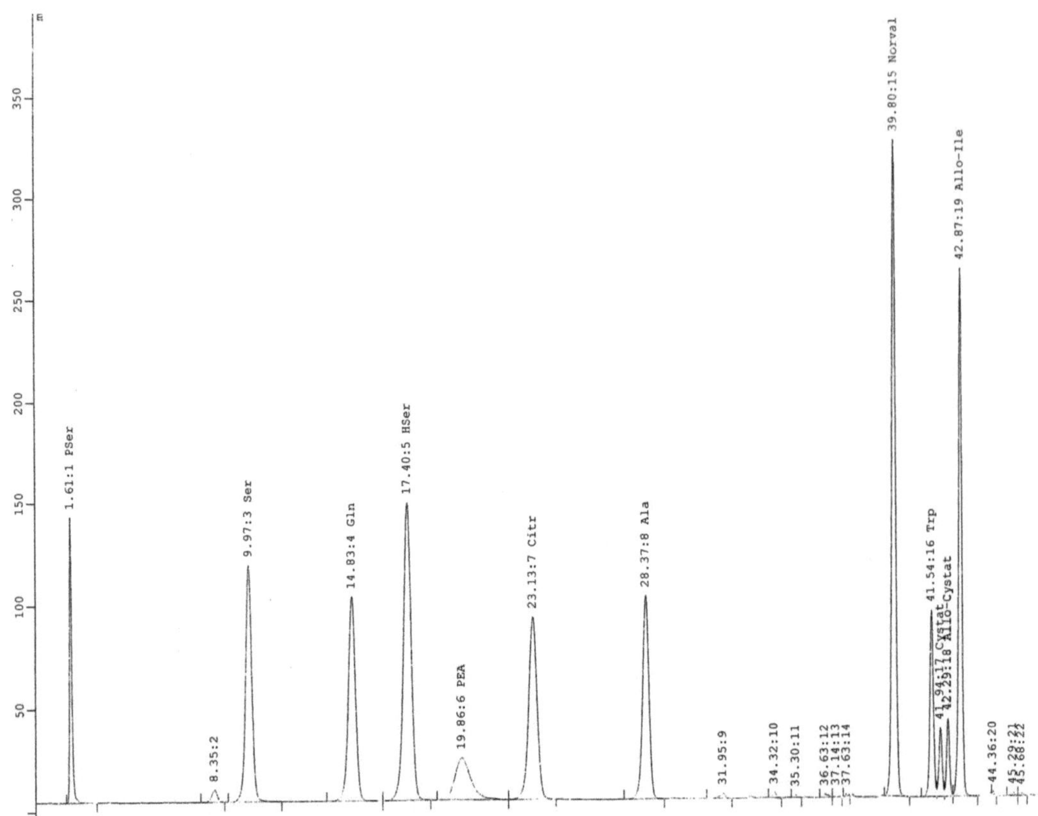

Fig. 5. Chromatogram of 125 μM tryptophan standard solution (see Subheading 2.3, item 2) after 110 injections. For details, see legend to Fig. 2.

An example of a chromatogram of the physiological calibration solution after approx. 100 injections is shown in Fig. 6. When we compare this Fig. 6 with Fig. 2, it can be seen that Citr is now completely separated from Carn and Anser (see Note 19). After approx. 250 injections, the resolution between His and Gln, 1-Me-His and Thr, and beta-alanine (b-Ala) and alanine (Ala) becomes worse (see Fig. 7 and Note 20).

The resolution between some amino acids during the lifetime of the analytical column may become unsatisfactory. The resolution may be increased by changing the time that solvent B starts or ends (see Note 21). Some amino acids, which are normally not present in human plasma, never have a satisfactory resolution with our method (see Notes 17–19). The calibration factors for Phe have to be adjusted during the lifetime of the column (see Note 22).

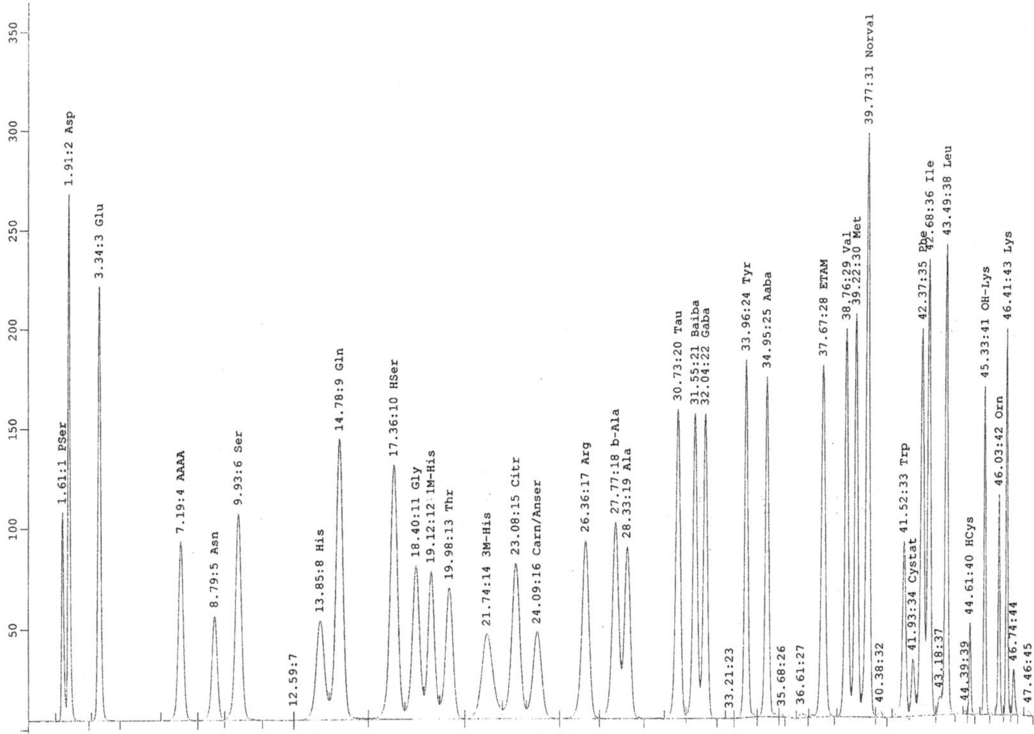

Fig. 6. Chromatogram of the physiological calibration solution after approx. 100 injections. For details, see legend to Fig. 2.

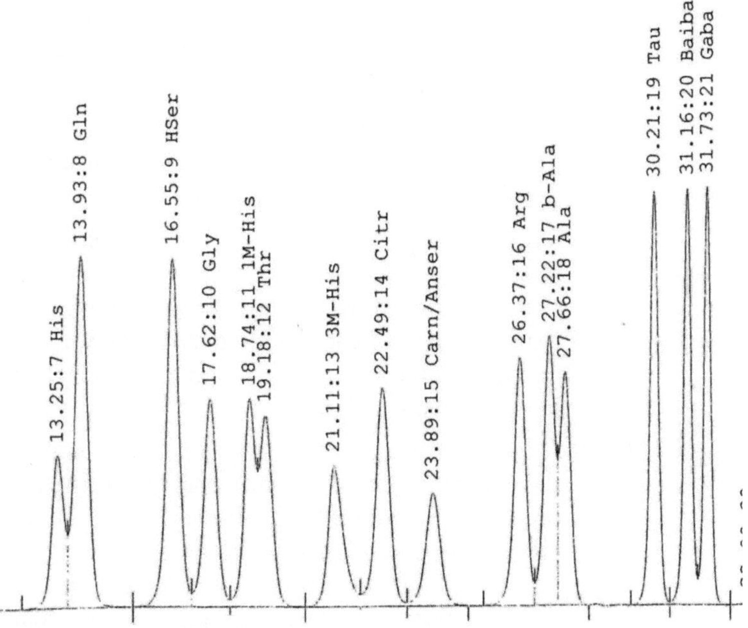

Fig. 7. Chromatogram of the physiological calibration solution after approx. 250 injections. For details, see legend to Fig. 2.

5. Notes

1. The amber glass jar we use is an empty bottle which contained the original OPA powder. For each preparation of OPA reagent, we clean the jar with HPLC water and methanol, respectively.
2. The metal-chelating agent nitrilotriacetic acid is added to the OPA reagent to reduce the oxidation of 2-mercaptoethanol and thus improve its stability (2, 5).
3. Brij-35 is added to the OPA reagent to increase the fluorescence responses of lysine and hydroxylysine.
4. Prepare the OPA reagent freshly every 2 weeks. When the reagent is older than 2 weeks, the fluorescence intensities of especially His, ornithine, and lysine may become very low.
5. These standards are diluted with solvent A to yield comparable fluorescence intensities with those of the samples.
6. For several reasons, we use a tryptophan (Trp) standard next to the physiological calibration solution: (a) due to instability and unknown concentration of Trp in the physiological calibration solution of Sigma; (b) to discover the retention times of PEA (see Subheading 3.4, second paragraph), Allo-Cystat, and Allo-Ile; and (c) to know the exact concentration of Citr, the amino acid which is not always resolved from Carn and Anser during the first 70 runs (see Subheading 3.4, second paragraph). We use four different concentrations of Trp because this amino acid is very important in our research (biological psychiatry). We also added some amino acids that are present in the calibration solution of Sigma to control for possible errors in making the standard.
7. Solvent A is used most during HPLC analysis. Therefore, we prepare 5 L of this solvent and store it at 4°C. When needed, this solution is filtered and used.
8. Since the Hypersil ODS guard column with a length of 20 mm is no longer available in the Netherlands, we now use a Security Guard cartridge C18, 5 µm, 4.0 mm × 2.1 mm from Phenomenex (part no. AJO-4286). This guard column is shorter (4 mm instead of 20 mm), cheaper, and has a longer lifetime than the Hypersil guard column.
9. In order to prevent blood platelets becoming activated with the resultant release of taurine and PEA, the use of siliconized tubes maybe advised (4). However, it is more important to prevent hemolysis (do not shake the blood tube, but gently invert it a few times) because this leads to false increases in the concentrations of aspartic acid, glutamic acid, taurine, PEA, and ornithine while the arginine levels may be decreased (4, 6).

10. The centrifuge speed and time are necessary to obtain platelet poor plasma (platelets contain high amounts of taurine and PEA). The temperature of 20°C may prevent the release of amino acids from the blood cells, heat inactivation of some amino acids (especially glutamine), and protein degradation.
11. Higher storage temperatures or longer storage times result in drastic changes in the concentrations of many amino acids due to, among others, hydrolysis of proteins (7). We found that storage of plasma for a period of 4 years at –30°C resulted in significant decreases in the levels of methionine, glutamine, and asparagine of 50, 40, and 25%, respectively. Immediate deproteinization of plasma with SSA and storage of this sample after buffering at pH 2.2 have been recommended to stabilize the amino acid concentrations (7). Although methionine and asparagine were stable under these conditions, we found that plasma levels of tryptophan, glutamine, and citrulline all decreased around 40% after storage of 4 years at –30°C.
12. In case of lower plasma amounts available, all quantities may be halved, i.e., deproteinize 250 μl of plasma with 50 μl of SSA solution. It is recommended not to use polypropylene microvials because sometimes these may be contaminated resulting in spurious peaks in the chromatogram. Instead, use clear 0.2-ml glass microinserts or 1.5-ml glass vials with integrated 0.2-ml microinserts.
13. A typical gradient program is shown in Table 1. This gradient program is not fixed, but changes during the lifetime of the column (see Notes 16 and 21).
14. Injection-to-injection time is routinely 58 min (run time: 48 min, equilibration time: 7 min, and derivatization time: 3 min). Under these circumstances, the coefficient of variation for the retention times was below 0.5%. Shorter times may be possible, but then retention times and peak widths of the first amino acids appearing in the chromatogram show a higher variation. The latter may result in less-reliable peak heights and consequently inaccurate levels of amino acids.
15. To ensure appropriate peak identification, we also include another internal standard, i.e., homoserine (Hse) in the calibration solution, which arises in the first part of the chromatogram (see Fig. 2).
16. During the first runs, 1-Me-His and threonine (Thr), 3-Me-His, and citrulline (Citr) may coelute. By changing the start of solvent B – starting solvent B a few minutes earlier results in less retention of both methylhistidines – the resolution between the mentioned amino acids becomes better.
17. The dipeptides carnosine (Carn) and anserine (Anser) are not resolved from Citr. In addition, allo-cystathionine (Allo-Cystat)

Table 1
Example of gradient conditions

	% Solvent		
Time (min)	**A**	**B**	**C**
0	100	0	0
11.5	100	0	0
12	99.6	0	0.4
12.5	99.2	0	0.8
13	98.8	0	1.2
13.5	0	98.4	1.6
23	0	90.8	9.2
24	90	0	10
28	80	0	20
34	62	0	38
38	48	0	52
40	44	0	56
44	15	0	85
45.8	0	0	100
46.2	0	0	100
46.5	100	0	0

coelutes with phenylalanine (Phe; see Note 22). In practice, this is not a problem since Carn, Anser, and Allo-Cystat normally are not present in high amounts in human plasma. When plasma does contain Carn and/or Anser, we are able to notice this by comparing the peak width of Citr in the calibration solution and the sample and report that the concentration of Citr is lower than the value found due to bad separation.

18. Although PEA is not present in the physiological calibration solution, which means that no calibration problems arise for PEA and the coeluting amino acids, PEA may be present in high amount in plasma. If this is the case, we notice this by the presence of a higher peak width of Hse, glycine (Gly), 1-Me-His, or Thr. During the first 70 runs, PEA coelutes with Hse (Fig. 4). Thereafter, the retention time of PEA increases and after 110 injections PEA elutes between 1-Me-His and Thr (Figs. 5 and 6).

19. Carn and Anser mostly never become resolved with our method. Fortunately, in psychiatric research, these amino acids

are not important. Although the separation between 3-Me-His and Citr is also better after 100 injections, the peak width of 3-Me-His may become larger. The resolution between His and glutamine (Gln) is fairly good due to the presence of tetrahydrofuran in solvents A and B (1), but deteriorates after approx. 250 injections (see Figs. 2, 6, and 7).

20. After more than 250 injections, quantification of His is no longer reliable, since the concentration of Gln in plasma is very high compared to that of His. This may result in inaccurate, higher plasma levels of His. Quantification of Ala due to bad resolution with b-Ala faces no problem, since this amino acid is also present in the Trp standards, which in this case can be used for calibration of Ala. When the resolution between arginine (Arg) and b-Ala also becomes worse, we routinely discard the column. It is possible to flush the column with acetonitrile or dichloromethane to restore the column performance (1), but this works only for 50–100 injections. In practice, we perform between 350 and 500 runs with the Spherisorb cartridge column. We replace the guard column when the back pressure increases significantly or when extra or double peaks arise in the chromatogram. Using a 70% acetonitrile solution and a flow rate of 1 ml/min, the back pressure of a new guard column is 14–17 Bar. When the back pressure is over 33 Bar, the guard cartridge is replaced.

21. In general, lowering the pH of solvent B (from 5.9 to 5.8 or lower) or starting solvent B earlier in the gradient program (see Table 1) results in a decrease in the retention times of His, 1-Me-His, and 3-Me-His, and in a better resolution between beta-aminoisobutyric acid and gamma-aminobutyric acid. Since there is no net effect on the retention times of Gln, Hse, and Gly, and a slight increase in the retention times of Thr and Arg, this action results in a better resolution between His and Gln, 1-Me-His and Thr, and 3-Me-His and Citr. However, the retention time of PEA increases, which may result in a problem with the quantification of Gly or Thr, and also the resolution between b-Ala and Ala and between Citr and Carn + Anser may become worse. A more gradual increase of solvent C in the gradient program generally results in higher retention times of the amino acids eluting between 18 and 30 min, i.e., from Gly until Ala.

22. The resolution between Allo-Cystat and Phe is not satisfactory. A small increase in the percentages of acetonitrile and methanol by 2 and 1%, respectively, may improve this resolution (1), but during column aging this resolution may become problematic again. Although Allo-Cystat normally is not present in plasma, this amino acid is added to the Physiological calibration solution of Sigma and therefore, problems with the calibration arise.

We solve this by comparing the peak heights of Phe and isoleucine (Ile). When the resolution between Allo-Cystat and Phe is good, the ratio of the peak height of Phe and Ile is 0.82 or lower and the concentration of Phe is 250 μM. If the ratio is 0.82–0.86, we calibrate Phe at 265 μM. A ratio between 0.87 and 0.92 gives a concentration of 280 μM, and a ratio of 0.92–0.94 yields a concentration of 295 μM Phe. This method is validated by performing amino acid analysis on plasma samples which have been analyzed earlier.

References

1. Fekkes D et al (1995) Validation of the determination of amino acids by high-performance liquid chromatography using automated pre-column derivatization with ortho-phthaldialdehyde. J Chromatogr B 669:177–186
2. Fekkes D (1996) State-of-art of high-performance liquid chromatographic analysis of amino acids in physiological samples. J Chromatogr B 682:3–22
3. Eriksson T, Voog L, Walinder J (1989) Diurnal rhythm in absolute and relative concentrations of large neutral amino acids in human plasma. J Psychiat Res 23:241–24
4. Perry TL, Hansen S (1969) Technical pitfalls leading to errors in the quantitation of plasma amino acids. Clin Chim Acta 25:53–58
5. Uhe AM et al (1991) Quantitation of tryptophan and other amino acids by automated precolumn ortho-phthaldialdehyde derivatization high-performance liquid chromatography: improved sample preparation. J Chromatogr B 564:81–91
6. Shih ES (1996) Amino acid analysis. In: Blau N, Duran M, Blaskovics ME (eds) Physician's guide to the laboratory diagnosis of metabolic diseases, first edn. Chapman & Hall, London
7. Schaefer A, Piquard F, Haberey P (1987) Plasma amino-acids analysis: effects of delayed samples preparation and of storage. Clin Chim Acta 164:163–169

Chapter 17

RP-LC of Phenylthiocarbamyl Amino Acid Adducts in Plasma Acetonitrile Extracts: Use of Multiple Internal Standards and Variable Wavelength UV Detection

Lionella Palego, Gino Giannaccini, and Antonio Lucacchini

Abstract

The measurement of physiological amino acids in body fluids and circulating cells can be relevant in the search for biological correlates of neuropsychiatric, neurological, body weight, and pain diseases. Several techniques are available for the quantitative analysis of free amino acids, including UV detection after precolumn derivatization. These systems have low specificity due to possible interferences at the analytical wavelength. Another problem linked to these methods is variations potentially occurring during extraction, derivatization, and chromatography of amino acids in biological matrices. We present here a modified reversed-phase LC of phenylthiocarbamyl amino acids in plasma deproteinated by the organic solvent acetonitrile. Specificity was monitored by UV-photodiode array detection and accuracy was controlled by a plasma spiking procedure with three internal standards. A dual-wavelength spectrophotometry (254, 283 nm) was used to quantify coeluting ornithine and tryptophan adducts. The method is simple and economical and enables the measure of most plasma amino acids for clinical research, also during therapeutic drug monitoring. Dual UV-fluorimetric detection solutions can improve its sensitivity.

Key words: Reversed-phase column liquid chromatography, Plasma amino acids, Phenyl isothiocyanate, Photodiode array UV detection, Amino acid analysis

1. Introduction

The quantitative analysis of physiological amino acids (AAs) in body fluids is a basic technique, applicable to many fields of biochemical, biomedical, pharmaceutical, and food sciences. Amino acids are indeed nutrients, act as pH and redox buffers, and are fundamental substrates of cell metabolism: they are, mostly, the structural units of protein synthesis, produce energy, and are pre-

Michail A. Alterman and Peter Hunziker (eds.), *Amino Acid Analysis: Methods and Protocols*, Methods in Molecular Biology, vol. 828, DOI 10.1007/978-1-61779-445-2_17,

cursors of neurotransmitters, polyamines, porphyrines, and nitric oxide (1). Moreover, some AAs, as glutamate, glycine, and the nonprotein γ-aminobutyric acid (GABA), act directly as neurotransmitters while some others are important intermediates of the detoxification and methyl transport systems (serine, sulfur-containing AAs). Branched-chain amino acids (BCAAs) influence the secretion of catecholamines and serotonin by competing with tyrosine and tryptophan for a common large–neutral amino acid (LNAA) transport system at the blood–brain barrier (2, 3). Additionally, AAs have been found directly linked to specific signaling pathways through the activation of metabotropic receptors (4, 5).

In human pathology, the profiling of physiological AAs in body fluids has acquired great importance in clinical evaluation: in the diagnosis and therapy of inherited diseases of AA metabolism (e.g., phenylketonuria and Maple syndrome disease); in biomedical research (neurological and neuropsychiatric diseases, bone and muscle disorders, and body weight and pain illnesses) (2, 6–10); in appraising nutritional states (11, 12). Pertaining to biomedical research, body fluid AA profiles may be particularly helpful to search biological indicators of disease and therapy success, such as proteins that mediate AA signaling, transport, absorption, and metabolism, representing a valid support to the screening of genetic polymorphisms and/or gene expression/protein changes in many pathologies. In some disorders, it can be useful to evaluate only a single AA (e.g., hydroxyproline, rheumatology) or a small group of AAs: the analysis of plasma tryptophan, aromatic AAs, and BCAAs has been and is currently employed to evaluate 5-HT metabolism in patients with different psychiatric disorders (6, 13–15). In many other instances, a full profiling of AAs is required.

Several analytical protocols are available to quantify physiological AAs: gas chromatography (GC) techniques, ion-exchange chromatography with postcolumn derivatization (AA autoanalyzer), combined GC–mass spectrometry (GC–MS) or MS–MS spectrometry, and high-performance liquid chromatography (HPLC) techniques. Each method presents disadvantages like time-consuming sample preparation, dedicated instrumentation, high coasts, or inability at quantifying the complete AA profile. Reversed-phase LC (RP-LC) is a versatile and economical technique. A number of RP-LC methods have been developed and described in the literature with a pre- or postcolumn AA derivatization. The separation of phenylthiocarbamyl (PTC) AA adducts (16) or a related technique (17) permits a reliable analysis of primary and secondary AAs by UV detection (254 nm).

There is also a number of references describing either advantages or limits of PTC-AA chromatography. A major advantage consists in the high stability of PTC-adducts, in terms of days, permitting the method's automation as well as an optimal sample organization. In addition, PTC-AA chromatography is a flexible method, as demonstrated by the development of many variants of the original analytical protocol adapted to different laboratory necessities (18–21). Nevertheless, the PTC-AA chromatography presents problems linked either to specificity, as observed for other UV detection methods, or matrix interferences (22). The low specificity is associated to possible coelution of drugs or other interfering substances absorbing at the analytical wavelength. Matrix interference alters the method accuracy and reproducibility. To circumvent these problems, a PTC-AA RP-LC coupled to electrochemical detection has been developed (23).

We report here a PTC method's modification to evaluate plasma AAs (24), where the original separation has been modified by using: (1) a plasma spiking procedure with more than one internal standard (IS) to limit analytical errors; (2) a variable wavelength UV detection system (photodiode array, PDA) to improve specificity.

The PDA detector enables the simultaneous measure of PTC-AAs at 254 nm and a scan view of chromatogram peaks at variable wavelength, permitting to solve analytical interferences or to separately estimate coeluting PTC adducts showing different spectral absorbance: by integrating the tryptophan–ornithine adduct's coelution peak at 254 and 283 nm, we could easily obtain their separate concentrations without changing gradient conditions (24).

2. Materials

2.1. Equipment

1. A model 126 Beckman Coulter (Palo Alto, CA, USA) high-pressure solvent binary delivery system and a model 168 UV PDA; a manual injection valve equipped by a 20-μl loop (see Note 1).
2. A Karat 32 Beckman Chromatography Software to manage the pump and PDA modules.
3. A 300×3.9 mm i.d., 4 μm particle, Pico-Tag C_{18} analytical column (Waters, Milan, Italy); an in-line filter and a 20×3.9 mm i.d., 4 μm particle Nova-pack C_{18} guard column (Waters) (see Note 2).
4. A column oven to maintain a constant column temperature (Tea, Ravenna, Italy).
5. An SC100 Speed-Vac concentrator attached to a Savant RT4104 refrigerated trap.

2.2. Reagents

1. A 18-MΩ cm resistivity water, treated by a Millipore Milli-Q purification equipment (Millipore, Milford, MA, USA) and filtered (0.2 μm) through a Sartorius vacuum filtration system (Stedim, Florence, Italy), was used for all buffers, mobile phases, and standard or sample solutions.

2. Methanol and acetonitrile used for mobile phases or column washing were of ultra-gradient grade (JT Backer-Mallinckrodt Italia, Milan, Italy).
3. Phenylisothiocyanate (PITC), picotag diluent (sample diluent for injection), and all glassware for derivatization (Waters).

2.3. Solutions

1. 10% acetic acid (*v*/*v*).
2. Mobile phase A (aqueous): 2.5% (*v*/*v*) acetonitrile in 0.07 M sodium acetate buffer (pH 6.5, buffered with 10% acetic acid).
3. Mobile phase B (organic): 45:15:40 (*v*/*v*) acetonitrile–methanol–water (16).
4. Column washing solutions: Aqueous 0.1% trifluoroacetic acid (TFA) or H_2PO_4 (*v*/*v*); organic acetonitrile–1% tetrahydrofuran (*v*/*v*).
5. Derivatization (16): Re-drying solution: Methanol, 1 M Na acetate, triethylamine (TEA), 2:2:1, *v*/*v*); PITC-coupling solution: Methanol–water–TEA–pure PITC (7:1:1:1, *v*/*v*).
6. Amino acid standard stock solutions: 10 mM stock solutions of each AA to be quantified (analyte AAs) or AA used as internal standards.

3. Methods

3.1. Preparation of AA Standard Working Solutions

1. Standard stock solutions of each amino acid (10 mM) were separately prepared in 0.1 M HCl.
2. All stock solutions were stored at 4°C for 2 weeks or at −20°C for a longer time, with the exception of glutamine, freshly made to avoid its conversion into glutamate under acidic conditions.
3. Five working calibration solutions of analyte AAs, 25, 50, 100, 200, and 400 μM were prepared by diluting stock solutions in 0.1 M HCl.
4. For homocystine, working solutions were 6, 12, 25, 50, and 100 μM.
5. Calibration solutions of ISs (β-alanine, α-amino-*N*-butyric acid, and norleucine) were also prepared in the same way.
6. All working calibration solutions, used to calculate analyte AA calibration lines, contained a fixed concentration (100 μM) of IS AAs (see Note 3).

3.2. Plasma Sample Preparation

1. For the analysis of plasma AA under basal conditions, 2–4 ml of peripheral venous blood samples were collected from overnight fasting subjects by venipuncture between 07:00 and

09:00 h to avoid differences due to circadian rhythms. The use of vacutainer blood collection tubes, containing EDTA or heparin as the anticoagulants, is recommended to circumvent hemolysis and subsequent AA enzyme degradation (see Note 4).

2. Collected blood was centrifuged as soon as possible at 2,600 × *g* for 15 min at 15°C to obtain cell-free plasma (see Note 5).

3.3. Deproteination

A chemical deproteination procedure was chosen: Plasma sample dilution by acetonitrile (see Note 6).

1. The deproteination procedure consisted in 1:1 (*v*/*v*) dilution of: (1) plasma samples for AA measurement in 0.1 mM HCl containing 100 μM ISs; (2) plasma "blanks" in 0.1 M HCl without ISs; (3) plasma samples for AA identification or recovery evaluation in 0.1 M HCl containing known amounts of analyte AAs.
2. Acetonitrile was immediately added to all these samples in the ratio plasma–0.1 M HCl solutions–acetonitrile 1:1:5.6 (*v*/*v*), yielding a final acetonitrile concentration of 73.8%.
3. Samples were then maintained on ice for 15–20 min to precipitate proteins and centrifuged at 12,000 × *g* for 15 min at 10°C (see Note 7).

3.4. Derivatization

1. As for plasma samples, acetonitrile were added to working standard solutions: AA solutions in 0.1 M HCl–water–acetonitrile in the ratio 1:1:5.6 (*v*/*v*). Derivatization blanks were also prepared by diluting 0.1 M HCl (without analyte AAs)–water–acetonitrile.
2. All solutions and plasma sample supernatants were filtered through a 0.2-μm filter.
3. A 75-μl aliquot of resulting filtrates was then transferred into a derivatization glass tube and dried under high vacuum for 35 min.
4. Derivatization was carried out accordingly to the original method, with small modifications (16): dried samples were resuspended in 20 μl redrying solution (to ensure alkaline reaction conditions) and dried again under vacuum for 20–25 min; samples were then suspended in 20 μl PITC-coupling solution and incubated at RT for 25 min: during this step, AAs react with PITC yielding PTC derivatives. Excess PITC reagent was removed soon after by drying samples for at least 1 h under high vacuum.
5. Dried samples were either directly injected into the chromatograph or stored at −20°C in silicon-capped tubes.
6. PITC pure solution is contained in glass vials of 1 ml each: since volumes used for derivatization are small, PITC can be stored in aliquots at −20°C for about 1 or 3 weeks (under N_2).

7. Dried PTC adducts in either standard solutions or plasma were stable for 1 month at −20°C. Before injection, plasma samples were resuspended and sonicated in 150 μl Waters picotag diluent (see Note 8).

3.5. Chromatography and PDA Instrument Setup

1. The column was maintained at a constant temperature of 47 ± 1°C.
2. The mobile phase was a time-programmed binary gradient prepared from 2.5% (*v/v*) acetonitrile in sodium acetate buffer (Table 1; Fig. 1) and 45:15:40 (*v/v*) acetonitrile–methanol–water.
3. Flow and gradient program has been modified and extended for a better AA resolution in our conditions (see Note 9).
4. Injection volume was 20 μl, full-loop injection.
5. The PDA detector was set up at the wavelength range of 220–400 nm (scan interval, 1 nm). The signal detection threshold was 0.0005 AU/min, peak width was 0.5 min, and data rate was 1 Hz.

Table 1
Flow and binary gradient conditions for PTC-AA chromatographic analysis

PTC-AA separation gradient

Time	Flow		Composition			
min	ml/min	min	%A	%B	min	Gradient[a]
Start	0.8		100.0	0.0		
0.0	0.8		100.0	0.0		
20.0	1.0	0.5[b]	97.0	3.0	0.70	1
27.0	1.0		94.0	6.0	6.0	5
33.0	1.0		91.0	9.0	10.0	3
43.0	1.0		78.0	22.0	10.0	3
63.0	1.0		62.0	38.0	15.0	0
80.0	1.0		0.0	100.0	3.0	0

Washing and reequilibration gradient

Time	Flow		Composition			Gradient
105.0	1.0		100.0	0.0	7.0	0
112.0	0.8	0.5[b]	100.0	0.0		
120.0	0.8		100.0	0.0		

[a]1: Step gradient; 5: Concave gradient; 3: Convex gradient; and 0: Linear gradient
[b]Time for flow changes, 0.8–1.0 and 1.0–0.8 ml/min

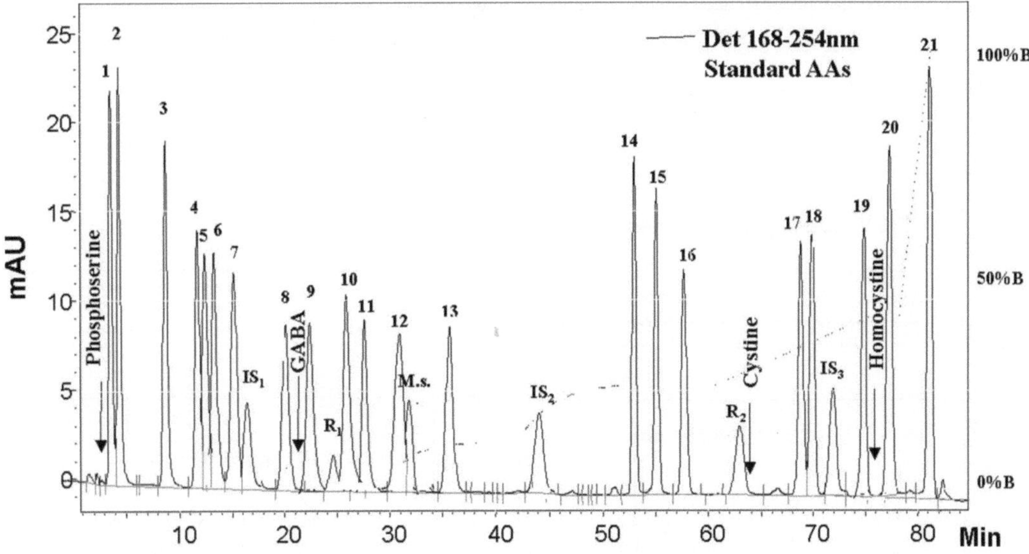

Fig. 1. Elution profile at 254 nm of a mixture of standard analyte AAs, 400 μM (pmoles injected: 520) and IS AAs, 100 μM, pmoles injected: 130 (IS_1: β-alanine; M.s.: methionine sulfone, tested as IS; IS_2: α-amino-*N*-butyric acid; IS_3: Nor-leucine), derivatized as indicated in Subheading 3. Numbers identify peaks corresponding to each analyte amino acid (*1–21*). *1*: aspartate; *2*: Glutamate; *3*: OH-Proline; *4*: Serine; *5*: Asparagine; *6*: Glycine; *7*: Glutamine; *8*: Taurine; *9*: Histidine; *R1*: Peak of reagent 1; *10*: Threonine; *11*: Alanine; *12*: Arginine; *13*: Proline; *14*: Tyrosine; *15*: Valine; *16*: Methionine; *R2*: Peak of reagent 2; *17*: Isoleucine; *18*: Leucine; *19*: Phenylalanine; *20*: Tryptophan + Ornithine; and *21*: Lysine. This chromatogram does not show peaks of some AAs, phosphoserine, GABA, cystine, and homocystine, Hcy, Det 168–254 nm: absorbance at 254 nm by UV-PDA detector 168.

6. Data collection and integration functions were managed by the Scan Mix View (220–400 nm) module of the Karat 32 software. Chromatograms were integrated at two wavelength, 254 nm (all IS and analyte AA peaks) and 283 nm (IS norleucine and tryptophan–ornithine peaks); the Scan Mixed View reports the 220–400 nm spectrum of selected peaks of the chromatogram (Fig. 2, scan view zoomed at 220–300 nm, and Fig. 4, zoomed at 225–375 nm); a 3D vision of selected parts of the chromatogram can be also obtained (Fig. 3) (see Note 10).

3.6. Column Washing

When required (e.g., analytical pressure increase, ≥3 Kpsi, during chromatographic runs), usually after a week of routine analyses, the analytical column was cleaned by a linear gradient formed by a solution C, 0.1% TFA, or H_2PO_4 (*v*/*v*) in water and a solution D, 100% acetonitrile, containing 1% tetrahydrofuran, flow rate 0.6 m/min; the gradient duration was 1 h from 0% solution C to 100% solution D; then, column was maintained at 100% solution D for 10–15 min; after changing solution C with mobile phase B, the column was reconditioned at 100% solution B (45:15:40, acetonitrile: methanol:water) through a linear gradient for at least 20–30 min, monitoring baseline (see Note 11).

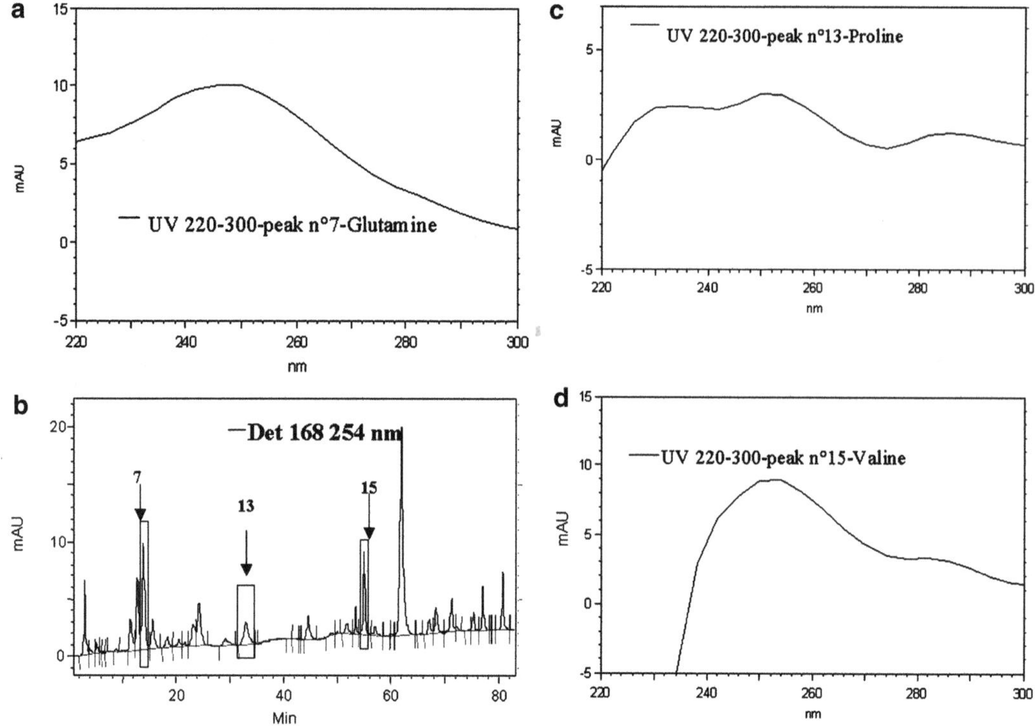

Fig. 2. Scan mixed view of a 254-nm plasma chromatogram of a subject with fibromyalgia syndrome. (**a**) Selection of a single peak and visualization of the corresponding spectrum (e.g., UV spectrum 220–300 nm): (**b**) peak no. 7: glutamine; (**c**) peak no. 13: proline; and **(d)** peak no. 15, valine.

3.7. Calculation of Plasma AA Concentrations

Calculation of analyte AA concentrations is obtained by the following equations:

1. μM analyte AAs eluting between 5 and 17 min = plasma $\text{slope}_{(254\text{nm})}$ (analyte AA, plasma peak area/IS_1, 100 μM β-alanine, plasma peak area)/external $\text{slope}_{(254\text{nm})}$ (AA standard analyte peak areas/IS_1, 100 μM β-alanine standard peak area).
2. μM analyte AAs eluting between 21 and 58 min = plasma $\text{slope}_{(254\text{nm})}$ (analyte AA, plasma peak area/IS_2, 100 μM α-amino-*N*-butyric acid, plasma peak area)/external $\text{slope}_{(254\text{nm})}$ (AA standard analyte peak areas/IS_2, 100 μM α-amino-*N*-butyric acid, standard peak area).
3. μM analyte AAs eluting between 68 and 78 min = plasma $\text{slope}_{(254\text{nm})}$ (analyte AA, plasma peak area/IS_3, 100 μM norleucine, plasma peak area)/external $\text{slope}_{(254\text{nm})}$ (AA standard analyte peak areas/IS_3, 100 μM norleucine, standard peak area).
4. To separately measure tryptophan and ornithine adducts, which coelute at the end of the chromatogram, we applied a dual-wavelength (254 and 283 nm) peak integration, based on respective tryptophan and ornithine UV spectra (see Note 12)

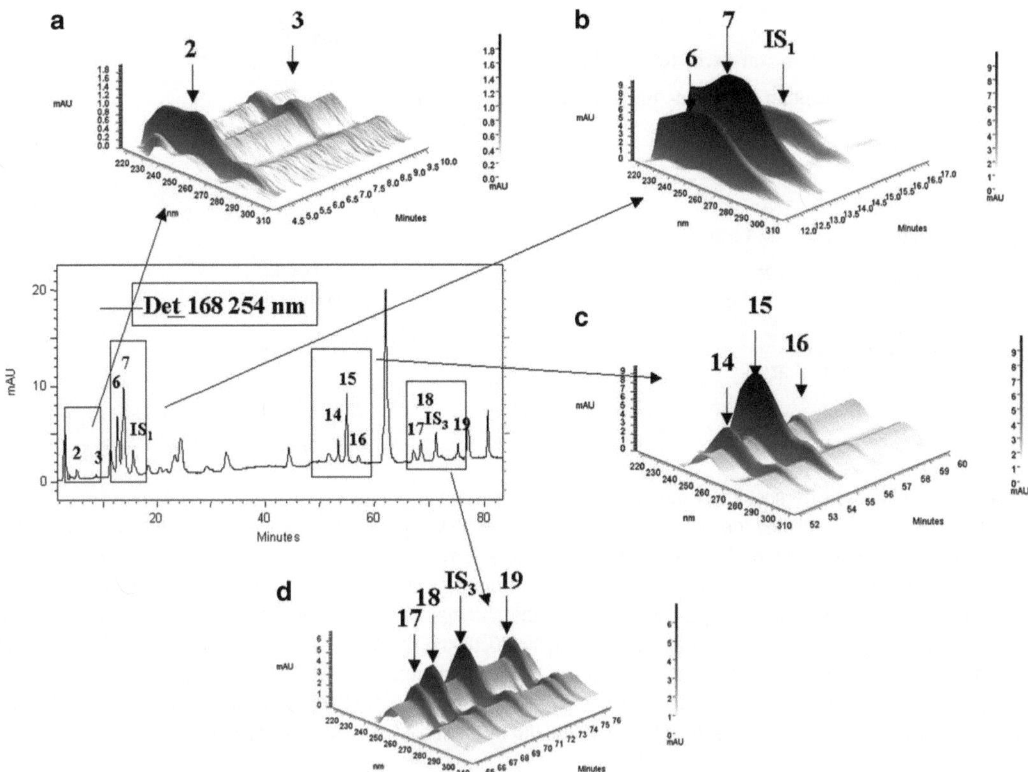

Fig. 3. 3D scan view of groups of peaks in a 254-nm chromatogram of a subject affected by fibromyalgia (same subject in Fig. 2); (**a**) peaks 2 and 3 (glutamate and hydroxy-proline); (**b**) peaks 6 and 7 (glycine, glutamine, and IS$_1$, β-alanine); (**c**) peaks 14–16 (tyrosine, valine, and methionine); and (**d**) peaks 17–18 (leucine and isoleucine), IS$_3$ norleucine, 19, phenylalanine.

and additivity of the Lambert–Beer principle; the following system of linear equation for two unknowns has been used:

(a) μM tryptophan (trp) (x_1) = [plasma trp-orn slope$_{(254nm)}$ (trp-orn, plasma peak area/IS$_3$, 100 μM norleucine, plasma peak area) − orn external slope$_{(254nm)}$ (standard orn peak areas/IS$_3$, 100 μM norleucine, standard peak area)$_{(254nm)}$ (μM orn) (x_2)]/trp external slope$_{(254nm)}$ (trp, standard peak areas/IS$_3$, 100 μM norleucine, standard peak area)$_{(254nm)}$.

(b) [μM trp (x_1) + μM orn (x_2)] = plasma trp-orn slope$_{(283nm)}$ (trp-orn plasma peak area/IS$_3$ 100 μM norleucine plasma peak area)/external trp-orn slope$_{(283nm)}$ (standard trp or orn peak areas/IS$_3$ 100 μM norleucine peak area) (Fig. 4).

5. Lysine slopes differ from those obtained for other AAs: in fact, this AA forms double-PTC adduct. For the optimal lysine measurement, a proximally eluting IS, also producing a double adduct, could be used: for instance – hydroxy-lysine. In this AA analysis protocol, we employed norleucine as the IS for

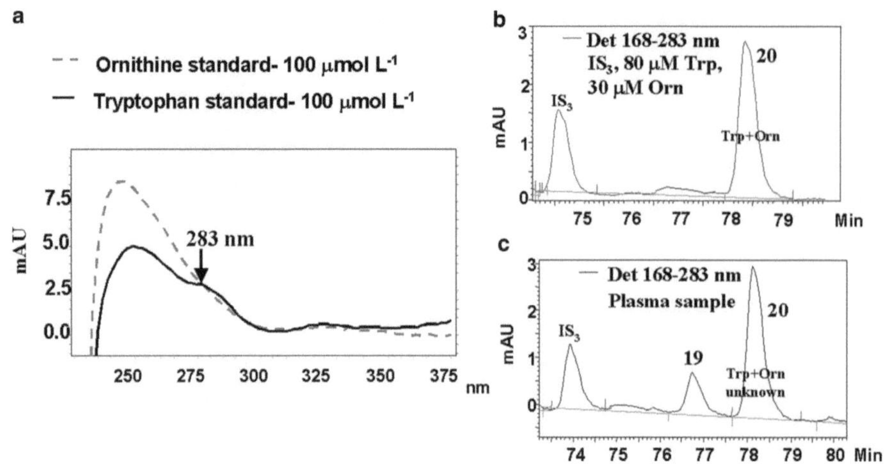

Fig. 4. (**a**) Superimposed UV spectra of the adducts of tryptophan and ornithine standards at the same concentration, 100 μM. Standards were diluted, derivatized as indicated in Subheading 3.4.1-5, and injected into the chromatograph; spectra were visualized by use of the PDA-mixed view module of the software. (**b**) Enlarged view of the chromatogram obtained from standards at 283 nm. Peak 20 was obtained from 30 μM ornithine and 80 μM tryptophan. IS_3 peak was obtained from 100 μM norleucine. (**c**) Enlarged view of the chromatogram obtained from plasma (healthy subject) at 283 nm of coeluting tryptophan + ornithine adducts (peak no. 20, unknown amounts). Peak IS_3 was obtained from 100 μM norleucine (*spiked*). Peak 19 is phenylalanine adduct. In (**c**), the dual-channel (254–283 nm) detection of peak 20 and calculation by use of the system of linear equations gave tryptophan and ornithine concentrations of 110 and 40 μM, respectively. Det 168 283 nm denotes absorbance at 283 nm by the model 168 UV-PDA detector.

lysine response factors and calculations, as reported above for AAs eluting between 68 and 78 min.

6. Table 2 presents the AA retention times in plasma with relative standard deviations, RSDs. At 254 nm, external calibration lines of analyte or IS AAs gave a good linearity ($r^2 > 0.993$) and displayed different linear slopes with regard to different groups of PTC-AAs, indicating diverse response factors (24). Reagent peaks 1 (R_1) and 2 (R_2) are peaks obtained by the chromatography of derivatization blanks, without plasma or AA standards. In plasma, variation in AA recoveries depends upon PTC-AA elution times, suggesting that response factors obtained using a single IS are not suitable to monitor analytical errors and matrix effects. Moreover, other authors have observed that, by this method, AA recoveries depend on matrix effects not only during the deproteination procedure, but also during the derivatization reaction step (25). As well, the spectral and 3D visualization of PTC-AAs shows slight difference depending on the chromatographic run position (Figs. 2 and 3) (see Note 10).

3.8. Conclusions

1. *AA concentrations in plasma*: By this PTC-chromatography variant, we could quantify 23 AAs in plasma, including homocystine, which elutes after the phenylalanine adduct. To our knowledge, this PTC-AA separation and UV detection method is the first that reports homocystine plasma levels, even if in some samples only (24). Most studies quantify homocystine by

Table 2
PTC-AA retention times in plasma and relative standard deviations (RSDs), obtained from five subjects

Amino acids	Retention time, min (RSD)
Aspartate	4.12 (5.8)
Glutammate	5.28 (2.0)
Hydroxy-proline	8.89 (8.0)
Serine	11.62 (2.1)
Asparagine	11.82 (2.4)
Glycine	13.01 (2.1)
Glutammine	13.94 (1.9)
β-Alanine	15.75 (2.2)
Taurine	18.70 (2.2)
Histidine	20.27 (1.9)
Reagent peak 1	21.64 (1.8)
Threonine	23.77 (2.0)
Alanine	24.83 (0.51)
Arginine	28.48 (0.84)
Proline	33.15 (4.0)
α-Amino-*N*-but. acid	45.03 (4.4)
Tyrosine	53.95 (1.9)
Valine	56.03 (2.1)
Methionine	58.46 (2.0)
Reagent peak 2	63.95 (2.8)
Isoleucine	70.09 (2.1)
Leucine	71.37 (1.9)
Nor-leucine	74.86 (2.4)
Phenylalanine	77.70 (1.6)
Homocystine	78.70 (0.71)
Tryptophan–ornithine	79.02 (0.73)
Lysine	82.91 (0.25)

electrochemical or fluorimetric detectors. In general, we obtained AA concentrations comparable to the previously published in literature, except relatively higher levels of taurine, tyrosine, methionine, phenylalanine, and tryptophan (Table 3).

Table 3
Comparison of AA plasma concentrations (μM) reported in literature and measured by the technique described in this chapter (24)

AAs	(28) (*n*=30) the USA, women	(20) (*n*=4) Italian	(18) (*n*=10) German	(21) (*n*=10) the USA, men	(11) French	(29) (*n*=34) Indian	(24) (*n*=14) Italian, women
Aspartate	–	24±5	–	4.3±0.7	3±1	16.9±11	12.6±4.1[a]
Glutammate	37±27	49±7	12±4	45.1±4	24±15	40.8±31	24.5±6.8
OH-proline	15.3±6	30±7	–	14±0.6	13±10	–	17.3±7.8
Serine	105±23	127±23	102±13	109±3.5	114±19	116.4±38.5	99.1±18
Asparagine	85.2±25	71±1	43.8±9.5	64.3±1.5	41±10	52.7±15.8	44.3±16.2
Glycine	232±68	202±14	191.8±23.4	239±6.2	230±52	266±76.1	198.2±78
Glutamine	–	450±31	493.1±56.7	632±15	586±84	–	415.3±61.4
Taurine	47.3±21	63±6	59.4±16.8	183±10	55±13	84.2±49.8	72.0±9.9
Histidine	64.9±16	93±9	79.8±20.3	74.2±2.4	82±10	68.5±34.4	96.0±19.6
Threonine	106.6±28.1	106±9	116.6±20.8	106±12	140±33	121.5±49.5	118.8±38.2
Alanine	341.1±79.8	285±14	339.9±51.5	460±16.5	333±74	339±113.3	371.7±72.2
Arginine	97.9±29.3	90±12	77.5±11.6	203±4	80±20	73.5±22.9	89.0±16.2
Proline	191.6±75.6	202±17	240.8±48.1	213±4	168±60	–	245.5±73.6
Tyrosine	54.7±15.8	64±7	54.7±5.0	65±3	59±12	61.2±26.1	89.3±23.9
Valine	206.2±45.8	187±18	222±28.3	–	233±43	168.9±31	255.9±66.7
Methionine	25.9±15.9	40±5	25.9±2.8	35.8±3.6	25±4	25±6.2	46.8±4.75
Isoleucine	54.3±15.3	67±5	68.7±10.3	71.8±4.1	62±14	60±14.1	70.8±15.7

Leucine	94.5 ± 30.9	106 ± 10	141.6 ± 16.8	151 ± 5.2	123 ± 25	77.4 ± 18.2	124.8 ± 30.1
Phenylalanine	52.2 ± 11.5	65 ± 12	60.3 ± 4.8	42.6 ± 1.5	57 ± 9	63.2 ± 21.6	75.3 ± 19.5
Homocystine	–	–	–	–	–	–	8.5 ± 2.3[b]
Tryptophan	50.6 ± 15.1	–	–	59.6 ± 2.1	44 ± 7	54.7 ± 15.7	73.2 ± 28.2
Ornithine	46.1 ± 12.2	–	–	72.8 ± 2.4	55 ± 16	65.6 ± 31.5	44.9 ± 22.1
Lysine	–	180 ± 10	202.7 ± 43.6	264 ± 15	188 ± 32	156.5 ± 41.8	161.6 ± 56.4

(28): Pico-Tag method: blood collected in heparine, ultrafiltration; (20): modified Pico-Tag method: extraction with 6% sulphosalicylic acid (SSA), shorter gradient elution program (13 min); (18): modified Pico-Tag method, shorter gradient elution time (30 min); (21): blood collected in EDTA, ultrafiltration, Pico-Tag method; (11): ion-exchange chromatography; (29): blood collected in heparin, extraction with methanol, *O*-phtalaldeide method

[a]Values obtained in ten subjects

[b]Values obtained in five subjects

At the same time, levels of both glutamate and glutamine observed were lower. We are trying to enlarge the number of healthy volunteers evaluated for their plasma AA concentrations for a better comparison. By this method, we have measured AA plasma levels in subjects with fibromyalgia syndrome, reporting lower levels of some plasma AAs than in non-fibromyalgic control subjects (8), quite accordingly to other authors (30).

2. *Limitations*: The major limitation consists in the method's sensitivity. Low-level plasma AAs, as aspartate, GABA, and homocystine (see previous section), can be missed sometimes, having concentrations lower than LOQ (24). Also, AA resolution problems can occur, for instance, in case of serine, asparagine, cystine, and methionine sulfone separation. For serine and asparagine, the issue can be resolved by the Karat integration program; regarding cystine and methionine sulfone, slight gradient changes should be applied. In our experience, addition of small volumes of TEA was not sufficient to improve resolution of these AAs. To increase the method's sensitivity, we plan to adapt the PDA UV detection module in line with a fluorimetric detector. To this end, tandem pump systems are suitable to enable a simultaneous acquisition of the chromatogram by two detection systems.

4. Notes

1. We obtained good results using these instrument modules, and a manual injection valve; the system can be improved by employing a refrigerated autosampler, which permits overnight runs and programmed column washing; the application of a "bidimensional" pump system also enables an optimal protection of the analytical column.
2. In our experience, the pico-tag column (Waters) is the most effective in AA separation; it requires particular care and protection using an in-line filter and a 4 μm particle Nova-pack C_{18} guard column (Waters), as reported here; do not use an organic phase composed by 100% methanol: in this case, it is preferable to use a 100% acetonitrile eluent; avoid rapid solvent changes during the column washing procedure.
3. As indicated in our paper (24), we used three ISs, β-alanine, α-amino-butyric acid, and norleucine, for plasma sample analyses; methionine sulfone was also tested as IS, but this AA adduct partially coeluted (Fig. 1), sometimes completely, with arginine; methionine sulfone could be used as additional IS for arginine and proline by slightly modifying gradient conditions.

4. To prevent sample hemolysis, a syringe and a needle not excessively small should be employed; a butterfly needle is suitable, being aware that blood can clot more likely under this condition: to avoid clotting, it is recommended to dampen needles with the chosen anticoagulant (EDTA or heparin).
5. This is a critical centrifugation step: at lower *g* values, there was a cell contamination of plasma samples, which can significantly interfere with subsequent sample preparation steps, altering final AA concentrations. The best blood centrifugation condition is obtained at least at 2,500 × *g*; temperature should be between 10 and 15°C.
6. Chemical acetonitrile deproteination presents the following advantages: a much easier processing of lipemic plasma samples; best recoveries in taurine and tryptophan (25–27); moreover, this simple procedure can be applied to other body fluids, as saliva or urines, or to tissues and cells; on the other side, a multiple ISs spiking plasma or urine procedure together UV-PDA detection can be applied also to a physical deproteination by ultrafiltration.
7. This is another critical step during plasma sample preparation: deproteination at 5,000–6,000 × *g* can result in poor precipitation efficiency; centrifugation at 12–15,000 × *g* is recommended.
8. Under the here-described instrumental conditions, the diluent volume to resuspend dried samples before injection into the chromatograph should be comprised between 75 and 300 μl; volumes lower than 75 μl may provoke sample loading excess, more than 300 μl significant loss of PITC-AA response.
9. Gradient flow conditions have been changed to improve separation of first AAs (aspartate–glutamine) and AAs eluting between 23 and 28 min; separation of these AA adducts can vary and worsen after several chromatographic runs: by prolonging the chromatographic AA separation, we delayed the loss in efficiency of the analytical column; thus, if the method is time consuming, on the other side it improves column maintenance; moreover, the method's automation by an autosampler together its improvement by bidimensional pump systems can completely counteract the problem; it should be also mentioned that slight additional retention time variations may occur when mobile phases are manually prepared, without, however, substantially affecting PTC-AA elution order.
10. This PDA setup needs a Karat software installation in PCs with adequate RAM characteristics to ensure optimal storage of data: an insufficient PC memory capacity can result in loss of data acquisition. Concerning the visualization of PTC-AA

spectra in plasma, this depends upon AA concentration and chromatographic gradient position; PTC adducts of low-concentrated AA in plasma, as aspartate, glutamate, and hydroxy-proline, display low signal and resolution of spectra (Fig. 3); sensitivity is a problem, as already described (24); on the other hand, any change in spectral response can be easily monitored by PDA.

11. After each solvent change, pumps were disconnected and conditioned with the new solvent. Pumps were then started again slowly to avoid rapid solvent changes for the analytical column.

12. Tryptophan and ornithine adducts have the same UV response at wavelength approx. >280 nm; 283 nm is the wavelength, where, by separate standard injections, we obtained the best trp and orn calibration line fit.

References

1. Kaspar H et al (2009) Advances in amino acid analysis. Anal Bioanal Chem 393:445–452
2. Gessa GL et al (1974) Effect of the oral administration of tryptophan-free amino acid mixtures on serum tryptophan, brain tryptophan and serotonin metabolism. J Neurochem 22:869–870
3. Gibson CJ, Wurtman RJ (1978) Physiological control of brain norepinephrine synthesis by brain tyrosine concentration. Life Sci 22:1399–1405
4. Wellendorph P et al (2005) Deorphanization of GPRC6A: a promiscuous L-alpha-amino acid receptor with preference for basic amino acids. Mol Pharmacol 67:589–97
5. Conigrave AD, Hampson DR (2010) Broad-spectrum amino acid-sensing class C G-protein coupled receptors: molecular mechanisms, physiological significance and options for drug development. Pharmacol Ther 17:398–407
6. Lucca A et al (1995) Neutral amino acid availability in two psychiatric disorders. Prog Neuro Psychopharmacol Biol Psychiatry 19:615–626
7. Richardson MA et al (1999). Branched chain amino acids decrease tardive dyskinesia symptoms Psychopharmacology (Berl) 143:358–364
8. Bazzichi L et al (2009) Altered amino acid homeostasis in subjects affected by fibromyalgia. Clin Biochem 42:1064–1070
9. Lee M et al (2011) Decreased plasma tryptophan and tryptophan/large neutral amino acid ratio in patients with neuroleptic-resistant schizophrenia: Relationship to plasma cortisol concentration. Psychiatry Res 185: 328–333
10. Osher Y et al (2004) Elevated homocystine levels in euthymic bipolar disorder patients showing functional deterioration. Bipolar Disorders 6: 82–86
11. Cynober LA (2002) Plasma amino acid levels with a note on membrane transport: characteristics, regulation, and metabolic significance. Nutrition 18:761–766
12. Badawy AAB, Morgan CJ, Turner JA (2008) Application of the Phenomenex EZ:faastTM amino acid analysis kit for rapid gas-chromatographic determination of concentrations of plasma tryptophan and its brain uptake competitors. Amino Acids 34:587–596
13. Tortorella A et al (2001) Plasma concentrations of amino acids in chronic schizophrenics treated with clozapine. Neuropsychobiology 44:167–171
14. Bailara KM et al (2006). Decreased brain tryptophan availability as a partial determinant of post-partum blues. Psychoneuroendocrinology 31:407–413
15. Bruce KR et al (2009). Impact of acute tryptophan depletion on mood and eating-related urges in bulimic and nonbulimic women. J Psychiatry Neurosci 34:376–382
16. Cohen SA, Strydom DJ (1988) Amino Acid Analysis utilizing phenylisothiocyanate derivatives. Anal Biochem 174:1–16.
17. Woo KL (2003) Determination of amino acids in foods by reversed-phase HPLC with new precolumn derivatives, butylthiocarbamyl, and

benzylthiocarbamyl derivatives compared to the phenylthiocarbamyl derivative and ion exchange chromatography. Mol Biotechnol 24:69–88

18. Furst P et al (1990). Appraisal of four pre-column derivatization methods for the high-performance liquid chromatographic determination of free amino acids in biological materials. J Chromatogr 499:557–569
19. Buzzigoli G et al (1990) Characterization of a reversed-phase high-performance liquid chromatographic system for the determination of blood amino acids. J Chromatogr 507:85–93
20. Fierabracci V et al (1991) Application of amino acid analysis by high-performance liquid chromatography with phenyl isothyocyanate derivatization to the rapid determination of free amino acids in biological samples. J Chromatogr 570:285–291
21. Feste AS (1992) Reversed-phase chromatography of phenylthiocarbamyl amino acid derivatives of physiological amino acids: an evaluation and a comparison with analysis by ion-exchange chromatography. J Chromatogr B 574:23–34
22. Fernandez-Figares I, Rodriguez LC, Gonzales-Casado A (2004) Effects of different matrices on physiological amino acids by liquid chromatography: evaluation and correction of the matrix effect. J Chromatogr B 799:73–79
23. Sherwood RA (2000) Amino acid measurement in body fluids using PITC derivatives. In: Cooper C, Packer N, Williams K (eds) Amino Acid Analysis Protocols (Methods in Molecular Biology). Humana Press, NJ
24. Palego L et al (2010) Modified RP-LC of phenylthiocarbamyl amino acid adducts in plasma acetonitrile extracts using multiple internal standards and photo-diode UV detection. Chromatographia 71:291–297
25. Davey JF, Ersser RS (1990) AA analysis of physiological fluids by high-performance liquid chromatography with phenyl-isothiocyanate derivatization and comparison with ion-exchange chromatography. J Chromatogr 528:9–23
26. Aristoy MC, Toldrà F (1991) Deproteinization technique for HPLC amino acids analysis in fresh pork muscle and dry-cured ham. J Agric Food Chem 39:1792–1795
27. Mou S, Ding X, Liu Y (2002) Separation methods for taurine analysis in biological samples. J Chromatogr B 781:251–267
28. Yunus M et al (1992) Plasma tryptophan and other amino acids in primary fibromyalgia: a controlled study. J Rheumatol 19:90–94
29. Suresh Babu SV et al (2002) HPLC method for amino acids profile in biological fluids and inborn metabolic disorders of aminoacidopathies. Ind J Clin Biochem 17:2–26
30. Russell IJ et al (1989) Serum amino acids in fibrositis/fibromyalgia syndrome. J Rheumatol S19:158–16

Chapter 18

Quantification of Underivatised Amino Acids on Dry Blood Spot, Plasma, and Urine by HPLC–ESI–MS/MS

Giuseppe Giordano, Iole Maria Di Gangi, Antonina Gucciardi, and Mauro Naturale

Abstract

Enzyme deficiencies in amino acid (AA) metabolism affecting the levels of amino acids and their derivatives in physiological fluids may serve as diagnostically significant biomarkers for one or a group of metabolic disorders. Therefore, it is important to monitor a wide range of free amino acids simultaneously and to quantify them. This is time consuming if we use the classical methods and more than ever now that many laboratories have introduced Newborn Screening Programs for the semiquantitative analysis, detection, and quantification of some amino acids needed to be performed in a short time to reduce the rate of false positives.

We have modified the stable isotope dilution HPLC–electrospray ionization (ESI)–MS/MS method previously described by Qu et al. (Anal Chem 74: 2034–2040, 2002) for a more rapid, robust, sensitive, and specific detection and quantification of underivatised amino acids. The modified method reduces the time of analysis to 10 min with very good reproducibility of retention times and a better separation of the metabolites and their isomers.

The omission of the derivatization step allowed us to achieve some important advantages: fast and simple sample preparation and exclusion of artefacts and interferences. The use of this technique is highly sensitive, specific, and allows monitoring of 40 underivatized amino acids, including the key isomers and quantification of some of them, in order to cover many diagnostically important intermediates of metabolic pathways.

We propose this HPLC–ESI–MS/MS method for underivatized amino acids as a support for the Newborn Screening as secondary test using the same dried blood spots for a more accurate and specific examination in case of suspected metabolic diseases. In this way, we avoid plasma collection from the patient as it normally occurs, reducing anxiety for the parents and further costs for analysis.

The same method was validated and applied also to plasma and urine samples with good reproducibility, accuracy, and precision. The fast run time, feasibility of high sample throughput, and small amount of sample required make this method very suitable for routine analysis in the clinical setting.

Key words: Newborn screening, Blood spot, Underivatized amino acids, HPLC–ESI–MS/MS, Amino acid analysis

Michail A. Alterman and Peter Hunziker (eds.), *Amino Acid Analysis: Methods and Protocols,* Methods in Molecular Biology, vol. 828, DOI 10.1007/978-1-61779-445-2_18,

1. Introduction

Amino acids (AAs) and their derivatives are very important for biological processes, such as protein synthesis and metabolic pathways (1, 2). Enzyme deficiencies in amino acid metabolism lead to variation of their physiological concentrations causing clinical symptoms, which are often not a single-disorder specific. Therefore, it is of great importance to simultaneously determine the levels of free amino acids in physiological samples as one or several compounds may play the role of a biomarker for one specific or a group of metabolic disorders (3), for example, phenylalanine for phenylketonuria (PKU), ornithine, citrulline, and argininosuccinic acid (ASA) for disorders of urea synthesis cycle, allo-isoleucine and valine for maple syrup urine disease (MSUD), and glutamine for ornithine transcarbamylase (OTC) deficiency.

There are many chromatographic separation techniques for the analysis of amino acids, most of which require pre- or postcolumn derivatization of AAs in order to achieve proper chromatographic separation and detection. The still most widespread technique used is the automated amino acid analyzer based on ion-exchange chromatography (4–6). However, the procedure adopted for these methods is time consuming and may lead to artefacts and interferences. In order to avoid these problems, some authors have developed qualitative and quantitative analysis methods of amino acids without performing derivatization. First, this was achieved for only a few underivatized amino acids by using HPLC or CE coupled to electrochemical detectors (7, 8), but with the advent of more sensitive and specific MS/MS techniques a wider range of amino acids, even those in trace amounts, became detectable (9–11). Consequently, analysis of underivatized amino acids in complex biological matrices (plasma and urine) by fast, straightforward, and sensitive HPLC–electrospray ionization (ESI)–MS/MS method based on the use of a volatile ion-pair reagent was reported (12–14). Recently, a quantitative method for the analysis of AA in plasma and urine was developed by Waterval using UPLC–MS/MS (15).

A significant number of clinical laboratories have introduced Newborn Screening Programs for the semiquantitative analysis of amino acids on dried blood spots by flow injection into the tandem mass spectrometer (flow injection analysis, FIA) in order to provide an early diagnosis of metabolic diseases and thus prompting intervention (16). While FIA–MS/MS allows for high-throughput screening, it cannot distinguish isobaric compounds unless an appropriate liquid chromatography step is applied before the mass spectrometric analysis. However, in many published techniques used for neonatal screening, the samples undergo a derivatization step and the amino acid profile is then acquired in the form of a mass spectrum making discrimination of isomers impossible because they appear as a single peak. Such a phenomenon particularly occurs in case of leucine that has three major isomers (leucine, isoleucine, and allo-isoleucine) and is also isobaric with 3-hydroxyproline and propionyl-glycine. Correspondingly, that may lead to false-positive results if further analyses are not performed (17).

We have developed an HPLC–ESI–MS/MS method for rapid, sensitive, and specific analysis of 40 underivatized amino acids, including the key isomers, from a dried blood spot, plasma, and urine. For analysis of samples, two different gradients (10 or 5 min) have been developed, depending on the need to analyze the whole set of amino acids or only a part of it. In the case of the analysis of all amino acids, they have been divided into three groups within the same acquisition method according to their retention times. For those amino acids which play a significant role in detection of metabolic diseases by newborn screening, such as Val, Leu/Ile/a-Ile, Met, Phe, Cit, and Tyr, we used a shorter chromatographic run of 5 min. This allowed the simultaneous monitoring of a reduced number of amino acids, and thus maximized the detection sensitivity and productivity.

2. Materials

Prepare all solutions using ultrapure water (prepared by purifying deionized water to attain a resistance of 18 MΩ.cm at 25°C) and analytical grade reagents. Store all reagents at –20°C (unless indicated otherwise).

2.1. Unlabelled Amino Acids Standard

L-Glycine (Gly), β-Alanine (β-Ala), L-Alanine (Ala), Sarcosine (Sar), α-Amino-*n*-butyric acid (α-Abu), γ-Aminobutyric acid (GABA), L-Serine (Ser), L-Proline (Pro), L-Valine (Val), L-Threonine (Thr), Taurine (Tau), L-Leucine (Leu), L-Isoleucine (Ile), δ-Aminolevulinic acid (δ-ALA), L-Ornithine HCl (Orn), L-Asparagine (Asn), L-Aspartic acid (Asp), L-Glutamine (Gln), L-Glutamic acid (Glu), L-Lysine (Lys), L-Methionine (Met), L-Histidine (His), L-Phenylalanine (Phe), L-Arginine (Arg), L-Citrulline (Cit), L-Tyrosine (Tyr), L-Carnosine (Carn), L-Tryptophan (Trp), and L-Homocystine $(Hcy)_2$ (Fluka); Ethanolamine (EA), Creatinine, Guanidineacetic acid (GAA), L-allo-Isoleucine (a-Ile), Creatine, L-α-Aminoadipic acid (α-AAD), 3-Methyl-L-histidine (3-Me-His), 1-Methyl-L-histidine (1-Me-His), L-Homoserine (Hse), and L-ASA Disodium Salt (Sigma). 3-Hydroxyproline (3-Hyp) (Merck).

2.2. Preparation of Unlabelled Standard Solution (Table 1)

1. Weigh each amino acid (as indicated in column A) and dissolve in water as indicated in column C to prepare a primary solution for each amino acid (column D).

Table 1
Weights and volumes to prepare unlabelled standard solution for calibration curves (Subheading 2.2)

	A	B	C	D	E	F	G
Unlabelled amino acids	Weight (mg)	MW	Volume (ml)	Primary solution (mM)	Volume to draw/10 ml (µl)	Final solution (µM)	Range of calibration curves (µM)
1-Me-His	30	169.18	10	17.73	56.4	100	0–100
3-Me-His	30	169.18	10	17.73	56.4	100	0–100
3-Hyp	50	131.13	10	38.13	52.5	200	0–200
a-Ile	30	131.18	10	22.87	109.3	250	0–250
Ala	50	89.10	10	56.12	160.4	900	0–900
α-Abu	30	103.12	10	29.09	34.4	100	0–100
α-AAD	30	161.16	10	18.62	53.7	100	0–100
Arg	50	174.20	10	28.70	139.4	400	0–400
ASA[a]	5	334.20	10	1.50	2673.6	400	0–400
Asn	30	132.12	10	22.71	110.1	250	0–250
Asp	30	133.11	10	22.54	133.1	300	0–300
β-Ala	30	89.09	10	33.67	29.7	100	0–100
Carnosine	30	226.23	10	13.26	75.4	100	0–100
Cit	50	175.19	10	28.54	70.1	200	0–200
Creatine	30	131.13	10	22.88	43.7	100	0–100

Creatinine	50	113.12	10	44.20	67.9	300	0–300
δ-ALA	30	167.59	10	17.90	55.9	100	0–100
EA (ml)[b]		61.08		61.81	32.4	200	0–200
GAA	30	117.10	10	25.62	78.1	200	0–200
GABA	20	103.12	10	19.39	51.6	100	0–100
Gln	30	146.15	10	20.53	243.6	500	0–500
Glu	30	147.13	10	20.39	294.3	600	0–600
Gly	30	75.07	10	39.96	225.2	900	0–900
$(Hcy)_2$	30	268.35	10	11.18	89.5	100	0–100
His	50	155.16	10	32.22	124.1	400	0–400
Hse	30	119.12	10	25.18	39.7	100	0–100
Ile	30	131.18	10	22.87	109.3	250	0–250
Leu	50	131.18	10	38.12	131.2	500	0–500
Lys	50	146.19	10	34.20	263.1	900	0–900
Met	50	149.21	10	33.51	59.7	200	0–200
Orn	50	168.62	10	29.65	134.9	400	0–400
Phe	50	165.19	10	30.27	264.3	800	0–600
Pro	30	115.13	10	26.06	76.8	200	0–200
Sar	30	89.09	10	33.67	29.7	100	0–100
Ser	50	105.09	10	47.58	126.1	600	0–600
Tau	50	125.15	10	39.95	225.3	900	0–900

(continued)

Table 1 (continued)

	A	B	C	D	E	F	G
Unlabelled amino acids	Weight (mg)	MW	Volume (ml)	Primary solution (mM)	Volume to draw/10 ml (μl)	Final solution (μM)	Range of calibration curves (μM)
Thr	30	119.12	10	25.18	357.4	900	0–900
Trp	50	204.22	10	24.48	122.5	300	0–300
Tyr[c]	10	181.19	30	1.84	2717.9	500	0–500
Val	30	117.15	10	25.61	195.3	500	0–500

[a]Solubility in water: 0.456 mg/ml at 25°C
[b]Density: 1.012 g/ml at 25°C
[c]Solubility in water (g/100 g): 0.351 at 25°C. Acidify with HCl 0.1N until the powder dissolves

2. To prepare a final solution (for concentration, see column F), take the volumes indicated in column E for each amino acid and pipette in a volumetric flask of 10 ml. Add methanol up to the volume.
3. Use this solution to prepare five points of calibration curves at decreasing concentrations, obtained by serial dilution (1:1).

2.3. Labelled Amino Acids Standard

L-Gly-1,2-$^{13}C_2$, DL-Ala-3,3,3-d_3, L-Glu-2,4,4-d_3, L-Val-d_8, L-Met-methyl-d_3, L-Tyr-ring-d_4, L-Leu-5,5,5-d_3, L-Orn-3,3,4,4,5,5-d_6, L-Phe-ring-d_5, L-Arg-5-^{13}C-4,4,5,5-d_4 (Cambridge Isotope Laboratories); Creatinine-methyl-d_3 (CDN); GAA-2,2-d_2 (Isotec); L-Pro-d_7, L-Asp-2,3,3-d_3, L-Gln-2,3,3,4,4-d_5, L-Cit-5,5-d_2, GABA-2,2-d_2, DL-Ser-2,3,3-d_3, and $(Hcy)_2$-d_8 (generous donation from Mayo Clinic).

2.4. Preparation of Labelled Internal Standard Solution (Table 2)

1. Weigh 10 mg of each amino acids and dissolve in water as indicated in column C to prepare a primary solution for each amino acid (column D) (see Note 1).
2. To prepare an intermediate solution (column F), take the volumes indicated in column E for each amino acid and pipette in a volumetric flask of 10 ml. Add methanol up to the volume.
3. Prepare final solution (column G) by diluting 1:10 with methanol intermediate solution. Use 150 μl to prepare both calibration curves and human samples (column H showed pmol added).

2.5. Reagents and Solvents

1. Acetonitrile and Methanol HPLC grade; Tridecafluoroheptanoic acid (TDFHA) (Sigma–Aldrich) (see Note 2).
2. HPLC solutions:
   (a) Wash solvent: Solution of 70% acetonitrile in water.
   Measure 700 ml of acetonitrile into a 1,000-ml graduated cylinder and pour in a 1-L solvent reservoir. Measure 300 ml of water, add into the bottle, and mix the solution.
   (b) Purge solvent: Water solution 100%.
   Measure 1,000 ml of water into a 1,000-ml graduated cylinder and place it in the reservoir.
   (c) Mobile phase solvent A: 0.1% TDFHA solution in water.
   Measure 999 ml of water into a 1,000-ml graduated flask and add 1 ml of TDFHA. Mix the solution and place it in the reservoir solvent A.
   (d) Mobile phase solvent B: 0.1% TDFHA solution in acetonitrile.
   Measure 999 ml of acetonitrile into a 1,000-ml graduated flask and add 1 ml of TDFHA. Mix the solution and place it in the reservoir solvent B.

Table 2
Preparation scheme of labelled internal standard solution as described in Subheading 2.4

	A	B	C	D	E	F	G	H
Labelled amino acids	Weight (mg)	MW	Volume (ml)	Primary solution (mM)	Volume to draw/10 ml (μl)	Intermediate solution (μM)	Final solution (μM)	pmol added to sample
$(Hcy)_2$-d_8	10	276.35	20	1.81	553	100	10.0	1,500
DL-Ser-d_3	10	108.09	20	4.63	216	100	10.0	1,500
GAA-d_2	10	119.10	20	4.20	715	300	30.0	4,500
GABA-d_2	10	105.12	20	4.76	210	100	10.0	1,500
L-Ala-d_3	10	92.10	20	5.43	184	100	10.0	1,500
L-Arg-^{13}C-d_4	10	179.20	20	2.79	358	100	10.0	1,500
L-Asp-d_3	10	136.11	20	3.67	272	100	10.0	1,500
L-Cit-d_2	10	177.19	20	2.82	354	100	10.0	1,500
L-Gln-d_5	10	151.18	20	3.31	302	100	10.0	1,500
L-Glu-d_3	10	150.13	20	3.33	300	100	10.0	1,500
L-Gly-1,2-$^{13}C_2$	10	77.07	20	6.49	154	100	10.0	1,500
L-Leu-d_3	10	134.18	20	3.73	268	100	10.0	1,500
L-Met-methyl-d_3	10	152.23	20	3.28	304	100	10.0	1,500

L-Orn-d_6	10	174.62	20	2.86	349	100	10.0	1,500
L-Phe-ring-d_5	10	170.19	20	2.94	340	100	10.0	1,500
L-Pro-d_7	10	122.13	20	4.09	244	100	10.0	1,500
L-Tyr-ring-d_4[a]	10	185.19	30	1.80	556	100	10.0	1,500
L-Val-d_8	10	125.15	20	4.00	250	100	10.0	1,500

[a] Solubility in water (g/100 g): 0.351 at 25°C. Acidify with HCl 0.1N until the powder dissolves

2.6. HPLC/ESI–MS/MS System

1. HPLC System: Waters 2795 Alliance HT.
2. Analytical Column: Discovery C_{18} (50 mm×2.1 mm×5 μm) (Supelco). The column temperature was set at 40°C. The eluent from the LC column was split 1:10 after the column.
3. Mass Spectrometer: Micromass Quattro Ultima (Waters Corporation, Manchester, UK) with an ESI source.
4. Software data analysis: Mass Spectrometer Data were acquired with MassLynx 4.0 and processed for calibration and quantification of the analytes by QuanLynx.

2.7. Filter Paper

1. Guthrie filter paper for collection of blood spots (Schleicher and Schuell S&S 903).
2. PerkinElmer blood spot puncher (1296–071 Delfia) for punching discs from blood spots on filter paper.

3. Methods

3.1. Chromatography

Two different gradients have been established, depending on the need to analyze the whole set of amino acids or only a part of it.

1. In the first method, chromatographic separation took 10 min with a reequilibration time of 5 min in between the runs. Gradient composition was:

Time (min)	Solvent A (%)	Solvent B (%)	Flow (ml/min)	Curve
0	90	10	0.3	1
1	90	10	0.3	1
3	85	15	0.3	1
5	80	20	0.3	1
6	75	25	0.3	6
7	60	40	0.3	6
8.9	25	75	0.3	6
9	2	98	0.3	1
10	2	98	0.3	6

2. To monitor only those amino acids playing an important role in detection of metabolic diseases by newborn screening, such as Val, Leu/Ile/a-Ile, Met, Phe, Cit, and Tyr, another gradient has been created for a shorter chromatographic run with duration

time of 5 min and a reequilibration time of 1 min. Gradient composition was as follows:

Time (min)	Solvent A (%)	Solvent B (%)	Flow (ml/min)	Curve
0	80	20	0.3	1
1	80	20	0.3	1
3	70	30	0.3	1
3.1	10	90	0.3	1
5	10	90	0.3	1

3. The underivatized samples dissolved in 50 μl of water containing 0.1% TDFHA was injected into the HPLC analytical column for chromatography. Figures 1 and 2 show representative chromatograms of a plasma human sample spiked with standard amino acid solution (Fig. 1) and two DBS samples with diagnosis of classical MSUD (Fig. 2a) and PKU (Fig. 2b).

3.2 Calibration Curves

1. Calibration has been achieved by applying the stable isotope dilution method in which scalar concentrations of final standard solution of unlabelled amino acids (see column F Table 1, Subheading 2.2) were added to 150 μl of the labelled internal standard solution (column G Table 2, Subheading 2.4).
2. Two sets of calibration curves were determined:
   (a) In water, to access linearity and to validate method using the range of concentrations for each amino acid showed in Table 1.
   (b) In all matrices: Dried blood spot, plasma, and urine. Add 30 μl of standard solutions at increasing concentrations of unlabelled amino acids to each matrix and prepare the curves as described in Subheading 3.3. The endogenous concentrations of the analytes in all matrices were calculated using the intercept value to *y*-axis of calibration curves, using peak area ratios of the analytes to internal standard.

3.3. Samples Preparation

3.3.1. Dried Blood Spot Samples

1. Punch from every filter paper three blood spot discs (each with Ø = 3.2 mm) and add 150 μl of labelled final solution to extract the amino acids.
2. Sonicate the samples for 20 min.
3. Transfer into clean vial or in a 96-well plate.
4. Take to dryness under a nitrogen flow at 60°C for 15 min.
5. Dissolve the samples in 50 μl of water with 0.1% TDFHA and proceed with analysis.

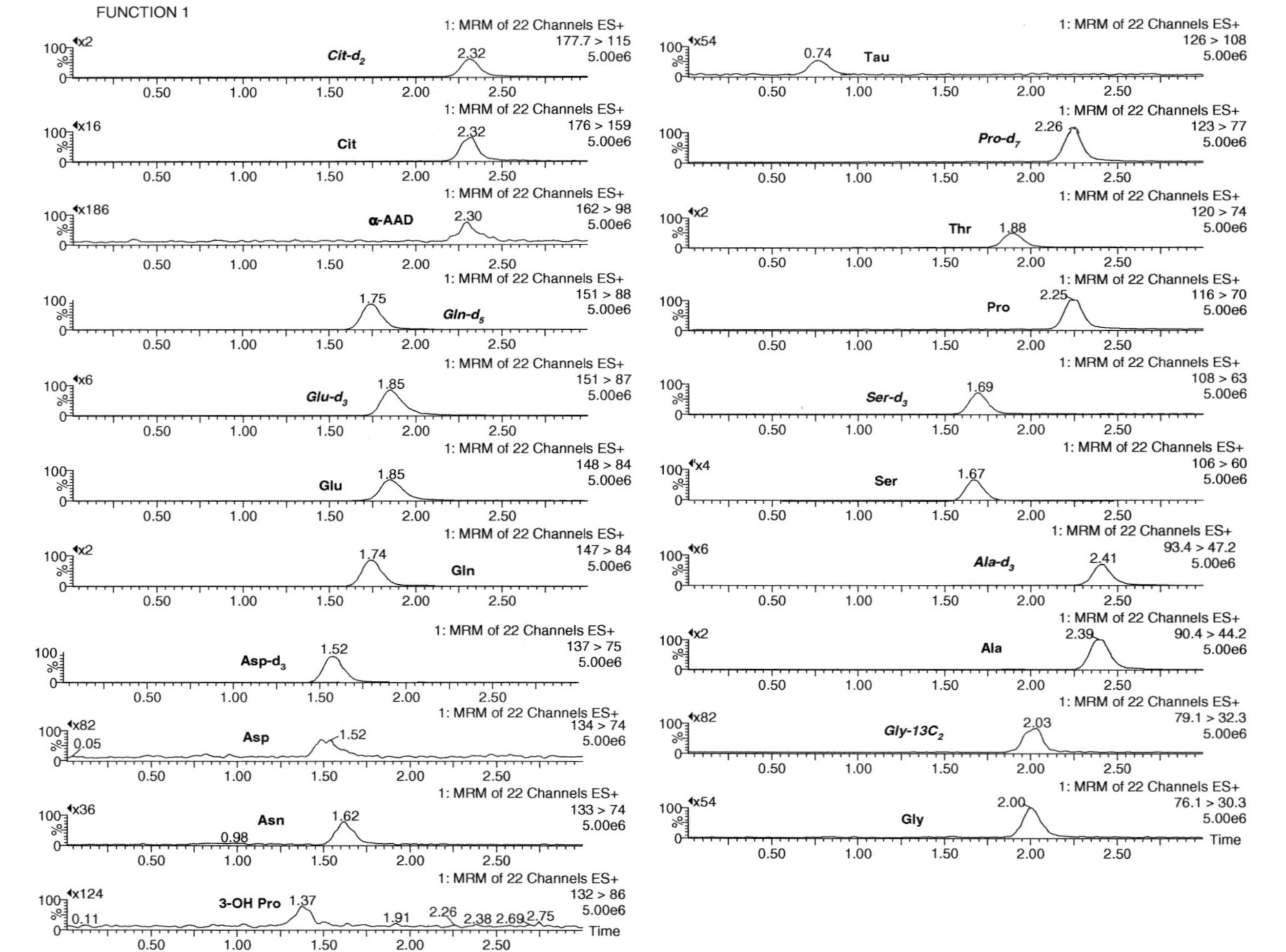

Fig 1. Reconstructed LC–ESI–MS/MS ion chromatograms in tree window of a plasma spiked with standard amino acids, with the addition of the labelled internal standard.

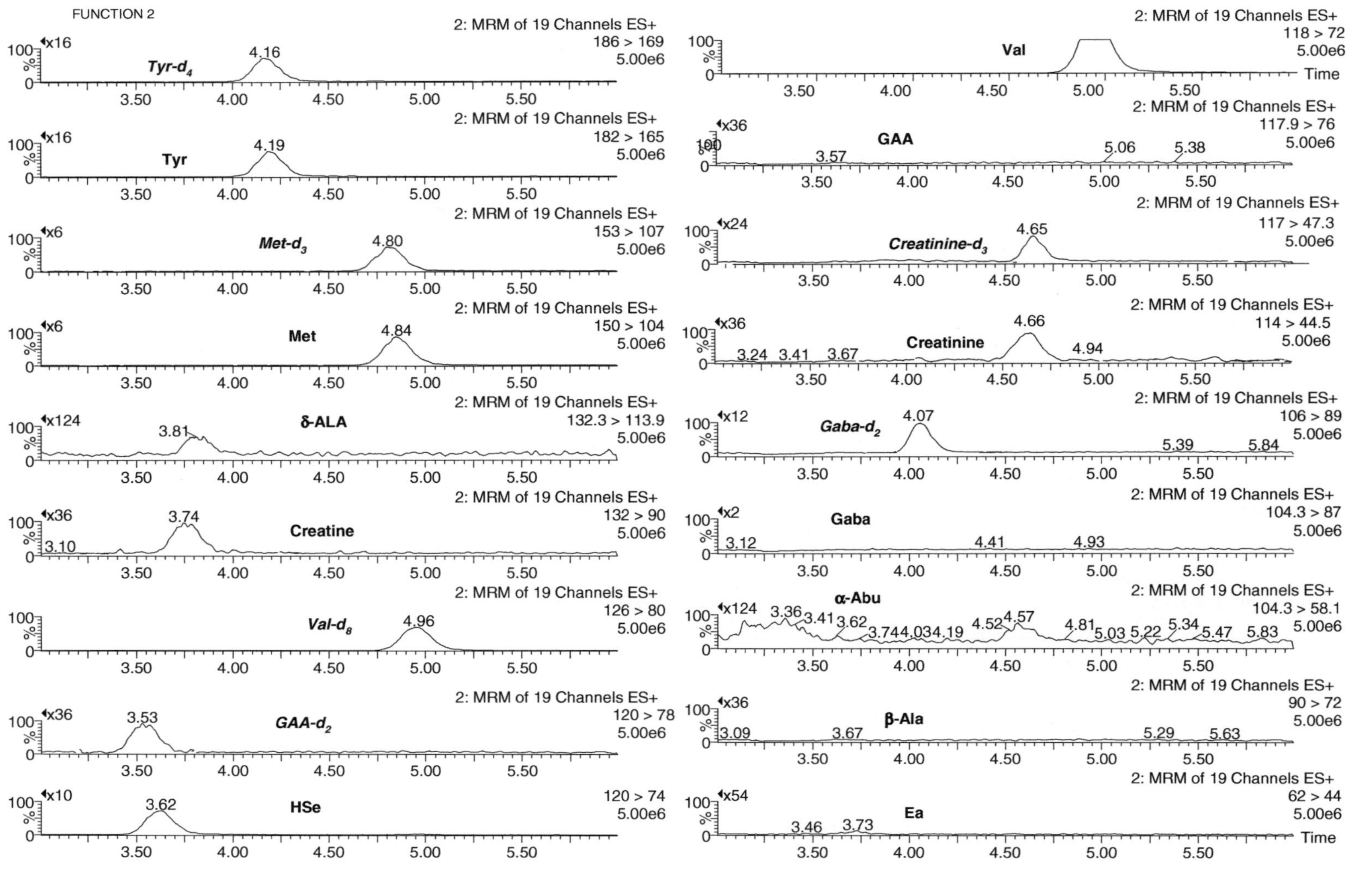

Fig 1. (continued)

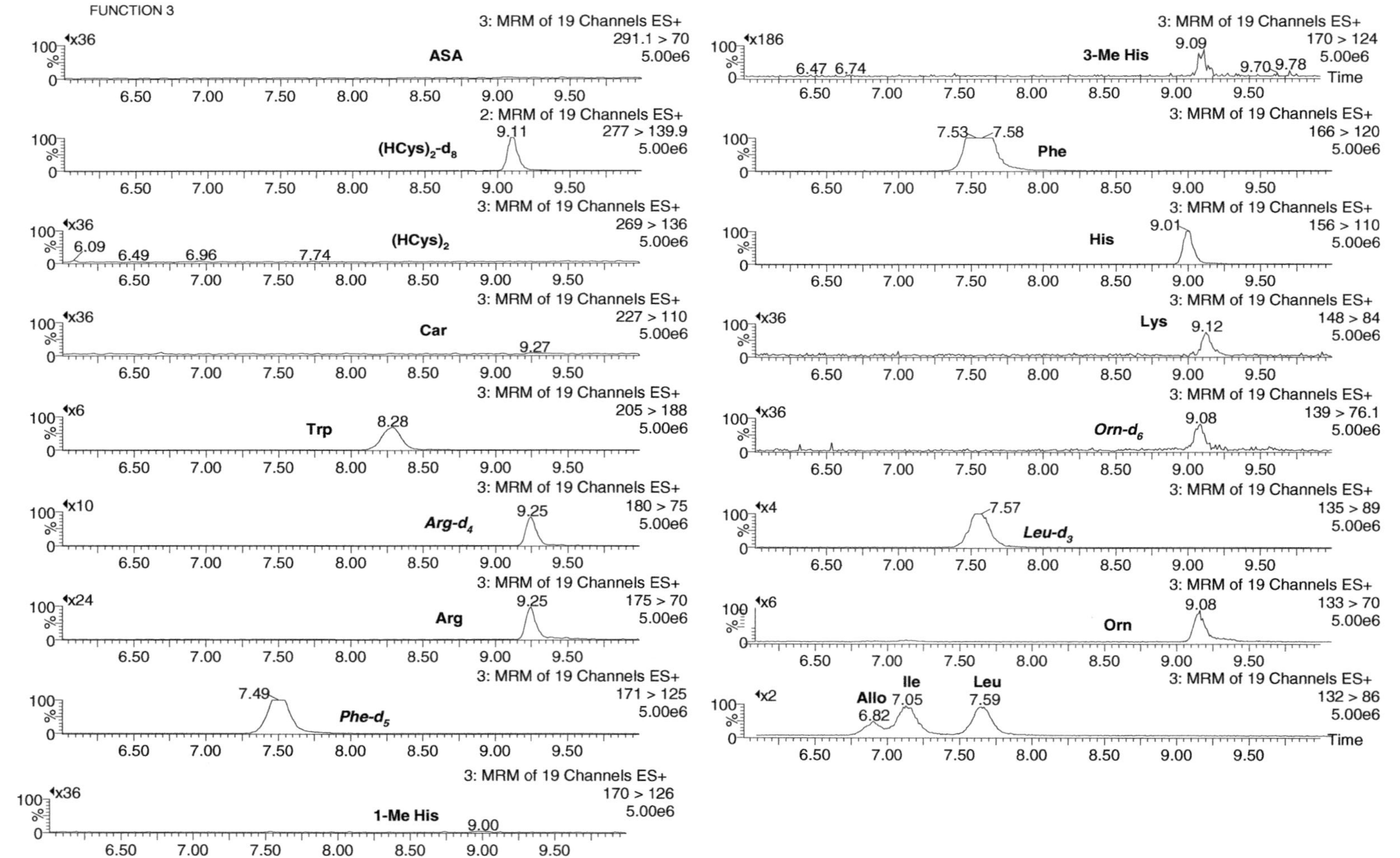

Fig 1. (continued)

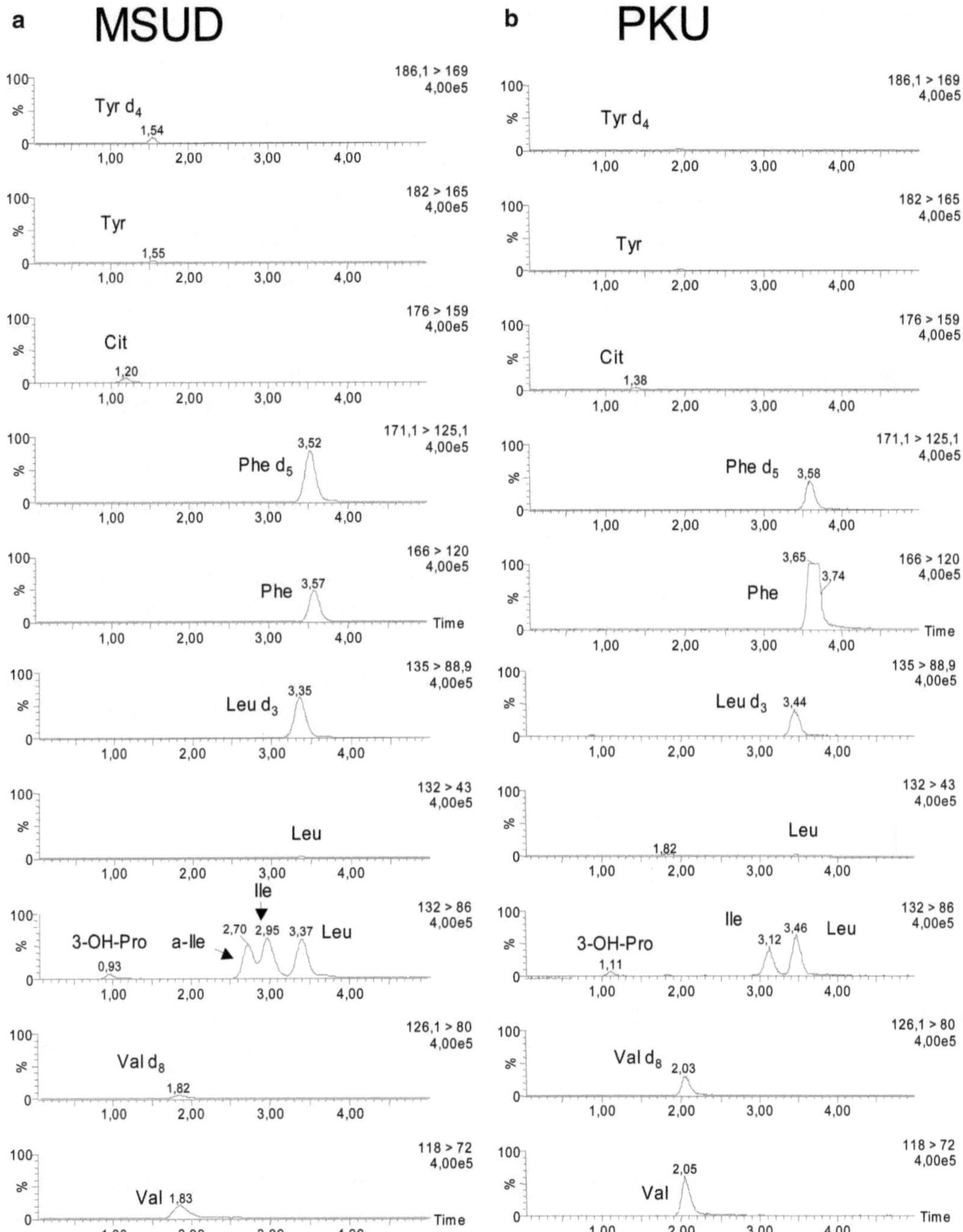

Fig. 2. Example of shorter reconstructed ion chromatogram (5 min), of two DBS samples with diagnosis of classical MSUD (**a**) and PKU (**b**).

3.3.2. Plasma Samples

1. Transfer 10 μl of human plasma into the propylene vial and add 150 μl of Internal Standards Solution.
2. Cap the vial, mix on the vortex for several seconds, and keep at −20°C for 10 min to precipitate proteins.
3. Centrifuge at 15,700 × *g* for 10 min to remove proteins.
4. Transfer into clean vial or in a 96-well plate.
5. Take to dryness under a nitrogen flow at 60°C for 15 min.
6. Dissolve the samples in 50 μl of water with 0.1% TDFHA and proceed with analysis.

3.3.3. Urine Samples

1. Prior to analysis, defrost urine samples by slowly warming up to room temperature and centrifuge it for 10 min at a rotor speed of 12,000 × *g* to remove possible particulates.
2. Determine the creatinine concentration using SPOTCHEM II Creatinine kit and SPOTCHEM Analyzer (Menarini Diagnostics) (see Note 3).
3. Transfer 10 μl of centrifuged urine into the propylene vial, add 150 μl of Internal Standards Solution, cap the vial, and mix on the vortex for several seconds.
4. Load the sample in the solid-phase extraction (SPE) cartridge previously activated with 1 ml of MeOH and washed with 1 ml of water.
5. Wash the cartridge with 1 ml of H_2O with 0.5% TDFHA and elute with 500 μl of MeOH.
6. Take to dryness the eluate under a nitrogen flow at 60°C for 15 min.
7. Dissolve the samples in 50 μl of water with 0.1% TDFHA and proceed with analysis.

3.4. Analyses

1. MS/MS analysis was done in positive ion mode: The ion transitions for unlabelled and labelled amino acids and the instrumental parameters for their MS/MS detection in MRM are showed in Table 3. The detection limits varied from 1 to 40 pmol depending on the nature of the amino acids injected.
2. Multiple reaction monitoring (MRM) measurements were performed using optimal cone and collision energy values derived from preliminary fragmentation studies with a continuous infusion of a 50 μM solution of each analyte. A list of exploited transitions, collision energy, and others acquisition parameters is reported in Table 3. Figure 3 shows a typical chromatogram of unlabelled standard solution, expressed as percent relative abundance versus time in minutes. To improve response of mass spectrometer detector, MRM analysis was separated on three windows of acquisition (function 1–3).

Table 3
Ion transitions for the unlabelled and labelled amino acids and instrumental parameters for their MS/MS detection ions in MRM mode

Compound	Ion transition	Dwell time (s)	Cone voltage (V)	Collision energy (eV)	Mean RT ± SD (min)
Function 1					
Gly	76.10 > 30.30	0.020	60	8	1.96 ± 0.1
Gly-$^{13}C_2$	79.10 > 32.30	0.020	60	8	2.03 ± 0.1
Sar	90.40 > 44.20	0.020	60	10	1.87 ± 0.1
Ala	90.40 > 44.20	0.020	60	10	2.40 ± 0.1
Ala-d_3	93.40 > 47.20	0.020	60	10	2.41 ± 0.1
Ser	106.00 > 60.00	0.020	70	15	1.69 ± 0.1
Ser-d_3	108.90 > 63.00	0.020	70	10	1.70 ± 0.1
Pro	116.00 > 70.00	0.020	70	15	2.25 ± 0.1
Thr	120.00 > 74.00	0.020	70	15	1.88 ± 0.1
Pro-d_7	123.00 > 77.10	0.020	70	20	2.25 ± 0.1
Tau	126.00 > 108.00	0.020	70	20	0.74 ± 0.1
3-Hyp	132.00 > 86.00	0.020	70	15	1.36 ± 0.1
Asn	133.00 > 74.00	0.020	70	15	1.62 ± 0.1
Asp	134.00 > 74.00	0.020	70	15	1.52 ± 0.1
Asp-d_3	137.10 > 75.00	0.020	50	15	1.53 ± 0.1
Gln	147.00 > 84.00	0.020	70	15	1.74 ± 0.1
Glu	148.00 > 84.00	0.020	70	15	1.85 ± 0.1
Glu-d_3	151.00 > 87.00	0.020	70	15	1.85 ± 0.1
Gln-d_5	151.90 > 88.00	0.020	70	15	1.74 ± 0.1
α-AAD	162.00 > 98.00	0.020	70	15	2.29 ± 0.1
Cit	176.00 > 159.00	0.020	70	15	2.32 ± 0.1
Cit-d_2	177.70 > 115.00	0.020	70	20	2.33 ± 0.1

(continued)

Table 3 (continued)

Compound	Ion transition	Dwell time (s)	Cone voltage (V)	Collision energy (eV)	Mean RT ± SD (min)
Function 2					
EA	62.00 > 44.00	0.020	70	15	3.73 ± 0.1
β-Ala	90.00 > 72.00	0.020	60	10	3.68 ± 0.1
α-Abu	104.30 > 58.10	0.020	70	10	3.66 ± 0.1
GABA	104.30 > 87.00	0.020	70	15	4.07 ± 0.1
GABA-d_2	106.00 > 89.00	0.020	70	8	4.07 ± 0.1
Creatinine	114.20 > 44.50	0.020	70	20	4.65 ± 0.1
Creatinine-d_3	117.00 > 47.30	0.020	70	15	4.65 ± 0.1
GAA	117.90 > 76.00	0.020	70	10	3.53 ± 0.1
Val	118.00 > 72.00	0.020	70	15	4.93 ± 0.1
Hse	120.00 > 74.00	0.020	70	15	3.62 ± 0.1
GAA-d_2	120.00 > 78.00	0.020	70	8	3.53 ± 0.1
Val-d_8	126.00 > 80.00	0.020	70	15	4.96 ± 0.1
δ-ALA	132.00 > 113.90	0.020	70	7	4.10 ± 0.1
Creatine	132.10 > 90.20	0.020	70	10	3.74 ± 0.1
Met	150.00 > 104.00	0.020	70	15	4.85 ± 0.1
Met-d_3	153.00 > 107.00	0.020	70	15	4.80 ± 0.1
Tyr	182.00 > 165.00	0.020	70	15	4.16 ± 0.1
Tyr-d_4	186.00 > 169.00	0.020	70	15	4.16 ± 0.1
Function 3					
Leu	132.00 > 86.00	0.020	70	15	7.46 ± 0.1
Ile	132.00 > 86.00	0.020	70	15	6.95 ± 0.1
a-Ile	132.00 > 86.00	0.020	70	15	6.74 ± 0.1

Orn	133.00>70.00	0.020	70	15	9.03 ± 0.1
Leu-d_3	135.00>89.00	0.020	70	15	7.57 ± 0.1
Orn-d_6	139.00>76.10	0.020	70	30	9.08 ± 0.1
Lys	148.00>84.00	0.020	70	15	9.10 ± 0.1
His	156.00>110.00	0.020	70	15	9.02 ± 0.1
Phe	166.00>120.00	0.020	70	15	7.53 ± 0.1
3-Me-His	170.00>124.00	0.020	70	15	9.04 ± 0.1
1-Me-His	170.00>126.00	0.020	70	15	8.99 ± 0.1
Phe-d_5	171.00>125.00	0.020	70	15	7.49 ± 0.1
Arg	175.00>70.00	0.020	70	15	9.23 ± 0.1
Arg-^{13}C-d_4	180.00>75.00	0.020	70	15	9.25 ± 0.1
Trp	205.00>188.00	0.020	70	15	8.33 ± 0.1
Carn	227.00>110.00	0.020	70	15	9.27 ± 0.1
$(Hcy)_2$	269.00>136.00	0.020	70	15	9.11 ± 0.1
$(Hcy)_2$-d_8	277.10>139.90	0.020	70	10	9.11 ± 0.1
ASA	291.10>70.00	0.020	150	40	9.04 ± 0.1

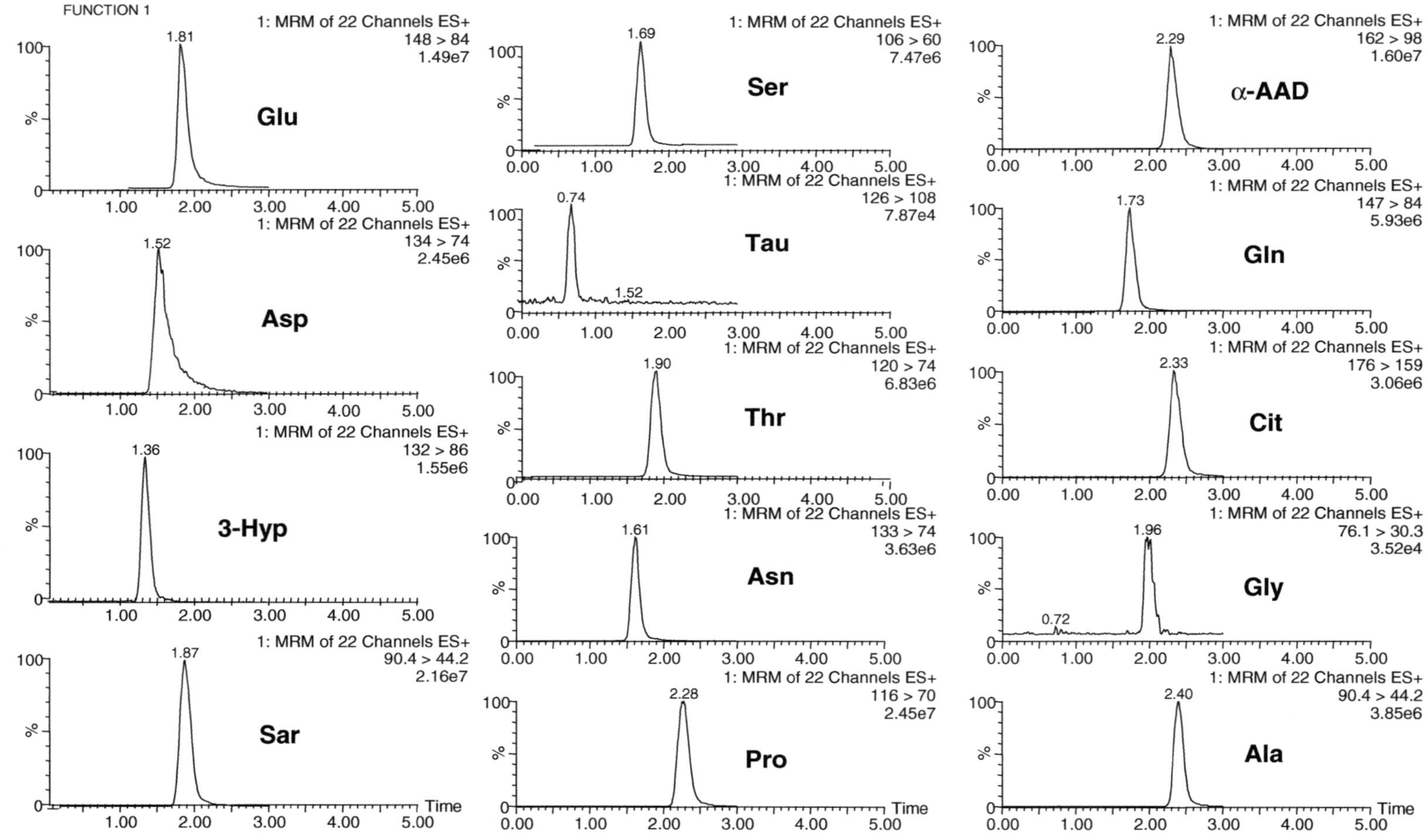

Fig. 3. Reconstructed LC–ESI–MS/MS ion chromatograms in three windows of unlabelled standard amino acids in positive-ion mode with the separation within 10 min. Reprinted with permission from Elsevier (18).

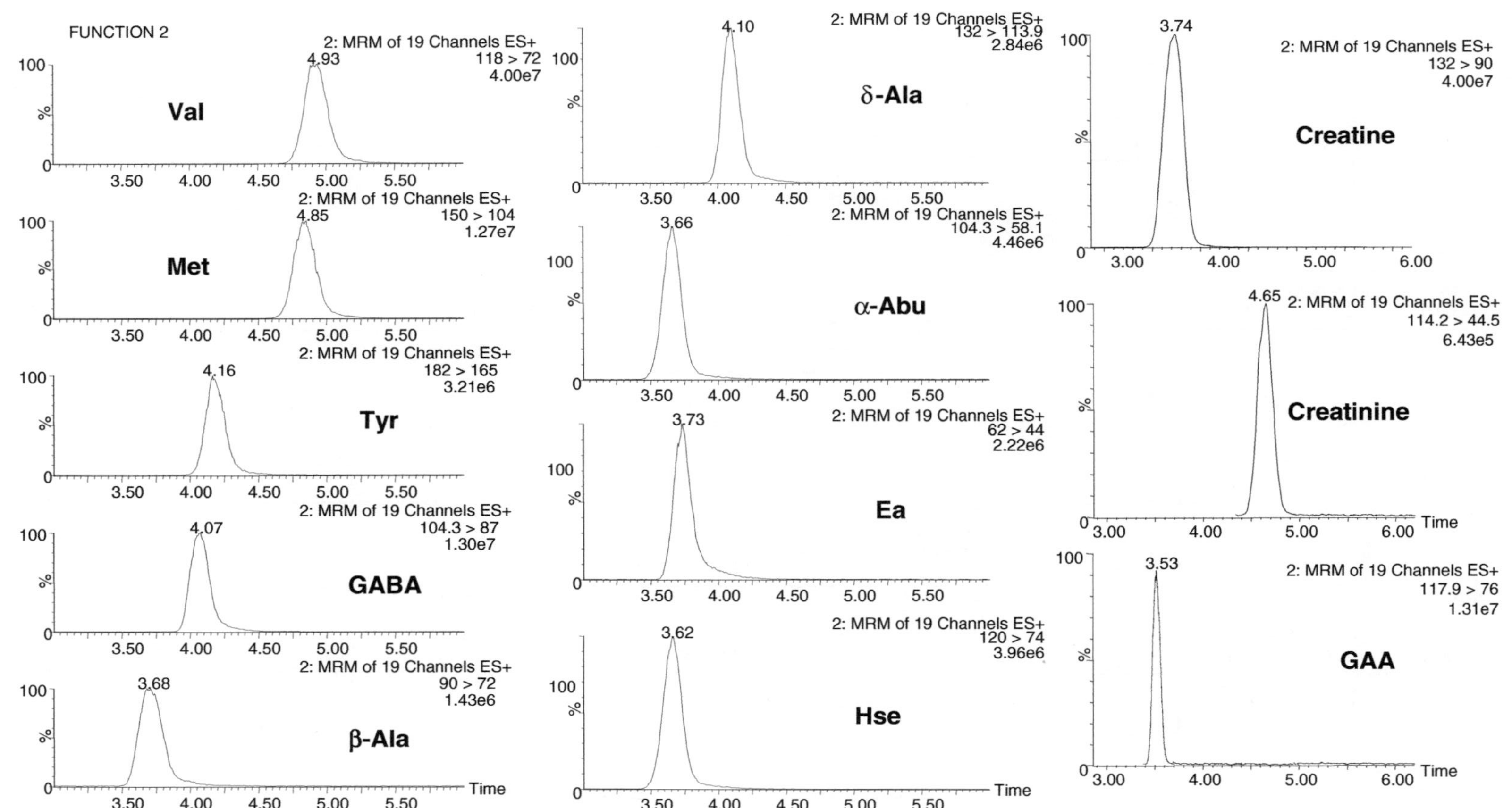

Fig 3. (continued)

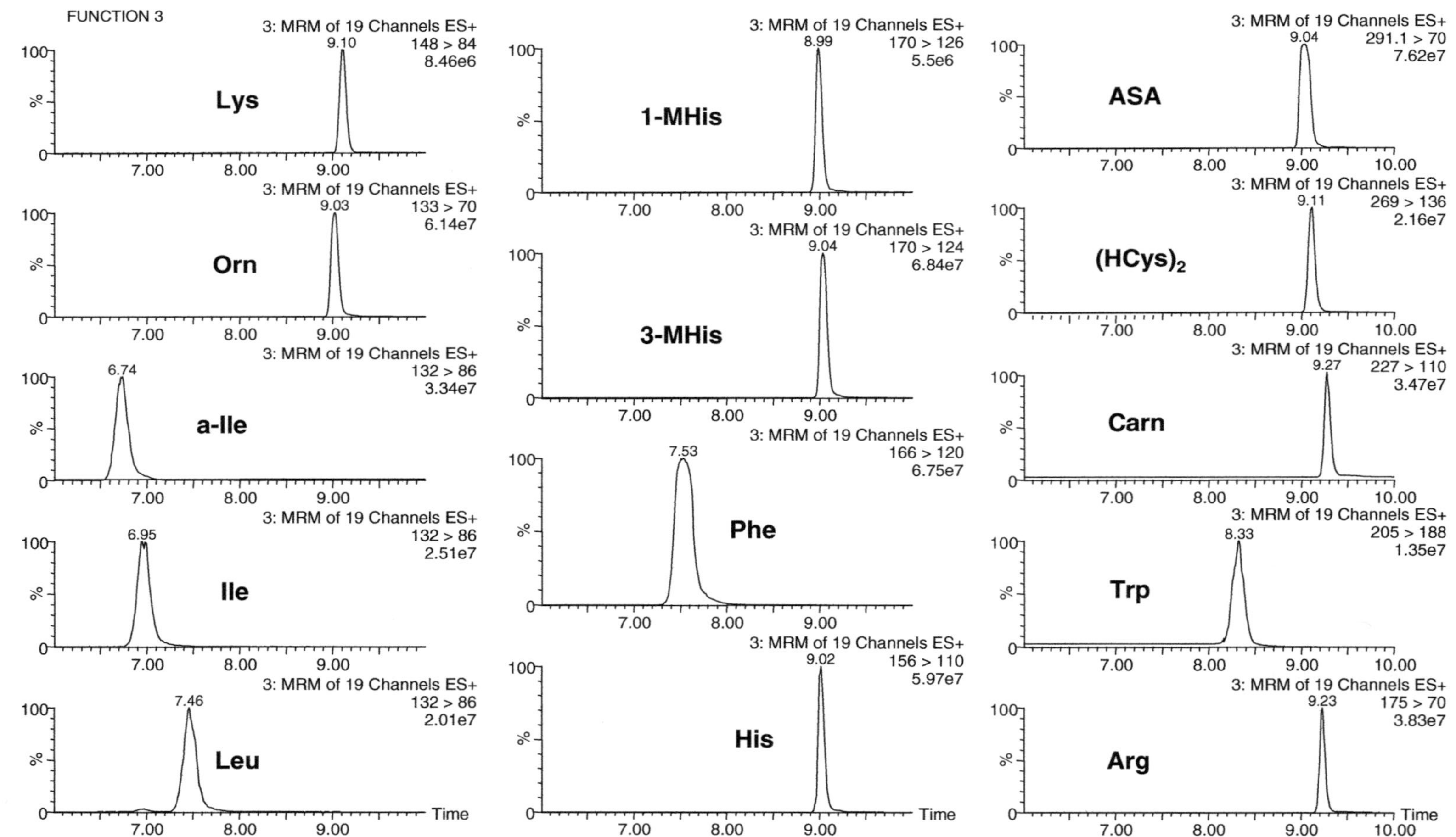

Fig 3. (continued)

4. Notes

1. Gln and Trp are unstable when stored in solution for long periods of time; ASA is strongly hygroscopic and stable only in its anhydrous form. In aqueous solution, anhydrides can be formed.
2. The TDFHA solution has to be heated at 35°C to dissolve completely. Density: 1.792 g/ml at 25°C; melting point 30°C.
3. Dilute 1:5 urine samples with 6% BSA water solution into the propylene tube. The quantification was performed by spectrophotometric measurement.

Acknowledgements

We acknowledge Dr. Piero Rinaldo and Dr. Dietrich Matern from Mayo Clinic for the gentle donation of some labelled amino acids to make our study possible.

References

1. Qu J et al (2002) Validated quantitation of underivatized amino acids in human blood samples by volatile ion-pair reversed-phase liquid chromatography coupled to isotope dilution tandem mass spectrometry. Anal Chem 74:2034–2040
2. Kopple J (1983) Amino acid metabolism in chronic renal failure. In: Blackburn GL, Grant JP, Young VR (eds) Amino Acids: Metabolism and Medical Applications, John Wright, Boston
3. Guthrie R (1968) Screening for "inborn errors of metabolism" in the newborn infant-a multiple test program. Birth Defects Orig Art Ser 4:96–98
4. Spackman DH, Stein WH, Moore S (1958) Automatic Recording Apparatus for Use in Chromatography of Amino Acids. Anal Chem 30:1190–1206
5. Robinson GW (1978) Detection Systems for Amino Acids in Ion-exchange Column Effluents. In: Blackburn S (ed) Amino Acid Determination – Methods and Techniques, Marcel Dekker, New York
6. Reilly AA, Bellissario R, Pass KA (1998) Multivariate discrimination for phenylketonuria (PKU) and non-PKU hyperphenylalaninemia after analysis of newborns' dried blood-spot specimens for six amino acids by ion-exchange chromatography. Clin Chem 44:317–326
7. Ye JN, Baldwin RP (1994) Determination of Amino Acids and Peptides by Capillary Electrophoresis and Electrochemical Detection at a Copper Electrode. Anal Chem 66:2669–2674
8. Pei J, Li X (2000) Determination of underivatized amino acids by high-performance liquid chromatography and electrochemical detection at an amino acid oxidase immobilized CuPtCI6 modified electrode. Fresenius J Anal Chem 367:707–713
9. Chace DH et al (1995) Rapid diagnosis of maple syrup urine disease in blood spots from newborns by tandem mass spectrometry. Clin Chem 41:62–68
10. Chace DH (2001) Mass spectrometry in the clinical laboratory. Chem Rev 101:445–477
11. Rashed MS (2001) Clinical applications of tandem mass spectrometry: ten years of diagnosis and screening for inherited metabolic diseases. J Chromatogr B Biomed Sci Appl 758:27–48
12. Piraud M et al (2005) Ion-pairing reversed-phase liquid chromatography/electrospray ionization mass spectrometric analysis of 76 underivatized amino acids of biological interest: a new tool

for the diagnosis of inherited disorders of amino acid metabolism. Rapid Commun Mass Spectrom 19:1587–1602

13. Qu J et al (2002) Rapid determination of underivatized pyroglutamic acid, glutamic acid, glutamine and other relevant amino acids in fermentation media by LC-MS-MS. Analyst 127:66–69
14. Piraud M et al (2003) ESI-MS/MS analysis of underivatised amino acids: a new tool for the diagnosis of inherited disorders of amino acid metabolism. Fragmentation study of 79 molecules of biological interest in positive and negative ionisation mode. Rapid Commun Mass Spectrom 17:1297–1311
15. Waterval WA et al (2009) Quantitative UPLC-MS/MS analysis of underivatised amino acids in body fluids is a reliable tool for the diagnosis and follow-up of patients with inborn errors of metabolism. Clin Chim Acta 407:36–42
16. Chace DH, Kalas TA, Naylor EW (2002) The application of tandem mass spectrometry to neonatal screening for inherited disorders of intermediary metabolism. Annu Rev Genomics Hum Genet 3:17–45
17. Oglesbee D et al (2008) Second-tier test for quantification of alloisoleucine and branched-chain amino acids in dried blood spots to improve newborn screening for maple syrup urine disease (MSUD). Clin Chem 54:542–549
18. Zoppa M, Gallo L, Zacchello F, Giordano G (2006) Method for the quantification of underivatized amino acids on dry blood spots from newborn screening by HPLC–ESI–MS/MS. The Journ Chrom B 831: 267–273

Chapter 19

Capillary Electrophoresis of Free Amino Acids in Physiological Fluids Without Derivatization Employing Direct or Indirect Absorbance Detection

Gordana D. Žunić, Slavica Spasić, and Zorana Jelić-Ivanović

Abstract

Whole blood and/or plasma amino acids are useful for monitoring whole body protein and amino acid metabolism in an organism under various physiological and pathophysiological conditions. Various methodological procedures are in use for their measurement in biological fluids. From the time when capillary electrophoresis was introduced as a technology offering rapid separation of various ionic and/or ionizable compounds with low sample and solvent consumption, there were many attempts to use it for the measurement of amino acids present in physiological fluids. As a rule, these methods require derivatization procedures for detection purposes.

Here, we present two protocols for the analysis of free amino acids employing free zone capillary electrophoresis. Main advantage of both methods is an absence of any derivatization procedures that permits the analysis of free amino acid in physiological fluids. The method using direct detection and carrier electrolyte consisting of disodium monophosphate (10 mM at pH 2.90) permits determination of compounds that absorb in UV region (aromatic and sulfur containing amino acids, as well as some peptides, such as carnosine, reduced and oxidized glutathione). The other method uses indirect absorbance detection, employing 8 mM *p*-amino salicylic acid buffered with sodium carbonate at pH 10.2 as running electrolyte. It permits quantification of 30 underivatized physiological amino acids and peptides. In our experience, factorial design represents a useful tool for final optimization of the electrophoretic conditions if it is necessary.

Key words: Amino acids, Capillary electrophoresis, Plasma, Blood, Amino acid analysis

1. Introduction

Free amino acids represent constant constituents of all biological fluids (blood, cells, plasma, and cerebrospinal fluid) at relatively low concentrations. Their presence in biological fluids is useful for monitoring whole body protein and amino acid metabolism in

Michail A. Alterman and Peter Hunziker (eds.), *Amino Acid Analysis: Methods and Protocols*, Methods in Molecular Biology, vol. 828, DOI 10.1007/978-1-61779-445-2_19, © Springer Science+Business Media, LLC 2012

an organism under various physiological and/or pathophysiological conditions (1–9). Because quantity of free amino acids in biological fluids represent good indicator of various metabolic disorders, interest in their measurement in different body fluids increases continuously (1, 2, 10–12).

Free aromatic amino acids (AAAs), such as phenylalanine (Phe), tyrosine (Tyr), tryptophan (Trp), and histidine (His), particularly reflect disturbances in protein anabolism/catabolism ratio in an organism (2, 8). The plasma level of Phe is considered the most important determinant of its catabolism occurring predominantly in the liver (13), while plasma molar ratio of Phe to Tyr reflects net protein catabolism in the peripheral tissue (3, 8, 14). On the other hand, valine (Val), leucine (Leu), and isoleucine (Ile), known as branched-chain (BCA) amino acids catabolized in the peripheral tissues represent regulators of insulin signaling (15). Circulating BCA versus AAA represents a good indicator of liver energy status (10). Sulfur containing free amino acids, such as cysteine (Cys) or homocysteine (Hcy) has important roles in various metabolic conditions (16). Glutamine (Gln) and alanine (Ala), as main nontoxic transporting forms of ammonia from peripheral tissues to liver, are good indicators of ammonia metabolism in the peripheral tissues (17). Plasma molar ratios between glycine (Gly) to BCA and Ala to BCA reflect nutritional status of an organism (12). On the other hand, urinary excretion of 3-methylhistidine (3Me-His), a constituent of skeletal muscle proteins, present in negligible quantities in the other tissues, has become a valuable catabolic marker of skeletal muscle proteins (1, 18–21). In addition, recent finding have shown that certain amino acids, particularly Gln and arginine (Arg), have important roles in healing processes effecting immunity and protein metabolism (1, 19–21).

Thiols, such are reduced glutathione (GSH), Cys, and Hcy, have important roles in overall metabolism, including cell homeostasis, radioprotection, and antioxidant defense (22). Their oxidation and/or interactions produce amino acids with disulfide bounds, such as homocystine (HcySS), cystine (CySS), mixed disulfides (Hcy-Cys), or oxidized glutathione (GSSG). Presence of other amino acids, such as Ala, asparagine (Asn), aspartic acid (Asp), beta-alanine (βAla), citrulline (Cit), gama-amino butyric acid (GABA), glutamic acid (Glu), Gly, hydroxylysine (Hyl), lysine (Lys), methionine (Met), ornithine (Orn), proline (Pro), serine (Ser), threonine (Thr), and some peptides including carnosine or anserine, are observed in biological fluids, too (3, 4).

Various methodological procedures were developed for amino acid measurement in biological fluids. They usually employ gas, liquid, or ion-exchange chromatography (18, 23). The classic ion-exchange separation followed by postcolumn derivatization with ninhydrin was considerably improved since its initial inception, particularly with the availability of modern dedicated amino acid analyzers (18, 24, 25). It has to be pointed out that, as amino acids lack chromophores, their precolumn or postcolumn derivatization is usually necessary for their detection. Chromatographic methods usually allow separation of 35–40 different amino acids, while high-pressure liquid chromatography provides for the identification of even larger number of these compounds.

Ever since capillary electrophoresis (CE) was introduced as a technology offering rapid separation of various ionic and/or ionizable compounds with low sample and solvent consumption, there were attempts to use it for amino acid studies (26, 27). Much of the early development of CE involved derivatized amino acids as test solutes (28). Also, aromatic (Tyr, Phe, and Trp) and sulfur (Cys, CySS, Hcy, HcySS, and Met) containing amino acids, absorbing light in the ultraviolet region could be separated by CE without any derivatization procedure (29).

However, due to the large number of the free amino acids present in physiological fluids, similarities in their electrochemical characteristics and particularly due to the fact that amino acids generally possess very limited chromophores, CE of free amino acids present in physiological fluids, still represent a challenging problem. Poinsot et al. (30) summarized recent advances in clinical studies and neuroclinical applications of amino acid analysis by CE.

The reported CE methods usually analyze amino acids derivatized with different reagents (11, 26, 31). These methods permit separation of 20 or fewer amino acids in a single run. The derivatization processes can be very time consuming while introducing possibilities for changing the native electrophoretic mobility of the analytes (32). It was shown that the limitations of capillary electrophoretic methods are rather due to the labeling efficiency than sensitivity of the detector (30). Derivatization procedures prior to CE could be eliminated by various strategies for amino acid detection, such as direct or indirect UV absorbance, indirect fluorescence, and electrochemical detection (32, 33).

The simultaneous separation and quantification of the analytes within a minimum analysis time and a maximum resolution and efficiency are the main objectives in the development of a CE method. Factorial design (34) is especially useful at the beginning of an experimental study, where the most influential factors, their ranges of influence and factor interactions are not yet known. Factorial experiments allow experiments to take place over the whole range of the factor space (34). They show a high degree of precision in exchange for a minimum experimental effort and they enable factor interactions to be revealed. Our experience confirms factorial design as suitable procedure in screening important variables leading to the best separations of the compounds present in a mixture (29).

Here, we present two different protocols for the direct analysis of amino acids in biological fluids without any derivatization prior to CE.

One method relies on direct UV detection at 200 nm (29). Within less than 30 min and without any derivatization procedures in a single run with good precision and linear relationship between peak area and concentrations, it permits separation and quantifications of His, Hcy, Trp, Phe, Tyr, CySS, as well as three peptides (carnosine, GSH, and GSSG). It can be used to analyze human capillary blood (29).

The other method use indirect UV detection (35). We have used *p*-amino salicylic acid, as one of the most suitable carrier buffer and background providers that have an effective mobility close to the mobilities of the most amino acids at alkaline pH (32). Using the same background electrolyte, the other authors recommended cationic surfactants as buffer additives, but separated not more than 19 amino acids. We developed the method providing separations of 30 underivatized physiological amino acids and peptides from the same sample, in a single run, within 30 min. It can be used for analyzing plasma samples, while use of indirect absorbance detection permits analyzing amino acid both with or without ultraviolet absorbing properties and without any prior derivatization.

2. Materials

Material and several steps are similar for both methods.

2.1. Instrumentation and Equipment

1. P/ACE 5010 capillary electrophoresis system (Beckman, Palo Alto, CA, USA) equipped with P/ACE system software controlled by an IBM computer is used for separations.
2. P/ACE 5010 system contains built-in 200-, 214-, 254-, and 280-nm narrow-band filters for online detection and quantification. The instrument detects direct or indirect absorbance and operates at thermostated temperatures that allows a minimum capillary temperature of 15°C, but operate on higher temperature if necessary.
3. Assemble capillary in the cartridge with a 100×800-μm aperture and the detection window locate at 6.5 cm from the capillary outlet.
4. Normal or inverse polarity can be set at this instrument.

2.2. Chemicals and Solutions

Water and reagents: Prepare all solutions with low conductivity deionized water and analytical grade reagents, storing them at room temperature. We used water obtained from the Milli-Q water purification system (Waters/Millipore, USA). Diligently follow all waste disposal regulations when disposing waste materials. We do not add sodium azide to the reagents.

2.3. Individual Amino Acid Stock Solutions

1. Prepare individual solutions of the following compounds: Ala, Arg, Asn, Asp, β-Ala, Cit, CySS, GABA, Glu, Gln, Gly, His, hCySS, Hcy, Hyl, Ile, Leu, Lys, Met, Orn, Phe, Pro, Ser, Thr, Trp, Tyr, Val, GSH, GSSG and carnosine in water at 100 mM concentration level. Place them in dark bottles and keep in refrigerator.
2. Prepare individual stock solutions of Hcy, Cys and GSH daily.
3. *Standard stock solution mixture S1 (solution S1)*: In a 10 ml volumetric flask put 100 μl of individually prepared stock solutions of all compounds listed in item 1 and fill it with water (concentration level 1 mM of each compound).
4. *Standard stock solution mixture S2 (solution S2)*: In a 10 ml volumetric flask put 100 μl of individually prepared stock solutions listed in item 2 and fill it with water (concentration level 1 mM of each compound). Prepare it daily.
5. *Working standard solution* prepared at 50, 100, 150 and 200 μM of individual compounds daily by mixing solutions S1 and S2. We test linearity in this concentration range and reproducibility at 100 μM concentration level.
6. *Carrier electrolyte 1* (10 mM disodium monophosphate, pH 2.85 ± 0.05) for method with direct absorbance detection: To prepare carrier electrolyte solution, use analytical grade chemicals and low conductivity deionized water. First, prepare stock solution of disodium monophosphate at 100 mM concentration, titrated with phosphoric acid to pH 2.85 ± 0.05. Running electrolyte at concentration of 10 mM is prepared daily by diluting stock solution with low conductivity deionized water following pH check.
7. *Carrier electrolyte 2* (8 mM *p*-aminosalicylic acid, 2 mM sodium carbonate, pH 10.2 ± 0.1) for method with indirect absorbance detection: We used *p*-aminosalicylic acid as the sodium salt (PAS, MW = 211.2), sodium carbonate (Na_2CO_3) and sodium hydroxide (NaOH). Prepare electrolyte with analytical grade chemicals and low conductivity deionized water.

3. Methods

3.1. Capillary Conditioning and Sample Injections for Both Methods

1. *Capillary conditioning*: Start with flushing with 0.1 M sodium hydroxide (2 min), followed by deionized water (2 min) and finally with carrier electrolyte (2 min). At the end of each run flush capillary with 0.1 M sodium hydroxide (2 min), followed by deionized water (2 min).
2. *Sample injections*: Perform by 0.5 psi injection for 10 s (approximately an 112 nl volume injected).

3.2. Capillary Zone Electrophoresis with Direct Absorbance Detection (See Notes 1–8)

The procedures comprise equipment adjustments employing carrier electrolyte 1.

1. *Equipment adjustment*: Adjust instrument working parameters as follows:
   (a) Capillary – assemble uncoated fused-silica capillary tubing with total length of 47 cm, 40.5 cm from the window to inlet and 75 μm internal diameter in the cartridge.
   (b) Detection – set on 200 nm online direct absorbance detection.
   (c) Polarity – set on normal polarity (inlet positive and outlet negative).
   (d) Voltage – adjust at constant 15 kV voltage.
   (e) Temperature 20°C.
2. *Calibration* of the system is performed using the standard solution consisting of eight amino acids and three peptides.
3. *Capillary blood sample* (10 μl) is rapidly mixed with ice-cold water (40 μl). Immediately ultrafiltrate obtained hemolysate to remove proteins by using a micropartition device with molecular mass cut-off 10 kDa (minicentrifuge filter probes) by centrifugation at 21,000×*g*, 20 min at 4°C in a microlaboratory refrigerated centrifuge MPW-350R. Before analyzing, dilute blood hemolysate three times with low conductivity deionized water (see Note 9).
4. *Electropherograms* obtained with standard mixture and biological samples (human capillary blood sample) are presented in Figs. 1 and 2.

3.3. Capillary Zone Electrophoresis with Indirect Absorbance Detection (See Notes 10–11)

The procedures comprise equipment adjustments employing carrier electrolyte 2.

1. *Equipment adjustment* consists of the adjustment of instrument working parameters as follows:
   (a) Capillary – assemble uncoated fused-silica capillary tubing with total length of 87 cm, 80.5 cm from the window to inlet and 75 μm internal diameter in the cartridge.
   (b) Detection – set on 254 nm online indirect absorbance detection.
   (c) Polarity – set on normal polarity (inlet positive and outlet negative).
   (d) Voltage – adjusts at constant 15 kV voltages.
   (e) Temperature is 20°C.
2. *Calibration* of the system is performed with the standard solution consisting of 27 amino acids and three peptides.

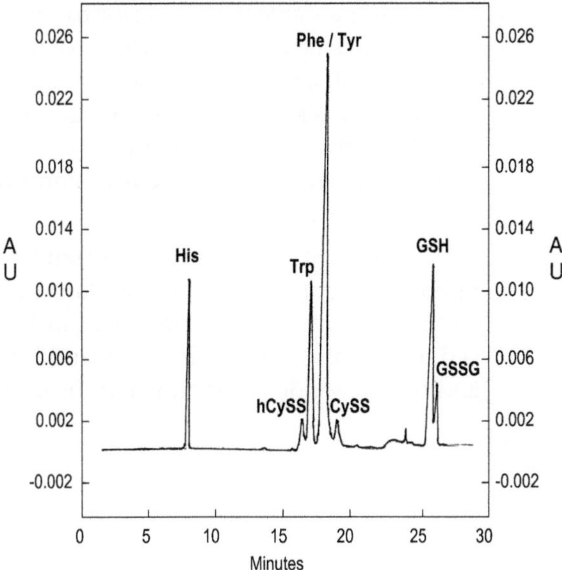

Fig. 1. Electropheregram of standard amino acid and peptide mixture. Carrier electrolyte is 10 mM disodium monophosphate, titrated to pH 2.90. The capillary has a total length 47 cm, effective length 40.5 cm, and I.D. 75 μm. Injection is performed by overpressure for 10 s. Voltage adjusted at 15 kV, temperature on 18°C, detection is direct UV at 200 nm. Concentration of His, HcySS, Trp, Phe, Tyr, CySS, and GSH are at 70 μM each while GSSG concentration is 7 μM. Figure reprinted from Žunić & Spasić (29) with permission.

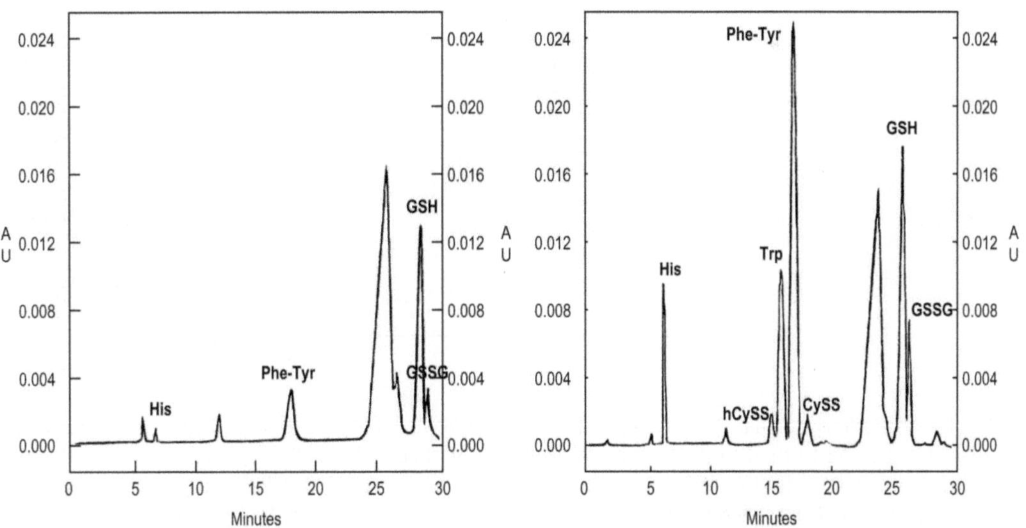

Fig. 2. Electropherogram of human capillary blood (diluted 12 times) without (*left*) and spiked (*right*) with standard mixture of examined compounds. Conditions used are as presented in Fig. 1. Figure reprinted from Žunić & Spasić (29) with permission.

3. *Plasma sample preparation*: Prior to the analysis plasma sample has to be deproteinized with absolute ethanol. Mix plasma (100 μl) with absolute ethanol (100 μl), centrifuge it for 1 min (Minifuge, Beckman) and fivefold dilute obtained supernatants with low conductivity deionized water. For recovery testing, dilute supernatant with standard solutions instead of water (see Notes 12–15).
4. *Electropherograms* obtained with standard mixture and plasma sample without and with the addition of standard amino acid and peptide mixture are presented in Figs. 3 and 4. This procedure permits separation and quantification of 27 amino acid and three peptide compounds in the single run.

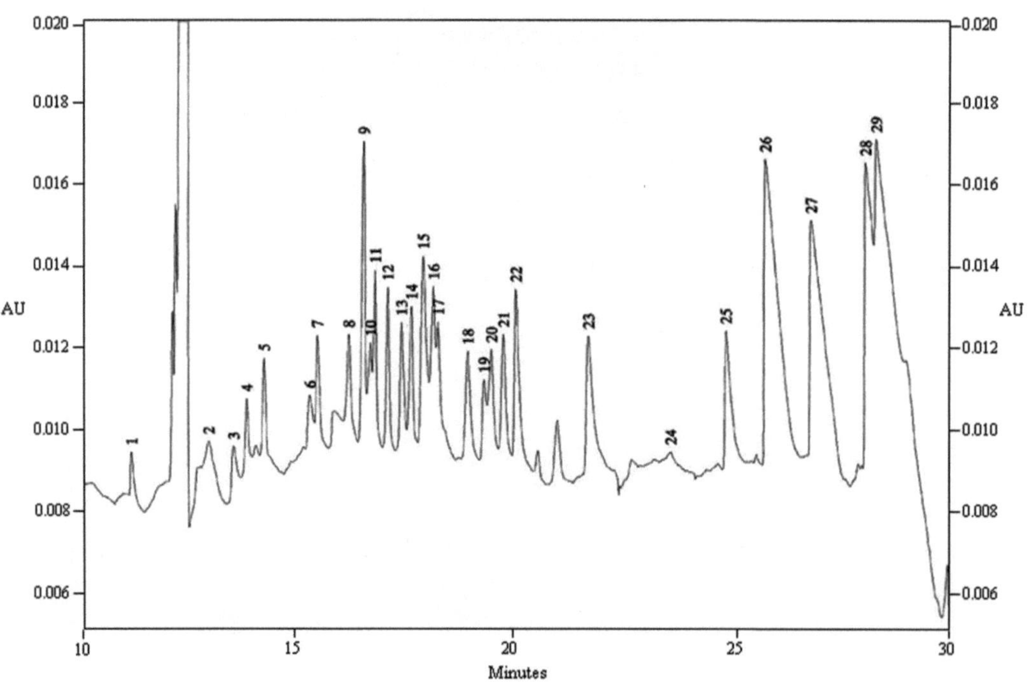

Fig. 3. A typical electropherogram of a separation of 30 amino acids and a peptide mixture. Carrier electrolyte is 8 mM PAS + 2 mM Na_2CO_3, pH 10.15. The capillary has total length 87 cm, effective length 80.5 cm, and I.D. 75 μm. Injection is performed by overpressure for 10 s. Voltage adjusted at 15 kV, temperature at 20°C, detection on indirect absorbance at 254 nm. AU (ordinate) = indirect absorbance at 254 nm. Migration peaks order: Arg (1), Lys (2), Orn (3), Pro (4), GABA (5), Hyl (6), β-Ala (7), carnosine (8), Leu + Ile (9), Trp (10), Cit (11), Val (12), Phe (13), Ala (14), His (15), Met (16), Gln (17), Thr (18), Asn (19), Gly (20), Ser (21), Tyr (22), HcySS (23), Hcy (24), Cys (25), GSSG (26), Glu (27), GSH (28), and Asp (29). Figure reprinted from Žunić et al. (35) with permission.

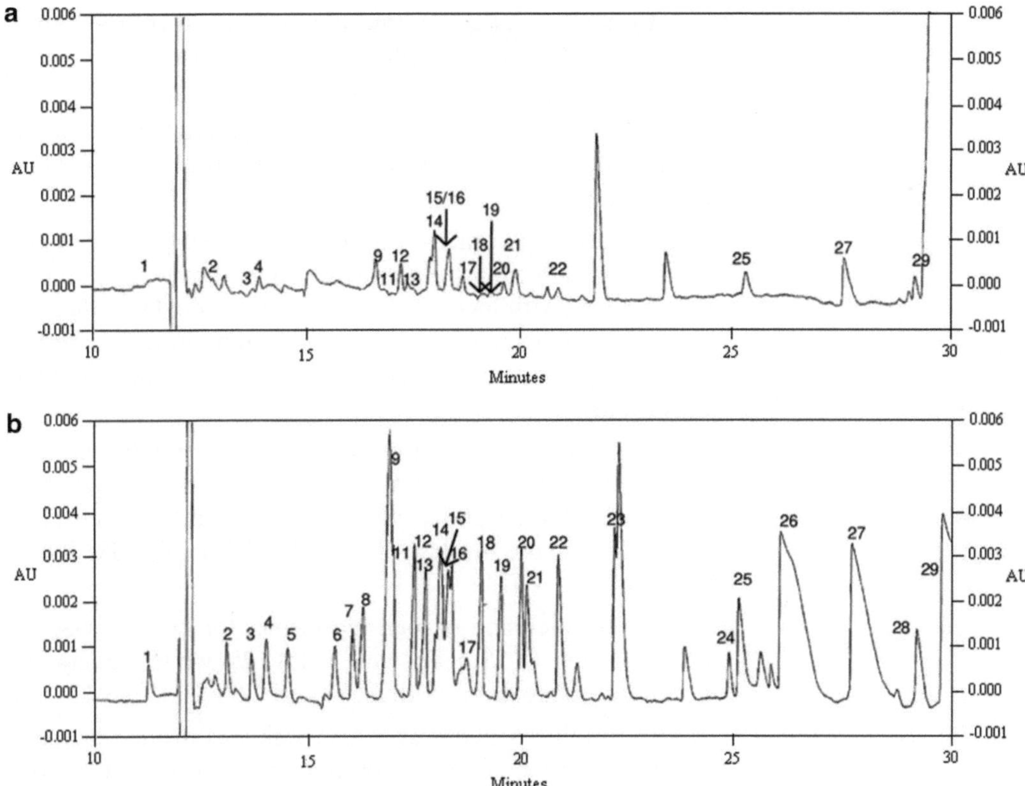

Fig. 4. Electropherograms of fivefold diluted deproteinized human plasma samples without (**a**) and spiked with (**b**) 100 μM standard amino acid and peptide mixture. Peak identification and capillary electrophoretic parameters are as in Fig. 3. Figure reprinted from Žunić et al. (35) with permission.

4. Notes

1. The method with direct detection needs no derivatization procedures, permitting analyses of native compounds in the blood samples.
2. This method permits measurement of GSH, GSSG as well as aromatic (i.e., His, Trp, and Phe + Tyr) and sulfur containing amino acids in blood hydrolysates of healthy subjects. Phe and Tyr co-migrated and can be measured as one peak, i.e., as Phe + Tyr (Fig. 1).
3. The pH of the carrier electrolyte has the largest impact on the resolution of carnosine and His within the examined experimental design. In fact, pH mainly effects His migration. Namely, faster migration of His compared to carnosine is at higher pH. His migrate before carnosine in running electrolyte with pH > 3.3 while at pH < 2.3 it migrates after carnosine. However, blood and plasma samples usually contain no carnosine. Thus,

considering resolutions of the other examined compounds we usually use running electrolyte with pH 2.85 ± 0.05.

4. The results obtained with factorial design suggest that Trp, Phe, Tyr, HcySS, and CySS are better separated in the running electrolyte with pH 2.04. However, at that pH GSH and GSSG co-migrate.
5. The greatest impact on the separation between GSH and GSSG has pH of the running electrolyte. Thus, their separations can be improved with increasing pH of the buffer. The other variables (i.e., temperature and concentration) have little effects on the separations between these peptides under examined conditions.
6. Sensitivity of the method evaluated by determining limits of detection (LOD) and limits of quantification (LOQ) is the highest for HcySS and CySS ranging around or even above 200 μM while the smallest LOD and LOQ are for GSSG ranging from 1 to 4 μM.
7. Within run precision of migration times for His, hCySS, Trp, Phe/Tyr, and CySS peaks were below 2% while higher variation were obtained for both GSH and GSSG.
8. Variations of corrected peak areas (peak area divided by migration time) for the examined compounds are usually below 10%.
9. Separation of examined compounds is tested by spiking physiological fluid with the standard mixture. It indicates that presented method permits the measurement of GSH, GSSG, as well as AAAs at physiological levels while HcySS and CySS as well as Hcy and Cys can be detected only at higher concentration levels.
10. The method with indirect absorbance detection under presented condition permits separation of 30 compounds (Fig. 3). Both separations and migration times of the examined compounds are significantly influenced by the running electrolyte composition, particularly by pH.
11. If the pH is higher than recommended, it is associated with increased migration times of all compounds, leading to co-migration of GSSG, GSH, Glu, and Asp, associated with disturbed separation between Leu/Ile, Trp and Cit, while separation of basic amino acids (Lys, Orn, and Pro) is improved.
12. When using only *p*-amino salicylic acid (titrated with NaOH to the desired pH) as the background electrolyte, there is fast pH decrease affecting both separation and migration times of examined compounds. Supplementation of the background electrolyte with sodium carbonate improved reproducibility.

13. The increase of voltage decreased migration times of all compounds, reducing their separations, while decrease of voltage to 10 kV almost doubled the time necessary for their separations.
14. As other ionizable compounds, usually present in physiological fluids, particularly proteins could influence capillary electrophoresis of amino acids, physiological fluids have to be deproteinized. We found absolute ethanol to be the most suitable reagent among those tested (acetone and sulfosalicylic acid).
15. Dilutions of the samples with water seem to be necessary for good separation of the examined compounds, probably due to high electrolyte contents in the physiological fluids. We used three- to fivefold dilution of deproteinized plasma, and even greater if necessary.

Acknowledgement

This work was supported by a grant of MMA (Project No. 06-10/B.5.)

References

1. Žunić G et al (2009) Increased nitric oxide formation followed by increased arginase activity induces relative lack of arginine at the wound site and alters whole nutritional status in rats almost within the early healing period. Nitric Oxide 20: 253–258
2. Žunić G et al (2000) Pulmonary blast injury increases nitric oxide production, disturbs arginine metabolism, and alters the plasma free amino acid pool in rabbits during the early posttraumatic period. Nitric Oxide 4: 123–128
3. Žunić G et al (1996) Early plasma amino acid pool alterations in patients with military gunshot/missile wound. J Trauma 40: S152–S156
4. Žunić G (1997) Free amino acids in the evaluation of protein metabolism. Vojnosanit Pregl 54: 615–620 (In Serbian)
5. Pinna A et al (2010) Plasma thiols and taurine levels in central retinal vein occlusion. Curr Eye Res 35: 644–650
6. Guzman-Ramos K et al (2010) Off-line concomitant release of dopamine and glutamate involvement in taste memory consolidation. J Neurochem 144(1): 226–236
7. Cherla G, Jaimes EA (2004) Role of L-arginine in the Pathogenesis and treatment of renal disease. J Nutr 134: 2801S–2908S
8. Sevaljević Lj, Zunic G (1996) Acute-phase protein synthesis in rats is influenced by alterations in plasma and muscle free amino acid pools related to lower plasma volume following trauma. J Nutr 126: 3136–3142
9. Cernak I, Savić J, Zunic G et al (1999). Recognizing, scoring and predicting blast injuries. World J Surg 23: 44–53
10. Ikai I et al (1989) Significance of hepatic mitochondrial redox potential on the concentration of plasma amino acids following hemorrhagic shock in rats. Circ Shock 27: 63–72
11. Zinellu A et al (2010) Quantification of neurotransmitter amino acids by capillary electrophoresis laser-induced fluorescence detection in biological fluids. Anal Bioanal Chem 398: 1973–1978
12. Oberholzer VG, Briddon A (1990) A novel use of amino acid ratios as an indicator of nutritional status. In: Lubec G, Rosenthal, GA (eds) Amino Acids, Chemistry, Biology and Medicine, Escom Science Publishers, Leiden
13. Rafii M et al (2008) In vivo regulation of phenylalanine hydroxylation to tyrosine, studied using enrichment in apoB-100. Am J Physiol Endocrinol Metab 294: E475–E479

14. Wannemacher RW et al (1976) The significance and mechanism of an increased serum phenylalanine-tyrosine ratio during infection. Am J Clin Nutr 29: 997–1006
15. Kawaguchi T et al (2007) Branched-chain amino acids improve insulin resistance in patients with hepatitis C virus-related liver disease: report of two cases. Liver Int 27: 1287–1292
16. Bolander-Gouaille C (2001) Focus on homocysteine and the vitamins involved in its metabolism. Springer, Verlag, France
17. Ruderman NB, Berger M (1974) The formation of glutamine and alanine in skeletal muscle. J Biol Chem 249:5500–5506
18. Žunić G, Stanimirović S, Savić J (1984) Rapid ion-exchange method for the determination of 3-methylhistidine in rat urine and skeletal muscle. J Chromatog 311:69–77
19. Peranzoni E et al (2007) Role of arginine in the metabolism, immunity and immunopathology. Immunobiology 212:795–812
20. Wischemeyer PE (2008) Glutamine: role in critical illness and ongoing clinical trials. Curr Opin Gastroenterol 24:190–197
21. Shi HP et al (2007) Supplemental L-arginine enhances wound healing following trauma/hemorrhagic shock. Wound Repair Regen 15: 66–70
22. Zinellu A et al (2005) Quantification of thiol-containing amino acids linked by disulfides to LDL. Clin Chem 51:658–660
23. Ziegler F et al (1992) Plasma amino acid determinations by reversed-phase HPLC: improvement of the orthophthalaldehyde method and comparison with ion-exchange chromatography. J Automat Chem 14:145–149
24. Moore S, Spackman DH, Stein WH (1958) Chromatography of amino acids on sulfonated polystyrene resins. Anal Chem 30:1185–1190
25. Cynober L et al (1985) High performance ion-exchange chromatography of amino acids in biological fluids using Chromakon 500—performance of the apparatus. J Automat Chem 7:201–203
26. Kuhr WG, Yeung ES (1988) Indirect fluorescence detection of native amino acids in capillary zone electrophoresis. Anal Chem 60:1832–1834
27. LiuJetal(1991)Designof3-(4-carboxybenzoyl)-2-quinolinecarboxaldehyde as a reagent for ultrasensitive determination of primary amines by capillary electrophoresis using laser fluorescence detection. Anal Chem 63:408–412
28. Altria KD (1996) Additional application areas of capillary electrophoresis. In: Altria KD (ed) Capillary Electrophoresis Guide Book – Principles, Operations and Applications, Humana Press, Totowa, New Jersey
29. Žunić G, Spasić S (2008) Capillary electrophoresis method optimized with a factorial design for the determination of glutathione and amino acid status using human capillary blood. J Chromatogr B 873:70–76
30. Poinsot V et al (2010) Recent advances in amino acid analysis by CE. Electrophoresis 31:105–121
31. Zhai C et al (2010) Ultraviolet detection of amino acids based on their on-column conjugation with cupric cation using a disposable electrophoresis microdevice. Talanta 82:67–71
32. Lee Y-H, Lin T-I (1994) Capillary electrophoretic determination of amino acids with indirect absorbance detection. J Chromatogr A 680:287–297
33. Martin-Girardeu A, Renou-Gonnord MF (2000) Optimization of a capillary electrophoresis – electrospray mass spectrometry for the quantification of the 20 natural amino acids in children blood. J Chromatogr B 742:163–171
34. Deming SN, Morgan SL (1993) Experimental design: a chemometric approach. Elsevier, Amsterdam – London – New York – Tokyo, Netherlands
35. Žunić G et al (2002) Optimization of a free separation of 30 free amino acids and peptides by capillary zone electrophoresis with indirect absorbance detection: a potential for quantification in physiological fluids. J Chromatogr B 772:19–33

Chapter 20

Measurement of 3-Nitro-Tyrosine in Human Plasma and Urine by Gas Chromatography-Tandem Mass Spectrometry

Dimitrios Tsikas, Anja Mitschke, and Frank-Mathias Gutzki

Abstract

Reaction of reactive nitrogen species (RNS), such as peroxynitrite and nitryl chloride, with soluble tyrosine and tyrosine residues in proteins produces soluble 3-nitro-tyrosine and 3-nitro-tyrosino-proteins, respectively. Regular proteolysis of 3-nitro-tyrosino-proteins yields soluble 3-nitro-tyrosine. 3-Nitro-tyrosine circulates in plasma and is excreted in the urine. Both circulating and excretory 3-nitro-tyrosine are considered suitable biomarkers of nitrative stress. Tandem mass spectrometry coupled with gas chromatography (GC–MS/MS) or liquid chromatography (LC–MS/MS) is one of the most reliable analytical techniques to determine 3-nitro-tyrosine. Here, we describe protocols for the quantitative determination of soluble 3-nitro-tyrosine in human plasma and urine by GC–MS/MS.

Key words: Derivatization, Electron-capture negative-ion chemical ionization, Gas chromatography–tandem mass spectrometry, Plasma, Stable isotopes, Urine, Amino Acid Analysis, 3-Nitro-tyrosine quantitation

1. Introduction

Nitrating agents, such as the peroxynitrite system, i.e. $ONOO^-$/ONOOH, and other reactive nitrogen species (RNS), including nitryl chloride (NO_2Cl), preferentially react with the aromatic ring of free (soluble amino acid) and protein-incorporated tyrosine to form soluble 3-nitro-tyrosine (NO_2Tyr) and protein-incorporated 3-nitro-tyrosine (NO_2TyrProt), respectively (Fig. 1). NO_2Tyr and NO_2TyrProt are chemically fairly stable and are thought to be footprints left by RNS. However, nitration of tyrosine can also occur artifactually at low pH values, e.g. by acidification of biological samples (Fig. 1). Origin and metabolism of NO_2Tyr and NO_2TyrProt are poorly investigated and little understood. It is actually unknown

Michail A. Alterman and Peter Hunziker (eds.), *Amino Acid Analysis: Methods and Protocols*, Methods in Molecular Biology, vol. 828, DOI 10.1007/978-1-61779-445-2_20,

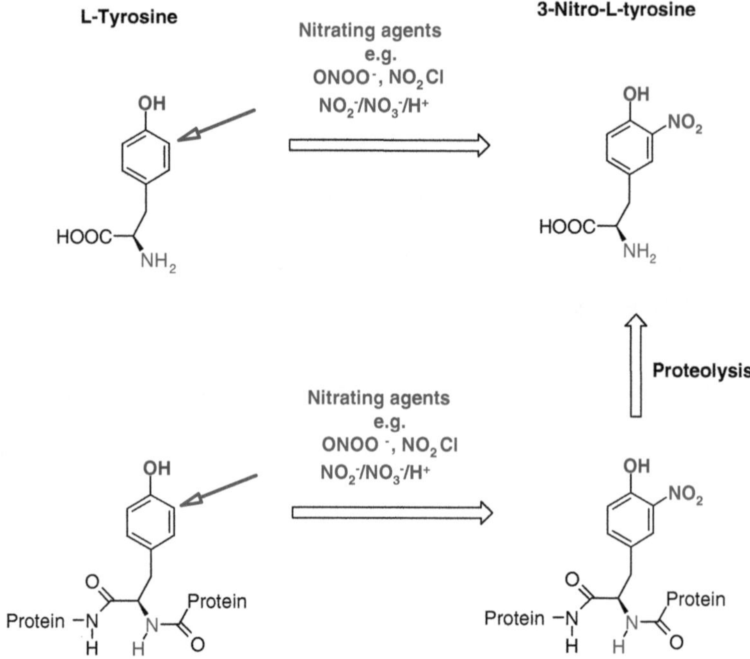

Fig. 1. Nitrating agents, such as peroxynitrite ($ONOO^-$) and nitryl chloride (NO_2Cl), attach the aromatic ring of soluble L-tyrosine and L-tyrosine residues in proteins to form soluble 3-nitro-L-tyrosine and protein-incorporated 3-nitro-L-tyrosine. The carbon atom attached is indicated by *arrow*. Protein-incorporated 3-nitro-L-tyrosine is hydrolyzed to soluble 3-nitro-L-tyrosine by regular proteolysis. Note that under acidic conditions, given the ubiquity of nitrite and nitrate, soluble 3-nitro-L-tyrosine and protein-incorporated 3-nitro-L-tyrosine can easily and abundantly be formed.

whether and to which extent NO_2Tyr derives from proteolyzed NO_2TyrProt or from nitrated soluble tyrosine.

Wide spectrum of physicochemical and immunological methods is currently available for NO_2Tyr and NO_2TyrProt in plasma, urine, and tissue (1–3). The 3-nitro-tyrosine concentration determined by these methods in plasma and urine varies over three orders of magnitude. In consideration of the unique specificity of the MS/MS methodology and of the sensitivity of modern GC–MS/MS and LC–MS/MS instruments, an NO_2Tyr concentration of the order of 1 nM appears a useful guide number for soluble 3-nitro-tyrosine (1).

NO_2Tyr and NO_2TyrProt are also present in urine (2). Healthy subjects excrete unchanged NO_2Tyr into the urine at mean excretion rates of about 0.5–1.4 nmol/mmol creatinine (4, 5).

Source and nature of the analytical problems, shortcomings and pitfalls associated with 3-nitro-tyrosine analysis include: (1) abundant artifactual formation of 3-nitro-tyrosine during sample treatment; (2) lack of sensitivity; and (3) lack of specificity. Given the pM-to-nM concentration of 3-nitro-tyrosine in biological samples, highest detection sensitivity is an indispensable prerequisite for accurate quantification of 3-nitro-tyrosine in plasma and urine. Highest sensitivity and specificity is provided by methods based on GC–MS/MS and LC–MS/MS. In consideration of the analytical challenge of 3-nitro-tyrosine quantification in biological samples, additional steps that ensure reliable 3-nitro-tyrosine results are necessary. Inclusion of a high-performance liquid chromatography (HPLC) step in GC–MS/MS minimizes artifactual formation of 3-nitro-tyrosine by separating 3-nitro-tyrosine from tyrosine, nitrite and nitrate, and enhances the method′s sensitivity (1).

2. Materials

Commercially available chemicals should be of the highest available chemical and isotopic purity. Where possible keep apparatus, biological samples and 3-nitro-tyrosine-containing samples cool (e.g. in an ice bath or at 4°C). Below, issues concerning plasma are indicated by a "P", those concerning urine by a "U" (see Note 1).

2.1. Chemicals

1. Use home-made doubly distilled water and commercially available distilled water of HPLC gradient grade.
2. All materials, chemicals and organic solvents, and water should not be contaminated with nitrite, nitrate, and tyrosine. Keep all samples closed in order to avoid contamination by nitrogen gases (e.g. NO_2) present in the laboratory air (6).

3P. For plasma, use ultrafiltration cartridges with a cut-off of 10 kDa (see Note 2), e.g. Vivaspin Hydrosart cartridges from Sartorius (Göttingen, Germany).

4. *Mobile phase*: 45 mM ammonium sulphate in 5 vol% methanol. Dissolve 11.3 g $(NH_4)_2SO_4$ (e.g. Merck, Darmstadt, Germany) in 1,900 ml distilled water and add methanol HPLC gradient grade (e.g. Baker, Deventer, The Netherlands) up to 2,000 ml. Do not adjust the pH value of the mobile phase which lies between 5.5 and 6.0.
5. *HPLC column*: Use a reverse phase packing material with a particle size of 5 μm, for instance Nucleodur 5 μm C_{18} Gravity or Nucleosil 100-5C_{18}AB (Macherey-Nagel, Düren, Germany), column length 250 mm and internal diameter 4 mm.
6. *Derivatization reagent #1*: 3 M HCl in *n*-propanol (see Note 3). Dilute concentrated hydrochloric acid in *n*-propanol (e.g. Merck) to reach a final HCl concentration of 3 M.
7. *Derivatization reagent #2*: 25 vol% pentafluoropropionic anhydride (PFPA) in ethyl acetate (see Notes 4 and 5). Dilute pure PFPA (e.g. Pierce, Rockford, IL, USA) in ethyl acetate (e.g. Merck).

8. *Derivatization reagent #3*: *N,O*-bis(trimethylsilyl)trifluoro acetamide (BSTFA). Use BSTFA (e.g. Pierce) pure, i.e. do not use solutions of BSTFA in organic solvents, such as heptane or pyridine.
9. *Borate buffer, 400 mM, pH 8.5*: Dissolve 6.183 g boric acid (e.g. Merck) in about 150 ml distilled water, adjust the pH to 8.5 by means of 2 M NaOH (e.g. Merck), and add distilled water up to 250 ml.
10. *Stable-isotope labelled 3-nitro-tyrosine*: If not commercially available, prepare stable isotope-labelled 3-nitro-tyrosine standard starting from commercially available stable isotope-labelled reference compounds of L-tyrosine, e.g. L-($^{2}H_{4}$)tyrosine or L-($^{15}N_{2}$,$^{13}C_{9}$)tyrosine (Isotec, Miamisburg, OH, USA), and from unlabelled or ^{15}N-labelled nitrate (e.g. Sigma-Aldrich, Steinheim, Germany) using concentrated sulphuric acid as a catalyst (7) (see Note 6).
11. *Stock solutions and dilutions of unlabelled and stable isotope-labelled 3-nitro-tyrosine and p-nitro-phenylalanine*: Make stock solutions (e.g. 1 mM) in distilled water and add 10 mM NaOH until complete dilution, make serial dilutions from stock solutions in distilled water, and store all samples frozen at −80°C. *p*-Nitro-L-phenylalanine (e.g. Sigma, Deisenhofen, Germany) is diluted to a final concentration of 100 nM in the mobile phase and used to determine the retention time of 3-nitro-tyrosine prior to analyze plasma and urine samples.

2.2. Sampling

1P. Draw blood (e.g. 3 ml) from the antecubital vein using monovettes containing EDTA (e.g. Sarstedt, Germany), shake gently, and put the sample into an ice bath. Centrifuge the blood sample immediately (5 min, 800×*g*, 4°C), transfer a 1-ml aliquot of the plasma into a safely pre-labelled 4-ml polypropylene tube, and store the sample at −80°C until further analysis.

2U. Collect spot urine (e.g. 10–50 ml), put the sample into an ice bath, transfer a 1-ml aliquot into a safely pre-labelled 4-ml polypropylene tube, and store the sample at −20°C (or −80°C) until further analysis (see Note 6).

2.3. HPLC Conditions

1. Keep the HPLC column well-thermostated (e.g. at 25°C) in order to hold the retention time of 3-nitro-tyrosine constant (see Note 7).
2. Perform isocratic analyses with a flow rate of 1 ml/min and inject plasma ultrafiltrate or diluted urine not less than 30 min apart (see Note 8).
3. If available use an autosampler and a fraction collector. Sample cooling in autosampler and fraction collector is not absolutely necessary. Close collected fractions as soon as possible.

4. Inject 200-μl aliquots of plasma ultrafiltrate or 200-μl aliquots of a urine sample previously diluted (1:1, *v/v*) with the mobile phase.
5. Collect an about 2-ml HPLC fraction in the retention time window of ±1 min of the retention time of synthetic 3-nitro-tyrosine (isocratic elution, 1 ml/min flow rate), or as appropriate for other HPLC conditions.

2.4. GC–Tandem MS Conditions

1. Use a GC–MS/MS instrument in the MS/MS mode, e.g. on triple-stage quadrupole mass spectrometer model TSQ 7000 (Finnigan MAT, San Jose, CA, USA). Keep constant temperatures of 280°C at the interface and 180°C at the ion-source. Other instruments may require different temperatures. Inject 1-μl aliquots of the final sample in the PTV-splitless mode. Hold the injection temperature at 70°C for 3 s and increase to 280°C at a rate of 10°/s for evaporation and transfer.
2. Use toluene for cleaning the glass syringe. Avoid contamination. Change toluene frequently for glass syringe cleaning.
3. Perform GC analyses on fused-silica capillary columns, e.g. Optima 5-MS, length 30 m, internal diameter 0.25 mm, film thickness 0.25 μm, or Optima 17-MS of the same dimensions from Macherey-Nagel.
4. Use helium (99.999%) as the carrier gas at a constant flow rate of 1 ml/min and an oven temperature programme, e.g. keep the initial column temperature at 90°C, then increase to 345°C at a rate of 25°C/min, and keep at this temperature for 1 min.
5. Perform electron-capture negative-ion chemical ionization (ECNICI) using methane (99.9995%) as the reagent (buffer) gas at a pressure of 530 Pa (or a flow rate of 1 ml/min) (see Note 9). Set electron energy (EE) and emission current (EC) to 200 eV and 300 μA, respectively. Other GC–MS/MS instruments may require differing EE and EC values.
6. For collision-induced dissociation (CID) use argon (99.999%) as the collision gas. Optimum values for collision energy and collision gas pressure are 6 eV and 0.27 Pa for the TSQ 7000 instrument, but other GC–MS/MS instruments may require other values.
7. Perform quantification in the selected-reaction monitoring (SRM) mode. Set the first quadrupole (Q1) alternately to *m/z* 396 for endogenous NO_2Tyr and *m/z* 399 for the internal standard 3-nitro-($^2H_3$)-tyrosine. Set the third quadrupole (Q3) alternately to *m/z* 379 for endogenous NO_2Tyr and *m/z* 382 for the internal standard 3-nitro-($^2H_3$)-tyrosine (see Notes 10 and 11). Set the dwell-time to 400 ms for each mass transition.

Set the detector, e.g. electron multiplier, voltage between 2,000 and 2,800 V.

8. *Calculation of concentrations*: Integrate manually our automatically the area of the peaks for endogenous 3-nitro-tyrosine and the internal standard, and calculate the peak area ratio (PAR) by diving the peak areas. The concentration of endogenous 3-nitro-tyrosin C (in nM-units) is calculated by multiplying the PAR (dimensionless) as estimated above by the known concentration of the internal standard C_{IS} (in nM-units) add to the sample:

$$C = \text{PAR} \times C_{\text{IS}}. \tag{1}$$

2.5. Small Equipment

1. Thermostat (e.g. model Bioblock Scientifics; sample tray capacity for 60 samples; Thermolyne Corp., Iowa, USA).
2. Nitrogen evaporator (e.g. TurboVap LV Evaporator; sample tray capacity for 50 vials; Zymark, Idstein/Taunus, Germany).
3. Glass ware (e.g. Macherey-Nagel). Crimp vials N11 flat (1.5 ml, 11.6 mm × 32 mm); Crimp vials N11 conical (1.1 ml, 11.6 mm × 32 mm); Crimp caps N11 with centre hole.
4. Autosampler/injector syringe (e.g. 5-µl SGE syringe, SK-5F-HP-0.63; SGE International Pty. Ltd., Australia).
5. Vortex mixer (e.g. Model Reax 2000; Heidolph, Germany).

3. Methods

If feasible, carry out 3-nitro-tyrosine analyses in a separate laboratory, where all steps of the method can be done. Arrange a set of pipettes and glass syringes for exclusive use in 3-nitro-tyrosine analyses. Avoid any contamination of working places and materials with nitrite, nitrate, and tyrosine. Do not acidify samples.

Derivatization should be performed preferably in safely and tightly closed glass vials by using metal-block thermostats in a well-functioning fume hood.

A schematic of the main analytical procedures and derivatization reactions of this protocol are shown in Figs. 2 and 3, respectively. A typical GC–MS/MS chromatogram from the quantitative determination of 3-nitro-tyrosine in plasma of a healthy subject by the current protocol is shown in Fig. 4.

1P. Add to an unthawed 600-µl plasma aliquot 6 µl of a 100 nM solution of the internal standard in distilled water (see Note 12P). The final concentration of the internal standard in the plasma sample is 1 nM ($C_{IS} = 1$ nM). Let the samples equilibrate for 10 min in a refrigerator.

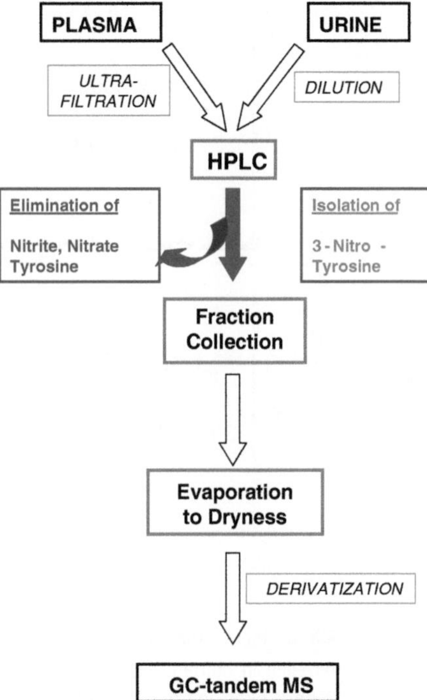

Fig. 2 Schematic of the main procedures of the GC–MS/MS (i.e. GC-tandem MS) method for 3-nitro-tyrosine quantification in human plasma and human urine. *HPLC* high-performance liquid chromatography. "Fraction Collection" means that the effluent of the HPLC fraction with the retention time of synthetic 3-nitro-tyrosine is collected.

Fig. 3. Schematic and chronological order of the three derivatization procedures for soluble 3-nitro-tyrosine and GC–MS/MS analysis. (1) *Esterification*: esterification is performed in 3 M HCl in *n*-propanol and yields the *n*-propyl ester. (2) *N-acylation*: the propyl ester of 3-nitro-L-tyrosine is further derivatized with pentafluoropropionic anhydride (PFPA) to the pentafluoropropionyl derivative. (3) *Etherification*: the last derivatization reaction is the trimethyl silylation of the hydroxyl group of 3-nitro-tyrosine with *N,O*-bis(trimethylsilyl) trifluoroacetamide (BSTFA). The nitro group of 3-nitro-tyrosine remains unaffected during all derivatization reactions.

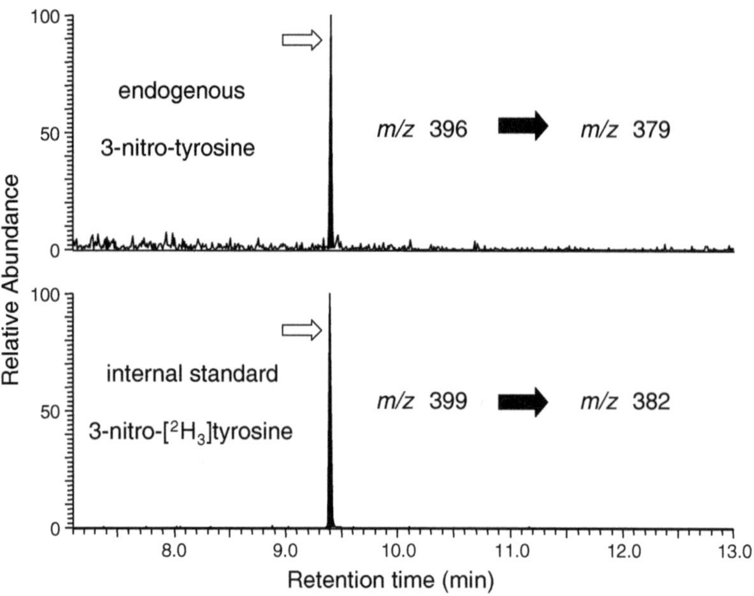

Fig. 4. Representative chromatogram from the GC–MS/MS analysis of soluble 3-nitro-L-tyrosine in human plasma. *Upper* and *lower panel* indicate the traces for endogenous 3-nitro-tyrosine (mass transition *m/z* 396→379) and the internal standard 3-nitro-($^2H_3$)-L-tyrosine (mass transition *m/z* 399→382), respectively. Endogenous 3-nitro-tyrosine in this plasma sample was measured to be 0.51 nM. The internal standard 3-nitro-($^2H_3$)-L-tyrosine was externally added to plasma at a final concentration of 5 nM. The propyl-PFP-TMS derivatives of 3-nitro-tyrosine and 3-nitro-($^2H_3$)-L-tyrosine elute at about 9.4 min (indicated by *unfilled arrows*).

In case of using frozen plasma samples, centrifuge the thawed plasma samples (800 × *g*, 5 min, 4°C), take a 600-µl aliquot and proceed as described above for unthawed plasma samples.

1U. Add to 1,000-µl urine sample a 10-µl aliquot of a 500 nM solution of the internal standard in distilled water (see Note 12U). The final concentration of the internal standard in the urine sample is 5 nM (C_{IS} = 5 nM).

In case of using frozen urine samples, centrifuge the thawed urine samples (800 × *g*, 5 min, 4°C), take a 1,000-µl aliquot and proceed as described above for unthawed urine samples.

2P. Prepare ultrafiltrate by centrifugation using for instance Vivaspin ultrafiltration cartridges (30 min, 8,000 × *g*, 4°C) (see Note 13P). Inject a 200-µl aliquot of the ultrafiltrate as soon as possible, or store frozen the whole ultrafiltrate sample at −80°C.

2U. Dilute a 500-µl aliquot of the spiked urine sample with a 500-µl aliquot of the mobile phase (45 mM ammonium sulphate/methanol, 95:5, *v*/*v*) and put the sample for 10 min in the refrigerator. Analyze the sample immediately or store at −80°C until HPLC analysis.

3. Determine the retention time of *p*-nitro-L-phenylalanine, calculate the retention time of 3-nitro-tyrosine and determine the HPLC fraction to be collected (see Note 7).
4. Inject a 200-µl aliquot of plasma ultrafiltrate or a 200-µl aliquot of a diluted urine sample (see Note 14). Injection of the next sample should occur not earlier than 30 min in case of isocratic elution (see Note 8).
5. Collect the HPLC fraction of 3-nitro-tyrosine in 12-ml polypropylene tubes, close with stoppers and put the samples in an ice bath in case of manual collection. In case of automated sample collection, set the temperature as low as possible.
6. Evaporate the collected HPLC fraction (commonly about 2 ml) to dryness under the stream of a nitrogen gas, preferably at 40°C. This step may take a considerable period of time.
7. Add to the solid residue 1,000 µl of ethanol (see Note 15), mix by vortexing, and centrifuge the sample (e.g. 800 × *g*, 2 min, 4°C).
8. Transfer the major fraction of the clear supernatant carefully into a 1.8-ml autosampler glass vial and evaporate to dryness under a stream of nitrogen at room temperature.
9. Perform esterification by using derivatization reagent #1 (3 M HCl in *n*-propanol). For this, add a 100-µl aliquot of this reagent to the samples, mix by vortexing, and close the vial tightly. Then, incubate the sample for 60 min at 80°C in a metal-block thermostat (see Note 16).
10. Subsequently, take off the sample, allow cooling to room temperature, evaporate to dryness under a nitrogen stream, and perform *N*-pentafluoropropionylation using derivatization reagent #2 (25 vol% PFPA in ethyl acetate). For this, add a 100-µl aliquot of this reagent to the sample, close the glass vial tightly and incubate the sample at 65°C for 30 min in a metal-block thermostat placed in a fume hood (see Note 17).
11. Take off the glass vial from the metal-block thermostat, let cool the sample to room temperature, and evaporate to dryness under a stream of nitrogen in a fume hood (see Note 17).
12. Sample washing/extracting with borate buffer/toluene. Perform this step twice without interruption (see Notes 18 and 19). Add to the residue a 200-µl aliquot of borate buffer immediately followed by addition of a 500-µl aliquot of toluene, by closing the glass vial and mixing by vortexing for not longer than 60 s (see Note 20). Centrifuge the sample immediately (800 × *g*, 5 min, 4°C) and transfer a large portion of the upper toluene phase into a cone glass vial (see Note 21). Repeat the step and transfer the second toluene aliquot into the toluene-containing cone glass vial.

13. Evaporate to dryness by means of a nitrogen stream, add 50 μl of BSTFA (see Note 22), close the glass vial tightly and incubate the sample for 60 min at 60°C in a metal-block thermostat. Alternatively, this derivatization can be performed by incubating the samples overnight at room temperature (see Note 23). The resulting sample is clear and colourless.
14. Analyze the sample by GC–MS/MS immediately or store at 4°C until analysis (see Note 24).
15. Perform GC–MS/MS analysis (see Note 25) in the SRM mode as described above (see Subheading 2.4).
16. Calculate the PAR from the measured peak areas of endogenous 3-nitro-tyrosine and the internal standard (see Note 26).
17. Calculate 3-nitro-tyrosine concentration by means of the Formula 1 (see Note 27).
18. For urine samples, divide the calculated concentration by the creatinine concentration measured in the same urine sample (see Note 28).
19. *Applications*: The 3-nitro-tyrosine concentrations measured in plasma of healthy humans by the GC–tandem MS method outlined above and by a similar GC–MS/MS method reported by Dr. Caidahl and his group (8, 9) revealed very similar values which may serve as a guidance for scientists wishing to use this technology and as reference values which should be achieved when using GC–MS/MS and LC–MS/MS methods for 3-nitro-tyrosine quantification in human plasma and urine samples (1). Dr. Caidahl´s groups eliminated interferences by reducing the nitro group of 3-nitro-tyrosine to an amino group (8) while our group used HPLC to eliminate potential interferences (4, 9, 10).

    The GC–MS/MS methods described above have been used by us for the quantitative determination of 3-nitro-tyrosine in vitro and in vivo studies in animals and humans (11–16). The major findings we observed and main conclusions that can be drawn from these studies are as follows:

    (a) Reliable quantitative determination of 3-nitro-tyrosine in human plasma and urine is highly challenging.
    (b) In plasma of healthy humans 3-nitro-tyrosine concentration at the basal state is of the order of 1 nM.

    Patients with end-stage renal diseases have slightly elevated 3-nitro-tyrosine concentrations in the plasma.

    It is assumed that 3-nitro-tyrosine concentration in other biological fluids, such as exhaled breath condensate will be below plasma levels (17).

    (c) In urine of healthy humans, 3-nitro-tyrosine concentration and excretion rate is of the order of 8 ± 10 nM or 0.5 ± 0.5 nmol/mmol creatinine and varies greatly.

Patients with chronic inflammatory rheumatic diseases excrete about three times more 3-nitro-tyrosine than healthy subjects.

(d) Circulating and excretory 3-nitro-tyrosine concentrations greatly deviating from those mentioned above should be considered incorrect. Analytical methods that do not provide – this applies for the majority of the methods reported so far – and newly developed methods that will not provide 3-nitro-tyrosine values of the order mentioned above should be regarded as invalid.

(e) Changes in circulating and excretory 3-nitro-tyrosine upon disease or pharmacological intervention are rather very small (3).

(f) Organic nitrates, such as isosorbide dinitrate (ISDN) and pentaerythrityl trinitrate (PETN) do not induce nitrosative and oxidative stress in healthy young subjects (12).

(g) Glucose at high concentration (i.e. 30 mM) increases drastically 3-nitro-tyrosine concentration in endothelial progenitor cells (14).

(h) Healthy rats have higher plasma 3-nitro-tyrosine concentration than healthy humans (i.e. 4 nM versus 1 nM).

4. Notes

1. In case of spot urine, 3-nitro-tyrosine concentration measured in the urine must be corrected for creatinine excretion. 3-Nitro-tyrosine excretion is expressed in nanomole 3-nitro-tyrosine per millimole creatinine. Otherwise, collect urine over a certain period of time (e.g. 12 or 24 h) and measure the total urine volume. 3-Nitro-tyrosine excretion is then expressed in nanomole 3-nitro-tyrosine per hour or per day.
2. Generate at least a 300-μl ultrafiltrate volume from a 600-μl plasma volume (see Note 13).
3. Esterification of 3-nitro-tyrosine can be performed by using 3 M HCl solutions in other alcohols, such as methanol and ethanol. Consider that methyl and ethyl esters are by 28 and 14 Da lighter than the propyl ester, respectively (7).
4. PFPA has an unpleasant smell. Keep all PFPA-containing samples tightly closed and work under well-working fume hood.
5. PFPA can be replaced by other perfluorated acid anhydrides, such as heptafluorobutyric anhydride (HFBA) (7, 8). Consider that HFB derivatives are by 50 Da heavier than PFP derivatives.

6. Chemical synthesis of stable isotope-labelled 3-nitro-($^2H_4$) tyrosine standards (7). Use aqueous solutions, e.g. dilutions of stock solutions of nitrate (e.g. $NaNO_3$) and ($^2H_4$)-tyrosine. Add to 30 μmol of ($^2H_4$)-tyrosine 60 μmol of $NaNO_3$, add up to 600 μl distilled water, and put the sample in an ice bath. Add small portions of ice cold concentrated H_2SO_4 under vortexing and put the sample again in the ice bath. Repeat this step until about 300 μl of H_2SO_4 were added in total. After 1 h of incubation in the ice bath, add to the sample a small portion of an ice cold 5 M NaOH and adjust the pH of the solution to about 3. Inject an aliquot of this solution in the HPLC system and collect the whole fraction of the peak eluting with the retention time of synthetic 3-nitrotyrosine. Determine by GC–MS (e.g. SIM of *m/z* 399 for 3-nitro-($^2H_3$)-tyrosine and *m/z* 396 for 3-nitro-($^2H_3$) tyrosine) the concentration of the prepared 3-nitro-($^2H_3$) tyrosine by using solutions of synthetic 3-nitro-tyrosine with known concentration as a calibrator. Store the stock solution, i.e. collected HPLC fraction, of 3-nitro-($^2H_3$)-tyrosine frozen at −80°C. Under these conditions, 3-nitro-($^2H_3$)-tyrosine is stable for several years.
7. Determine the retention times of 3-nitro-tyrosine (e.g. 8.363 min) and *p*-nitro-phenylalanine (e.g. 9.586 min) by injecting 200-μl aliquots of a mixture of these compounds in the mobile phase (each 100 nM) and setting the UV detector at 276 nm. If available use more sensitive detectors, such as electrochemical detectors. Calculate the retention of 3-nitro-tyrorine relative to that of *p*-nitro-phenylalanine (e.g. 0.872). When analyzing biological samples, multiply the currently measured retention time of *p*-nitro-L-phenylalanine by the previously experimentally obtained factor.
8. Proceed as described above, if the column has to be changed, and if gradient elution is possible or desired. In the latter case, use a second mobile phase with a higher methanol fraction (e.g. 30 vol%).
9. Electron ionization (EI) is not useful for quantitative 3-nitro-tyrosine analysis in plasma and urine.
10. Caution! Avoid contamination of the whole HPLC system by 3-nitro-tyrosine and stable isotope-labelled 3-nitro-tyrosine! Do not inject 3-nitro-tyrosine solutions with concentrations higher than 100 nM or amounts higher than 20 pmol of 3-nitro-tyrosine. Do not inject any 3-nitro-tyrosine prior to start with analysis of plasma or urine samples.
11. Tune the GC–MS/MS instrument and acquire full scan and product ion mass spectra in order to determine the precise *m/z* values for precursor and product ions, e.g. *m/z* 396.4, 399.4, 379.4, and 382.4, for use in SRM.

12P. As the 3-nitro-tyrosine concentration in human plasma is of the order of 1 nM, the stable-isotope labelled internal standard should be used at final concentrations of the same order. A higher concentration of the internal standard, e.g. 5 or 10 nM, in the plasma sample is advisable.

12U. As the 3-nitro-tyrosine concentration in human urine may vary in the lower nM-range, the stable-isotope labelled internal standard should be used at final concentrations of the same order. Thus, an internal standard concentration of 5 or 10 nM in the urine sample is advisable.

13P. Read and follow carefully the instructions of the manufacturer regarding physicochemical properties and utilization of ultrafiltration cartridges for plasma ultrafiltration.

14. Thawed plasma ultrafiltrate and urine samples may contain some precipitation. In such cases, centrifuge the samples prior to HPLC analysis.

15. Most of the precipitated solid material is ammonium sulphate which is poorly soluble in ethanol. Short centrifugation yields a powdery white precipitate and a clear supernatant.

16. Derivatization with reagent #1 yields *n*-propyl esters of amino acids. Because of the presence of HCl, acid-catalyzed hydrolysis of small proteins and peptides present in the HPLC fraction of 3-nitro-tyrosine to tyrosine and acid-catalyzed nitration of tyrosine by nitrite and nitrate, which are ubiquitous, would result in artifactual formation of 3-nitro-tyrosine. Therefore, all sources potentially contributing to tyrosine, nitrite, and nitrate and finally to 3-nitro-tyrosine should be avoided. Use glass vials, septa and stoppers with lowest nitrite and nitrate contamination. Ultrafiltration membranes of cut-off higher than 10 kDa would facilitate generation of plasma ultrafiltrate, but they would also increase the risk of artifactual 3-nitro-tyrosine formation.

17. Perform derivatization with PFPA in a well-functioning fume hood. Despite careful closing of glass vials, escape of unpleasant smelling PFPA and its volatile reaction products is not completely avoidable. This also applies to the evaporation of excess PFPA and its volatile reaction products (see below).

19. Derivatization with reagent #2 (PFPA) is not selective and specific but yields at least *N*-pentafluoropropionyl derivatives of primary and of secondary amines, as well as *O*-pentafluoropropionyl derivatives of aliphatic and aromatic hydroxyl groups. This also applies to 3-nitro-tyrosine (8).

20. Treatment of the sample with borate buffer and toluene aims at (1) eliminating acidic compounds which remain in the aqueous phase (borate buffer) and (2) extracting electrically neutral lipophilic compounds into the water-immiscible organic solvent toluene.

21. Unlike *N*-pentafluoropropionyl derivatives, *O*-pentafluoro propionyl derivatives are less resistant towards hydrolysis. A longer extraction time than 60 s with borate buffer/ toluene may lead to considerable hydrolysis of the *N*-pentafluoropropionyl group of 3-nitro-tyrosine and thus finally to lower sensitivity.
22. Use of cone glass vials at this stage saves costs because these vials can be used for the last derivatization step as well as for injection of samples. Cone glass vials allow use of volumes as small as 50 μl which are required for automated injection and highest analytical sensitivity for 3-nitro-tyrosine.
23. Silylating reagents, such as BSTFA undergo rapid hydrolysis in dependence of the humidity prevailing in the laboratory. Use preferably 1,000-μl BSTFA ampoules which suffice for derivatization of 20 samples.
24. Shorter derivatization time with BSTFA is possible for the chemically reactive hydroxyl group of 3-nitro-tyrosine and other acidic hydroxylic groups. Because the sample is likely to contain aliphatic hydroxyl groups which are non-acidic and much less reactive, a BSTFA derivatization time of 60 min at 60°C would ensure complete derivatization of all hydroxyl groups present in the sample. The benefit of this is a better gas chromatography and a shorter GC–MS/MS analysis time.
25. The *n*-propyl-*N*-pentafluoropropionyl-*O*-trimethylsilyl (propyl-PFP-TMS) derivative of 3-nitro-tyrosine is stable for several weeks and months when stored tightly in BSTFA at 4°C. Dispose the samples by letting open the BSTFA-containing glass vials in a well-functioning fume hood until no liquid is visible.
26. (a) Reliable quantification of 3-nitro-tyrosine in human plasma and urine samples by GC–MS in the selected-ion monitoring mode is not possible by this protocol (3, 4, 10, 11). (b) Avoid contamination of the GC–MS/MS system with unlabelled or stable-isotope labelled 3-nitro-tyrosine derivatives. For instance, do not analyze samples that contain more than 10 nM 3-nitro-tyrosine. Inject pure BSTFA to test 3-nitro-tyrosine contamination. (c) Analyze first plasma samples and then urine samples. (d) 3-Nitro-tyrosine analysis may be affected by preceding analysis of other derivatives, such as methyl ester pentafluoropropionyl derivatives of other amino acids, such as arginine. (e) Because of the very low analytically challenging concentrations of 3-nitro-tyrosine both gas chromatography (e.g. septum, glass liner, and column), including autosampler and mass spectrometry (e.g. ion-source, quadrupoles, and tuning) must be optimum.

27. (a) Under similar gas chromatography conditions, the retention time of the propyl-PFP-TMS derivative of 3-nitro-tyrosine is highly reproducible (e.g. 9.4 min; Fig. 4). (b) Note that propyl-PFP-TMS derivatives of deuterium-labelled but not of carbon-13- or nitrogen-15-labelled 3-nitro-tyrosine emerge from the gas chromatography column a few seconds earlier than that of unlabelled 3-nitro-tyrosine. (c) Shorten of the gas chromatography column may be required after the analysis of about 100 biological samples, and it will result in a little bit shorter retention times. Avoid injection of propyl-PFP-TMS derivatives of 3-nitro-tyrosine standards after maintenance of injector and gas chromatography column.
28. Expected 3-nitro-tyrosine concentrations in plasma samples of healthy and ill subjects are of the order of 1 nM (11).
29. Expected 3-nitro-tyrosine concentrations in urine samples of healthy subjects may range widely, e.g. between 1.6 and 33 nM or 0.05–1.07 nmol/mmol creatinine (4, 11).

References

1. Tsikas D, and Caidahl K (2005) Recent methodological advances in the mass spectrometric analysis of free and protein-associated 3-nitrotyrosine in human plasma. J Chromatogr B 814:1–9
2. Ryberg H, and Caidahl K (2007) Chromatographic and mass spectrometric methods for quantitative determination of 3-nitro-tyrosine in biological samples and their application to human samples. J Chromatogr B 851:160–171
3. Tsikas D (2010) Analytical methods for 3-nitro-tyrosine quantification in biological samples: the unique role of tandem mass spectrometry. Amino Acids, DOI 10.1007/s00726-010-0604-5.
4. Tsikas D et al (2005) Determination of 3-nitrotyrosine in human urine at the basal state by gas chromatography-tandem mass spectrometry and evaluation of the excretion after oral intake. J Chromatogr B 827:146–156
5. Kato Y et al (2009) Quantification of modified tyrosines in healthy and diabetic human urine using liquid chromatography/tandem mass spectrometry. J Clin Biochem Nutr 44:67–78
6. Tsikas D et al (2010) Evidence by GC–MS of ex vivo nitrite and nitrate formation from air nitrogen oxides (NOx) in human plasma, serum and urine samples. Anal Biochem 397:126–128
7. Schwedhelm E, Sandmann J, Tsikas D (1998) Synthesis of 3-nitro-L-($^2H_3$)tyrosine for use as an internal standard for quantification of 3-nitro-L-tyrosine by gas chromatography-mass spectrometry. J Labelled Cpds Radiopharm 41:773–780
8. Söderling AS et al (2003) A derivatization assay using gas chromatography/negative chemical ionization tandem mass spectrometry to quantify 3-nitro-tyrosine in human plasma. J Mass Spectrom 38:1187–1196
9. Schwedhelm E et al (1999) Gas chromatographic-tandem mass spectrometric quantification of free 3-nitro-tyrosine in human plasma at the basal state. Anal Biochem 276:195–203
10. Tsikas D et al (2003) Accurate quantification of basal plasma levels of 3-nitro-tyrosine and 3-nitrotyrosinoalbumin by gas chromatography-tandem mass spectrometry. J Chromatogr B 784:77–90
11. Tsikas D (2008) A critical review and discussion of analytical methods in the L-arginine/nitric oxide area of basic and clinical research. Anal Biochem 379: 139–163
12. Keimer R et al (2003) Lack of oxidative stress during sustained therapy with isosorbide dinitrate and pentaerythrityl tetranitrate in healthy humans: a randomized, double-blind crossover study. J Cardiovasc Pharmacol 41:284–292
13. Tsikas D (2006) Quantitative determination of free and protein-associated 3-nitro-tyrosine and S-nitrosothiols in the circulation by mass spectrometry and other methologies: a critical review and discussion from the analytical and review point of view. In: Dalle-Donne I,

Scaloni A, Butterfield DA (eds) Redox Proteomics: From Protein Modifications to Cellular Dysfunction and Diseases, John Wiley & Sons Inc., Hoboken

14. Thum T et al (2007) Endothelial nitric oxide synthase uncoupling impairs endothelial progenitor cell mobilization and function in diabetes. Diabetes 56: 666–674
15. Magné J et al (2009) Rapeseed protein in a high-fat mixed meal alleviates postprandial systemic and vascular oxidative stress and prevents vascular endothelial dysfunction in healthy rats. J Nutr 139:1660–1666
16. Pham VV, Stichtenoth DO, and Tsikas D (2009) Nitrite correlates with 3-nitro-tyrosine but not with the F(2)-isoprostane 15(S)-8-iso-PGF(2alpha) in urine of rheumatic patients. Nitric Oxide 21:210–215
17. Göen T et al (2005) Sensitive and accurate analyses of free 3-nitro-tyrosine in exhaled breath condensate by LC–MS/MS. J Chromatogr B 826:261–266

Chapter 21

Analysis of Hydroxyproline in Collagen Hydrolysates

Tobias Langrock and Ralf Hoffmann

Abstract

Hydroxyproline (Hyp) is an imino acid post-translationally formed by sequence-specific hydroxylases in the repeating collagen Gly–Xaa–Yaa triad present in all collagen types of all species. In both Xaa- and Yaa-positions, Pro is the most common residue, often oxidized to 4-Hyp in the Yaa- and rarely to 3-Hyp in the Xaa-positions. Here, we describe the qualitative and quantitative analysis of 3- and 4-Hyp-isomers by separating the free imino acids either with hydrophilic interaction chromatography (HILIC) or after derivatization with reversed-phase chromatography (RPC). In both cases, the compounds were detected by electrospray ionization mass spectrometry (ESI-MS).

Key words: Amino acid quantitation, Collagen, Hydroxyproline, Hydroxyproline isomers, Mass spectrometry, ESI-MS/MS, Hydrophilic interaction chromatography, Amino acid analysis

1. Introduction

Collagens are present in nearly all organs and tissues in mammals and form a major constituent of skin, bone, tendon, cartilage, blood vessels, and teeth. They constitute a quarter of the total protein content. The most abundant collagen types form fibrils with a 67 nm axial period, but there are also 14 "non-fibrillar" collagens. All form a tightly packed, right-handed triple-helix composed of three left-handed polyproline II (PPII)-like helical chains wound around each other. Gly-residues, with their small side chain, are present without distortion in every third position of the repeating sequence motif Gly–Xaa–Yaa (1–3). Both Xaa- and Yaa-positions contain mostly proline residues and can be hydroxylated in positions 3 or 4, respectively, by hydroxylases (4–8). In all collagen types 4-Hyp dominates over 3-Hyp, typically by a factor of ten or more.

Michail A. Alterman and Peter Hunziker (eds.), *Amino Acid Analysis: Methods and Protocols*, Methods in Molecular Biology, vol. 828, DOI 10.1007/978-1-61779-445-2_21, © Springer Science+Business Media, LLC 2012

Due to the limitations of Edman degradation in general and tandem mass spectrometry for isomeric and isobaric compounds, the analysis of Hyp residues still relies mostly on amino acid analysis (9). Including the stereocentres at the α-C atom and the hydroxylated C-atom, four stereoisomers exist for both 3- and 4-Hyp, i.e. a total of eight isomers. Although only the *trans*-isomers are formed from L-proline in collagen, the other stereoisomers can be formed during protein aging or acid hydrolysis in 6 M hydrochloric acid at elevated temperatures (typically above 100°C) and prolonged hydrolysis times (typically 4–24 h) (10, 11). Here, we describe two analytical strategies relying on high-performance liquid chromatography (HPLC) coupled online to electrospray ionization mass spectrometry (ESI-MS). The first technique separates the hydrolysates directly on a TSK-Gel Amide 80 column using hydrophilic interaction chromatography (HILIC) (12), i.e. a special type of normal phase chromatography applicable to polar compounds (13).

The second approach relies on derivatization of the protein hydrolysates with N^2-(5-fluoro-2,4-dinitrophenyl)-L-valine amide (L-FDVA) (14) followed by reversed-phase (RP-)HPLC (15).

2. Materials

Standard laboratory equipment, such as pipettes (1–1,000 μl), sample tubes (0.5 and 1.5 ml), analytical balance, centrifuge, and vortexer are needed in addition to the special materials described below.

2.1. Apparatus

1. An analytical gradient HPLC-system, ideally equipped with an autosampler, coupled to an UV-detector (to detect FDVA-derivatized amino acids) and an ESI-mass spectrometer.
2. To accomplish the most sensitive and selective detection, a tandem mass spectrometer, such as a triple quadrupole, is highly recommended.
3. A vacuum centrifuge (e.g. SpeedVac) is used to dry protein samples and hydrolysates.
4. For sample incubation, a thermo mixer is needed to hold 0.5 and 1.5 ml sample tubes.

2.2. Consumables

1. For reversed phase chromatography – an Aqua C_{18}-column (internal diameter 2.0 mm, length 150 mm, particle size 3 μm, pore size 12.5 nm; Phenomenx GmbH, Aschaffenburg, Germany).
2. For HILIC chromatography – a TSK gel Amide-80-column (internal diameter 2.0 mm, length 150 mm, particle size 5 μm, pore size 8 nm; TOSOH Bioscience GmbH, Stuttgart, Germany).

3. For gas phase hydrolysis, we used a simple and inexpensive setup with glass inserts for sample tubes (0.25 ml, SUPELCO, Bellafonte, USA) placed into screw cap tubes with rubber seals (Fischer Scientific, Leicestershire UK).

2.3. Chemicals

1. Use only reagents of the highest purity available, i.e. at least HPLC grade eluents and analytical grade reagents.
2. For online LC–MS coupling, LC–MS grade solvents, or even higher purity solvents, are necessary to provide stable spraying conditions with a low background during the whole gradient.
3. Prepare all solutions with ultrapure water (at least 18 MΩ·cm at 25°C) free of any organic contaminants that might bind to the columns or interfere with the UV-detection or mass spectrometry (see Note 1).
4. Formic acid should be the highest quality available, i.e. >99% for UV-detection and analytical grade for mass spectrometry.
5. Use only hydrochloric acid (6 M) specifically sold to conduct amino acid analysis (e.g. from Sigma-Aldrich, Steinheim, Germany).
6. FDVA is commercially available (Sigma-Aldrich) but can also be synthesized with reasonable yields and purities (14).

3. Methods

3.1. Gas Phase Hydrolysis

1. Dissolve the protein to be analyzed in an appropriate solvent in a defined concentration (ideally 1 mg/ml). Some collagens are water soluble; others have to be digested with collagenase.
2. Transfer the protein solution (150 μl, i.e. 150 μg protein) into a glass insert. Place the glass insert into a screw cap sample tube, but do not cap it.
3. Dry the sample in a vacuum centrifuge.
4. Pipet hydrochloric acid (6 M, 300 μl) into the sample tube, around the glass insert.
5. Screw the cap with the rubber seal onto the tube and close tightly. For safety reasons, you should place the sealed tube into a larger container (e.g. centrifuge tube or glass beaker) and put this into the drying oven. Make sure that no pressure can build up in the large container and that the tubes are not turned upside down.
6. Incubate samples at 110°C for 24 h.
7. Take the container out of the drying oven (wear heat protective gloves and protective glasses) and let the samples cool down to room temperature.

8. Take the glass inserts and clean all hydrochloric acid from the outside. To remove acid that might have condensed in the inserts, re-centrifuge them in the vacuum centrifuge (30 min).

3.2. HILIC–ESI-MS/MS (see Note 2)

1. Prepare HILIC eluent A by dissolving ammonium acetate (38.54 mg, 0.5 mmol) in water (100 ml). For eluent B, dissolve ammonium acetate (192.7 mg, 2.5 mmol) in water (400 ml).
2. Titrate both eluents to pH 5.5 with acetic acid.
3. Add acetonitrile to eluents A (900 ml) and B (600 ml).
4. Dissolve the amino acid samples in eluent A. If charged amino acids, such as Asp and Glu, are also to be analyzed, then use eluent B as solvent (see Note 3). For Hyp-analysis in proteins, dissolve the dried hydrolysates in eluent A (75 μl) and inject 10 μl (20 μg of the original protein).
5. Separate *trans*-4-Hyp, *trans*-3-Hyp and *cis*-4-Hyp (see Note 4), by using the following gradient, at a flow rate of 150 μl/min:

HILIC gradient

Time point (min)	Eluent B (%)
0	5
28.5	60
33.5	95
34	5
44	5

6. Depending on the type of your mass spectrometer, chose one of the following detection modes.
   (a) With a single quadrupole MS use the SIR-mode (*m/z* 132) to detect the isobaric Hyp-isomers and Ile/Leu or use a scan mode (*m/z* 70–300) to detect all amino acids.
   (b) With a triple quadrupole MS, you may enhance selectivity and sensitivity with MS/MS experiments. For detection of Hyp and Ile/Leu, use MRM experiments (*m/z* transition 132→86). In order to detect all amino acids selectively, apply a neutral loss scan (neutral loss of 46 u), as all amino acids tend to eliminate formic acid under MS/MS conditions (16). A chromatogram for Hyp and Ile/Leu is shown as an example in Fig. 1.
7. For quantitative analyses, prepare a standard containing all amino acids of interest and inject it several times to record the signal intensities for at least five different amounts ranging from 0.1 to 10 nmol of each amino acid. To further improve the accuracy, we add asparagine as an internal standard to the hydrolysates (see Note 5).

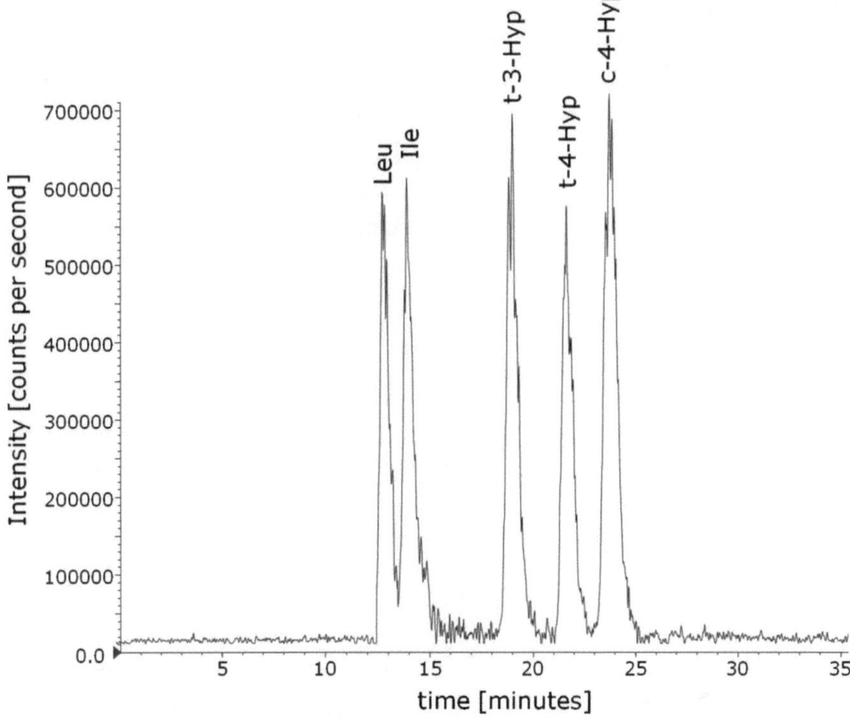

Fig. 1. Extracted ion HILIC–ESI-MS chromatogram at *m/z* 132 (hydroxyproline, isoleucine, and leucine). The sample was a mixture of 19 proteinogenic amino acids, *trans*-4-L-Hyp, *cis*-4-L-Hyp, and *trans*-3-L-Hyp using a neutral loss scan of 46 in positive ion-mode. All isobaric components were separated on the TSK Gel Amide 80 column with a linear gradient of 5–60% eluent B over 28.5 min at a flow rate of 150 μl/min.

3.3. Pre-column Derivatization with N^2-(5-Fluoro-2,4-Dinitrophenyl)-L-Valinamide for RP Chromatography

1. Dissolve FDVA (11 mg, 36.7 μmol) in acetone (1 ml) to obtain a 36.7 mM solution. Dissolve $NaHCO_3$ (840 mg) in water (10 ml) to obtain a 1 M solution. Prepare 1 M hydrochloric acid by diluting the 6 M hydrochloric acid (100 μl) six fold with water (500 μl).
2. Use a threefold molar excess of FDVA over the amino acids, as lower multiples will reduce the accuracy. For real samples, these numbers have to be judged from the individual amino acid amounts (see Note 6) or calculated from the estimated protein quantities (see Note 7).
3. In order to analyze Hyp isomers from collagen, dissolve the dried hydrolysate in water (15 μl). Transfer one third of this solution (5 μl) to a fresh tube and add water (16 μl) and $NaHCO_3$ buffer (8 μl, 1 M).
4. Vortex the solution shortly and add the FDVA solution (32 μl) prepared in step 1.
5. Incubate the samples (90 min, 40°C) and stop the reaction by adding hydrochloric acid (8 μl, 1 M). Add acetonitrile (200 μl) and water (731 μl) to obtain a final volume of 1 ml.

3.4. RP-HPLC

1. Prepare eluents by adding formic acid (1 ml) to water (1 L, eluent A) or acetonitrile (1 L, eluent B).
2. Separate the six Hyp-isomers (see Note 8) with the following gradient at a flow rate of 200 μl/min:

RP-HPLC gradient

Time point (min)	Eluent B (%)
0	20
30	28
45	36
58	46.5
68	65
73	95
83	95
84	20
105	20

3. This gradient was optimized for an Aqua C_{18}-column (column length 150 mm, internal diameter 2.0 mm, particle size 3 μm, and pore size 12.5 nm) originally operated with an aqueous acetonitrile eluent system containing 0.1% trifluoroacetic acid (TFA) as the ion pair reagent (see Note 9).
4. The amino acids can be detected and quantified with a UV-detector or a mass spectrometer, but ideally both systems are coupled online to the HPLC-system.
   (a) Record the absorption at a wavelength of 340 nm to detect all amino acids with a high sensitivity (Fig. 2).

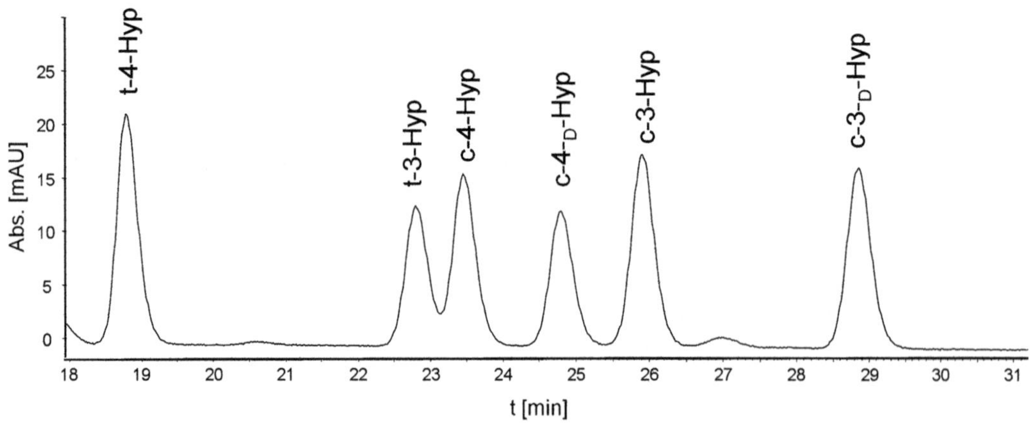

Fig. 2. A mixture of the six relevant hydroxyproline isomers, i.e. *trans*-4-Hyp, *cis*-4-Hyp, *trans*-3-Hyp, *cis*-4-D-Hyp, *cis*-3-D-Hyp, and *cis*-3-L-Hyp, was derivatized with N^2-(5-fluoro-2,4-dinitrophenyl)-L-valine amide and separated by RP-HPLC (C_{18}-Aqua column, absorbance recorded at 340 nm). Eluents were water (A) and acetonitrile (B), both containing 0.1% formic acid. The gradient used linear increments from 20 to 28% eluent B over 30 min, to 36% eluent B over 15 min, to 46.5% eluent B over 13 min, and to 65% eluent B over 10 min.

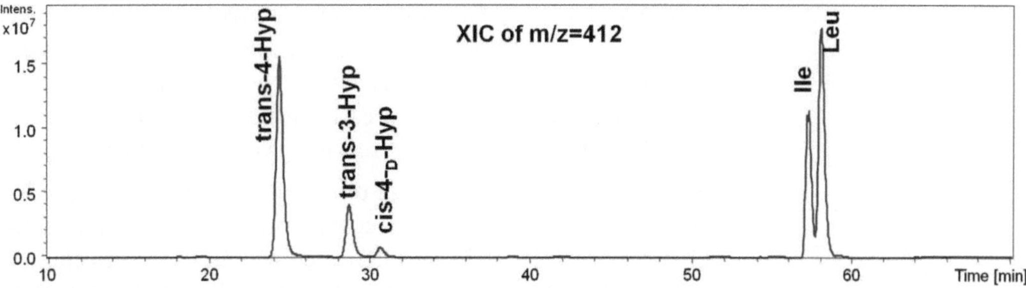

Fig. 3. Extracted ion chromatogram for the N^2-(5-fluoro-2,4-dinitrophenyl)-L-valine amide (FDVA) derivatives at *m/z* 412 (Hyp, Leu, and Ile) from the hydrolysate of the organic matrix of demineralized glass sponge spiculae. The high content of *trans*-3-Hyp (more than 20% of the total hydroxyproline) is characteristic for sponge collagen. The detected 4-*cis*-D-hydroxyproline most likely results from epimerization during acid hydrolysis.

(b) Use the full scan mode from *m/z* 300–800 to detect all derivatized amino acids, including the doubly labelled lysine. For best quantification, extract the chromatogram at *m/z* 412, which corresponds to derivatized Hyp, Ile and Leu (Fig. 3), or record only an SIR at *m/z* 412 (see Note 10). All Hyp-isomers, Ile and Leu can be directly quantified from this data set.

4. Notes

1. Reliable detection and, in particular, quantification, by LC–MS demands solvents of special purities. Even solvents sold for far-UV-applications in chromatography do not usually meet the high standards required for LC–MS coupling, as they yield a high chemical background, reducing the sensitivity of the analysis and contaminating the ionization source and the first section of the mass spectrometer.
2. Underivatized amino acids are retarded on the Amide-80 column in HILIC-mode and can be eluted by increasing the polarity of the eluent, e.g. by increasing the water content. As most amino acids have very low extinction coefficients in the UV region above 210 nm, alternative detectors have to be used, such as an online coupled mass spectrometer. Therefore, only isobaric amino acids have to be separated by HILIC. Cysteine was not included in the method development due to its fast oxidation and dimerization. Basic amino acids (His, Arg, and Lys) are eluted at high water contents (>50%), but the broad peaks are difficult to quantify and often show a strong tailing. All other proteinogenic amino acids can be analyzed by HILIC–ESI–MS, as the isomeric amino acids (Ile/Leu) and amino acids with similar masses (e.g. Asn/Asp and Gln/Glu) are well separated.

3. Asn, Asp, Gln, and Glu are not very soluble at high acetonitrile concentrations. Therefore, these amino acids were dissolved in eluent B, which can negatively affect the peak shape and chromatographic performance of early eluting amino acids, such as peak-broadening and in severe cases even peak-splitting.
4. All α-L-amino acids racemize during acid hydrolysis at the low percent level, forming the corresponding α-D-amino acids. Hydroxyproline with its two stereocentres can thus form epimers, this means that *trans*-4-L-Hyp, for example, epimerizes to *cis*-4-D-Hyp (11). As *trans*-4-Hyp is the main Hyp-isomer present in collagen, a significant amount of *cis*-4-D-Hyp will be formed during hydrolysis. The content of *cis*-4-D-Hyp is at the same level as *trans*-3-L-Hyp. It is important, therefore, that this epimer is separated in HILIC–ESI-MS. As the Amide-80 phase is non-chiral, HILIC cannot resolve *cis*-D-Hyp and *cis*-L-Hyp.
5. The limit of quantification and the linear range both strongly depend on the mass spectrometer used. The amino acid concentrations in collagen hydrolysates will typically range over more than two orders of magnitude. We found a linear quantification between 100 pmol and 10 nmol, with correlation coefficients better than 0.99. Relative standard deviations (RSD) between multiple injections were below 20%, which is still within an acceptable range for MS detection and can be further improved by using internal standards. We recommend Asn as an internal standard for collagen analysis, as it is not present in protein hydrolysates and the RSD is typically better than 10%. Therefore, dissolve Asn in water (100 pmol/μl) and use this solution to dissolve the collagen hydrolysate. Using this protocol, 1 nmol of Asn will be present in the injected hydrolysate (10 μl) as an internal standard. It is not unusual that amino acids show a non-linear response in ESI-MS. In such cases, a quadratic regression may be favourable.
6. To prepare standard solutions of derivatized amino acids, use stock solutions of 100 mM. Dilute an aliquot of the amino acid stock solution (10 μl) with water (40 μl) and add first $NaHCO_3$ buffer (20 μl, 1 M) and then FDVA solution (82.5 μl, 36.7 mM). Incubate on a shaker (90 min, 40°C) and neutralize the mixture with hydrochloric acid (20 μl, 1 M). Add acetonitrile (200 μl) and water (627.5 μl) to obtain a 1 mM solution of the derivatized amino acid (1 nmol/μl).
7. In order to derivatize the amino acids with the ideal three molar excess of FDVA, some assumptions for the derivatization of protein hydrolysates have to be made. As 150 μg of collagen are hydrolyzed and dissolved in 15 μl, the 5 μl to be derivatized contain 50 μg of protein. Assuming a molar mass of 100,000 g/mol for collagen, this amount reflects 0.5 nmol of protein. Assuming further, that a collagen molecule contains around

800 residues and is quantitatively hydrolyzed, 400 nmol of amino acids have to be derivatized. This requires 1,200 nmol FDVA, which corresponds to 32 μl of the FDVA stock solution.

8. A complete analysis of all Hyp-isomers potentially present in collagen hydrolysates demands the separation of at least six Hyp-isomers. Besides the four L-Hyp isomers potentially present in collagen, i.e. *trans*-4-, *trans*-3-, *cis*-4- and *cis*-3-Hyp, *cis*-4-D-, and *cis*-3-D-Hyp have to be separated as well, as these epimers are formed at a lower percentage from the corresponding *trans*-L-isomers during acid hydrolysis. Usually, this epimerization is accepted as an error in amino acid analysis. However, to be able to analyze existing low abundant *cis*-Hyp in collagen at exactly this level of concentration, these possible interferences must be eliminated.

9. During the optimization of the eluent conditions, we also used TFA instead of formic acid. When the first Aqua C_{18}-column was replaced by a new column, we realized that the *trans*-3-Hyp and *cis*-4-Hyp was only partially resolved. After flushing the column overnight with a solution containing 1% TFA, however, the separation was significantly improved, providing a baseline separation of both Hyp-isomers with the standard gradient in the presence of formic acid. This elution profile was stable afterwards.

10. For additional confidence or enhancement, the described RP-HPLC-method can be easily coupled online to a mass spectrometer. Only MS-experiments are useful, as the derivatized amino acids show a complex fragmentation pattern which hampers the detection in MS/MS-modes. Depending on the instrumental setup, MS-detection can improve the sensitivity tenfold compared to UV-detection.

References

1. Ramachandran GN, Kartha G (1955) Structure of collagen. Nature 176:593–595
2. Rich A, Crick FHC (1955) Structure of Collagen. Nature 176:915–916
3. Crick FHC (1954) A Structure for Collagen. J Chem Phys 22:347–348
4. Kivirikko KI, Myllyla R, Pihlajaniemi T (1989) Protein Hydroxylation – Prolyl 4-Hydroxylase, An Enzyme with 4 Cosubstrates and A Multifunctional Subunit. FASEB Journal 3:1609–1617
5. Pihlajaniemi T, Myllyla R, Kivirikko KI (1991) Prolyl 4-Hydroxylase and Its Role in Collagen-Synthesis. J Hepatology 13:S2–S7
6. Myllyharju J (2003) Prolyl 4-hydroxylases, the key enzymes of collagen biosynthesis. Matrix Biol 22:15–24
7. Gryder RM, Lamon M, Adams E (1975) Sequence position of 3-hydroxyproline in basement membrane collagen. Isolation of glycyl-3-hydroxyprolyl-4-hydroxyproline from swine kidney. J Biol Chem 250:2470–2474
8. Clifton IJ et al (2001) Structure of proline 3-hydroxylase. Evolution of the family of 2-oxoglutarate dependent oxygenases. Eur J Biochem 268:6625–6636
9. Crabb JW et al (1997) Amino Acid Analysis in Current Protocols in Protein Science (John Wiley & Sons, Inc.)
10. Bellon G et al (1984) Separation and Evaluation of the Cis and Trans Isomers of Hydroxyprolines – Effect of Hydrolysis on the Epimerization. Anal Biochem 137: 151–155

11. Dziewiatkowski DD, Hascall VC, Riolo RL (1972) Epimerization of Trans-4-Hydroxy-L-Proline to Cis-4-Hydroxy-D-Proline During Acid-Hydrolysis of Collagen. Anal Biochem 49:550–558
12. Langrock T, Czihal P, Hoffmann R (2006) Amino acid analysis by hydrophilic interaction chromatography coupled on-line to electrospray ionization mass spectrometry. Amino Acids 30:291–297
13. Alpert AJ (1990) Hydrophilic-Interaction Chromatography for the Separation of Peptides, Nucleic-Acids and Other Polar Compounds. J Chromatogr 499:177–196
14. Bruckner H, Keller-Hoehl C (1990) HPLC Separation of DL-amino acids derivatized with N^2-(5-fluoro-2,4-dinitrophenyl)-L-amino acid amides. Chromatographia 30:621–629
15. Langrock T, García-Villar N, Hoffmann R (2007) Analysis of Hydroxyproline by Reversed-Phase HPLC and Mass Spectrometry. J Chromatogr B 847:282–288
16. Kindt E et al (2003) Determination of hydroxyproline in plasma and tissue using electrospray mass spectrometry. J Pharm Biomed Anal 33:1081–1092

Chapter 22

Innovative and Rapid Procedure for 4-Hydroxyproline Determination in Meat-Based Foods

Maria Cristina Messia and Emanuele Marconi

Abstract

This report describes a rapid and innovative microwave procedure for protein hydrolysis coupled with high performance anion exchange chromatography and pulsed amperometric detection (HPAEC-PAD) to quantify the amino acid 4-hydroxyproline in meat and meat-based products. This innovative approach was successfully applied to determine collagen content (4-hydroxyproline × 8) as the index quality of meat material used in the preparation of typical meat-based foods.

Key words: 4-Hydroxyproline, Collagen, Microwave protein hydrolysis, HPAEC-PAD

1. Introduction

The addition of low value meat is generally considered to be the most frequent adulteration of meat-based products, such as sausages, fresh filled pasta, hamburger, etc. A suitable marker for identifying the quality of raw material used for meat-based preparations is collagen content which can be calculated by the concentration of the imino acid 4-hydroxyproline (1–3).

The method most commonly used for 4-hydroxyproline analysis is the colorimetric method based on hydrochloric acid or sulphuric acid hydrolysis of meat sample, oxidation of 4-hydroxyproline with chloramines-T and spectrophotometric measurement at 560 nm of red-purple colour formed. Although this method is very specific for 4-hydroxyproline, it is difficult to control oxidation, colour formation, and the hydrolysis step is very time consuming (16–24 h).

One of the most significant and recent developments in the performance of compositional food analysis is the use of microwave radiation energy for protein hydrolysis (4, 5). The procedure has been successfully used for fast protein hydrolysis (6) in determining

Michail A. Alterman and Peter Hunziker (eds.), *Amino Acid Analysis: Methods and Protocols*, Methods in Molecular Biology, vol. 828, DOI 10.1007/978-1-61779-445-2_22, © Springer Science+Business Media, LLC 2012

natural as well as non-natural single amino acids, such as tryptophan (7), lysine (8), furosine (9, 10), *meso*-diaminopimelic acid (11), and total amino acids (4, 12–15).

Several analytical techniques have been developed to detect total and single amino acids. An alternative to derivatization systems (16) is the electrochemical detection (17, 18), and in particular integrated amperometry (HPAEC-PAD) (19, 20) which allows the direct detection of amino acids on a platinum or a gold electrode after the separation by anion exchange.

In this chapter, we describe the use of microwave hydrolysis coupled to high performance anion exchange chromatography and pulsed amperometric detection (HPAEC-PAD) for 4-hydroxyproline analysis in meat-based products. This combined approach allows the reduction of protein hydrolysis time to 20 min and the direct 4-hydroxyproline detection without pre- or post-column derivatization.

2. Materials

2.1. Chemicals

1. Prepare all solutions using ultrapure water (prepared by purifying deionized water to attain a sensitivity of 18 MΩ cm at 25°C).
2. NaOH 50% (*p*/*v*) (Mallinckrodt Baker B.V., Deventer, Holland).
3. 4-Hydroxyproline standard (Sigma Chemical Co., St. Louis, MO, USA).
4. HCl: Approximately 6N HCl. Add 250 ml H_2O to 1-L volumetric flask. Slowly add, 500 ml of HCl (solution at 37.5%, density: 1.186). Cool to room temperature and dilute to volume with water.
5. Hydroxyproline standard solution: Stock solution (13 mg/ml). Dissolve 1.31 g of hydroxyproline in water in 100-ml volumetric flask. Dilute to volume with water. Solution is stable ca. 2 months at 4°C.
6. All other chemicals and reagents of HPLC grade.
7. Samples of meat-based foods are chopped and then homogenized with a steel blade homogenizer, for no longer than 15–20 s to minimize any increase in temperature.

2.2. Microwave Apparatus

Microwave Digestion System, Mod. MDS 2000 (CEM Corporation, Mattews, NC, USA). This microwave oven has a maximum power of 630 ± 50 Watts and a magnetron frequency 2,435 MHz. The system is equipped with probes to detect and control the pressure and temperature inside the sealed vessel. The oven is equipped with a 12-positioned turntable (3 rpm) and a mode stirrer to prevent the uneven distribution of microwaves in the hydrolysis vessels (Fig. 1).

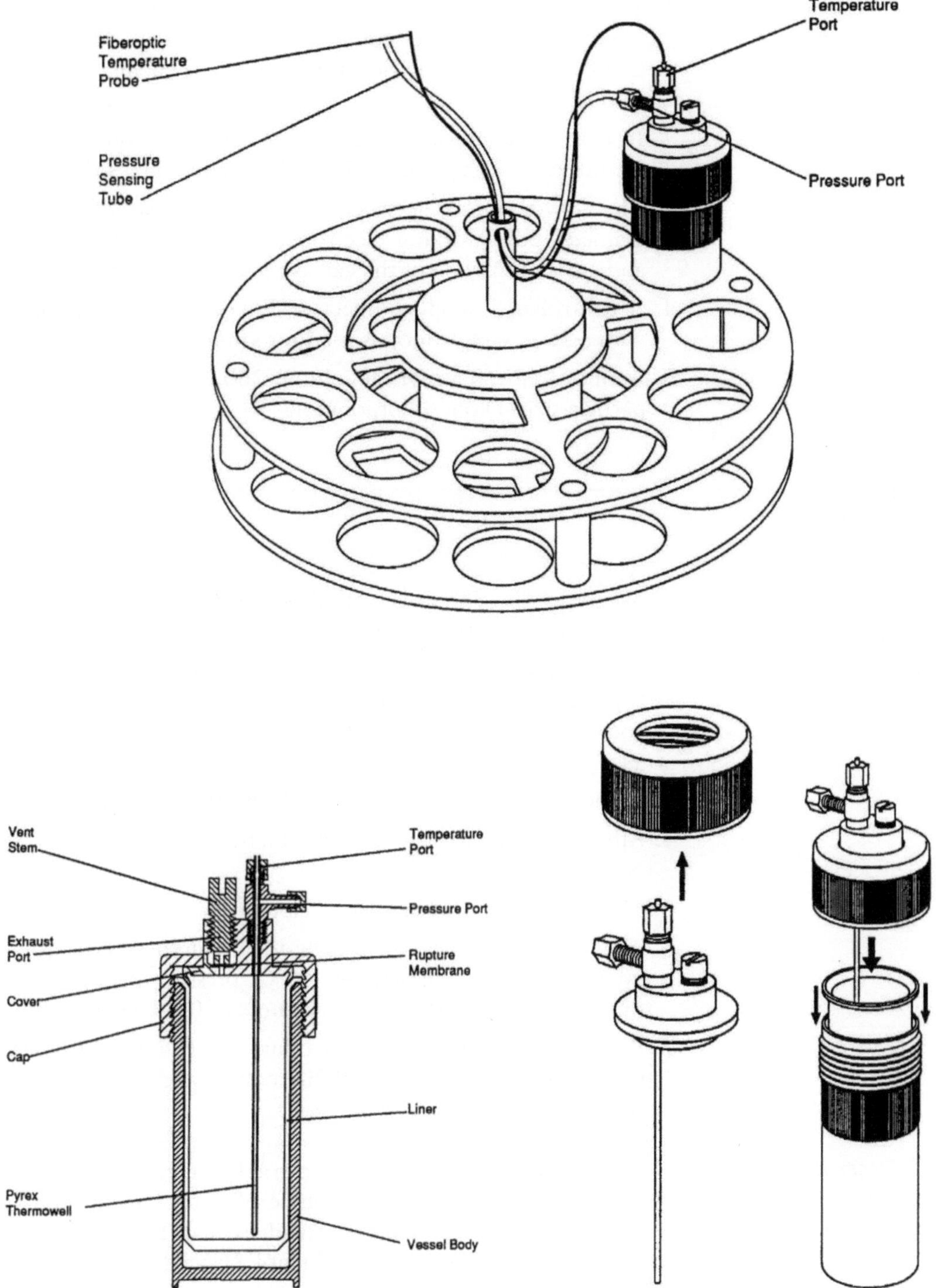

Fig. 1. Cross section of lined digestion vessel for pressure and temperature control and turntable collection vessel (MDS-2000 Operative manual, CEM Corporation, Mattews, NC, USA).

2.3. Chromatographic Apparatus

1. 1 Dionex DX500 Ion Chromatograph (Dionex Corporation, Sunnyvale, CA, USA) composed of a gradient pump (mod GP50) with an online degasser and electrochemical detector (model ED40).
2. Instrument control, data collection, and total quantification are managed using Chromeleon chromatography software.
3. The flow-through electrochemical cell consists of a 1 mm diameter gold working electrode, a pH reference electrode, and a titanium body of the cell as the counter electrode.
4. A controlled Rheodyne injector (Rheodyne L.P., CA, USA) with a 25 μl sample loop is used for sample injection.
5. Aminopac PA10 analytical column 250×2 mm, with 8.5 μm particle size (Dionex Corporation).

3. Methods

3.1. Microwave Protein Hydrolysis (see Ref. 21)

1. Microwave hydrolysis is carried out using a microwave digestion system designed for laboratory use.
2. The sample, corresponding to 25 mg of protein (see Note 1), is placed into four Teflon PFA digestion vessels (see Note 2) and 8 ml 6N HCl are added (see Note 3).
3. The vessel cup is screwed manually; the pressure and fibre optic probes are connected to the vessel with the triple ported cap.
4. After the irradiation cycles (Table 1), the vessels are cooled (see Note 4) and then removed.
5. The hydrolysates are filtered (see Note 5) by Whatman paper n.1.
6. The filtered samples are evaporated to dryness by a rotary evaporator (bath temperature 40°C, pressure 25 mbar) and then redissolved with a precise volume of 0.1N HCl.
7. Before the analysis, samples are diluted 1:50–1:100 with ultrapure water, filtered through 0.20 μm filter and then injected in the chromatographic system.

3.2. Chromatographic Separation (see Ref. 21)

1. Quantitative determination is carried out at a flow rate of 0.25 ml/min using a mobile phase of water (eluent E1) (see Note 6), 250 mM sodium hydroxide (eluent E2) (see Note 7) and 1.0 M of sodium acetate (eluent E3) (see Note 8) as shown in Table 2 and an optimized time-potential waveform as shown in Table 3.
2. 4-Hydroxyproline identification and quantification is carried out by means of external amino acid standard (see Note 9).

Table 1
Microwave protein hydrolysis conditions in meat-based foods

	1st Cycle	2nd Cycle
Power (% 630 Watts)	85	85
Time (min)	1	5
Temperature (°C)	100	155
Pressure max (psi)	100	130

Table 2
Gradient conditions for anion-exchange separation of 4-hydroxyproline

Time (min)	Eluent 1 water (%)	Eluent 2 NaOH (%)	Eluent 3 sodium acetate (%)
0.0	80	20	0
2.0	80	20	0
12.0	80	20	0
16.0	68	32	0
24.0	36	24	40
40.0	36	24	40
40.1	20	80	0
42.1	20	80	0
42.2	80	20	0
62.0	80	20	0

3. Figure 2 shows an HPAEC-PAD chromatogram of meat-based food hydrolysed by microwave procedure. The retention time of 4-hydroxyproline is 9 min. The same HPAEC-PAD chromatographic run can also permit the identification and quantification of all other amino acids.

3.3. Calculation of 4-Hydroxyproline Content

1. Calculate 4-hydroxyproline content (g/100 g) as follows:

$$\text{4-Hydroxyproline(g/100g)} = ((\text{SA} \times \text{Std})/\text{StdA}) \times \text{MW} \times \text{D} \times 100/\text{SW},$$

where: SA = 4-hydroxyproline sample area; Std = 4-hydroxyproline standard concentration; StdA = 4-hydroxyproline standard

Table 3
Integrated amperometry waveform used to detect 4-hydroxyproline

Time (s)	Potential (V)	Integration
0.00	+0.13	
0.04	+0.13	
0.05	+0.28	
0.11	+0.28	Begin
0.12	+0.60	
0.41	+0.60	
0.42	+0.28	
0.56	+0.28	End
0.57	−1.67	
0.58	−1.67	
0.59	+0.93	
0.60	+0.13	

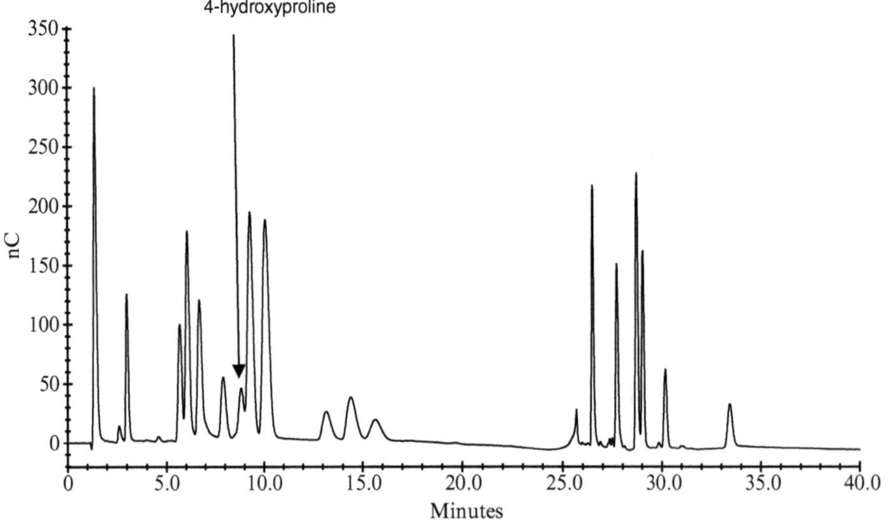

Fig. 2. HPAEC-PAD chromatogram of meat-based food after microwave hydrolysis.

area; MW = 4-hydroxyproline molecular weight; D = dilution factor which consider the final volume of hydrolysate; the volume of hydrolysate dried by rotary evaporator; the volume of 0.1 N HCl used to resume the sample after drying with rotary evaporator; the sample dilution (1:50 or 1:100) before the injection in the chromatographic system; and SW = sample weight.

2. The collagen content can be estimated by multiplying the 4-hydroxyproline content (g/100 g) by 8 as provided by AOAC method 990.26 (22) (see Note 9)

4. Notes

1. Protein content ($N \times 6.25$) was determined according AOAC method 928.08 (22).
2. The sealed vessels used consist of perfluoroalkoxy (PFA) sample liner (volume 50 ml) surrounded by an advanced composite sleeve.
3. Each cycle of hydrolysis has to be carried out using same number of vessels (four), same typology and quantity of samples in order to allow a uniform microwave distribution and absorption.
4. The turntable is cooled into an ice bath until the pressure is equal to atmospheric pressure (about 5 min).
5. The sample is filtered into 100-ml volumetric flask and diluted to mark with deionized water.
6. Eluent E1: Deionized water. Filter the pure deionized water through 0.2 μm nylon filters and then transfer into a DX500 eluent bottle. Seal the filtered water immediately. Remember, that atmospheric carbon dioxide adsorbs even into pure water, albeit at much lower levels than in alkaline solutions. Minimize the contact time of water surface with the atmosphere (Dionex DX500 Operative Manual).
7. Eluent E2: 250 mM Sodium hydroxide. The first step in the preparation of sodium hydroxide eluent is filtration of a water aliquot (typically 1.0 L). Hermetically seal the filtered water immediately after filtration while preparing a disposable glass pipette and a pipette filler. Using a pipette filler, draw an aliquot of 50% sodium hydroxide into pipette (13.1 ml), unseal the filtered water, and insert the full pipette approximately 1 inch. below the water surface and release the sodium hydroxide. If done properly and without stirring, most of the concentrated sodium hydroxide stays at the lower half of the container and the rate of carbon dioxide adsorption is much lower than that of a normally prepared 250 mM sodium hydroxide solution. Seal the container immediately after sodium hydroxide transfer is complete. Remember to put the screw cap back on the 50% hydroxide bottle immediately as well. Mix the content of the tightly sealed container holding the 250 mM hydroxide.

Unscrew the cap of the eluent bottle E2 attached to the chromatographic system. Allow the helium or nitrogen gas to blow out of the cap. Unseal the bottle holding 250 mM sodium hydroxide and immediately, without delay, start the transfer into the eluent bottle E2. Try to minimize the carbon dioxide absorption by holding the gas orifice of the bottle cap as close as possible to the 250 mM hydroxide during the transfer. With the inert gas still blowing, put the cap on the eluent bottle. Allow the pressure to build up inside the bottle and reopen the cap briefly several times, to allow trapped air to be gradually replaced by inert gas (Dionex DX500 Operation Manual).

8. Eluent E3: 1.0 M Sodium acetate. Dissolve 82.04 g of anhydrous sodium acetate in approximately 750 ml water. Seal the container during the dissolution step. Make up to 1.0 L with water and filter through a 0.2 μm nylon filter. Transfer the filtered sodium acetate eluent into the eluent bottle E3 of the chromatographic system, using the same procedure as for the sodium hydroxide (Dionex DX500 Operation Manual).
9. Collagen connective tissue contains 12.5% 4-hydroxyproline if the nitrogen-to-protein factor is 6.25.

References

1. Flores J (1980) Control de calidad de los productos carnicos. Rev Agroquim Tecnol Aliment 20:180
2. Vazquez-Ortiz FA, Gonzalez-Mendez NF (1996) Determination of collagen as a quality index in Bologna from Northwestern Mexico. J Food Comp Anal 9:269–276
3. Etherington DJ, Trevor SJ (1981) Detection and estimation of collagen. J Sci Food Agric 32:539–546
4. Chiou SH, Wang KT (1989) Peptide and protein hydrolysis by microwave irradiation. J Chromatogr 491:424–431
5. Fountoulakis M, Lahm HW (1998) Hydrolysis and amino acid composition analysis of proteins. J Chromatogr A 826:109–134
6. Huan YF et al (2009) PPESO3 monitoring the process of microwave enhanced enzymatic digestion on protein. Chem J Chinese Univ 30(1):54–56
7. Carisano A (1993) Digestione delle proteine in matrici alimentari con forno a microonde. 2-determinazione del triptofano. Ind Aliment 32:346–349
8. Marconi E et al (1996) Fast analysis of lysine in food using protein microwave hydrolysis and an electrochemical biosensor. Anal Lett 29 (7):1125–1137
9. Marconi E, Panfili G, Acquistucci R (1997) Microwave hydrolysis for the rapid analysis of furosine in foods. Italian J Food Sci 1 (9):47–55
10. Marconi E, Mastrocola L, Panfili G (1999) Improvement of microwave hydrolysis for fast analysis of furosine in milk and dairy products. Sci Tec Lattiero-Casearia 50(2): 127–137
11. Marconi E et al (2000) Rapid detection of *meso*-diaminopimelic acid in lactic acid bacteria by microwave hydrolysis. J Agric Food Chem 48(8):3348–3351
12. Chen ST et al (1987) Rapid hydrolysis of proteins and peptides by means of microwave technology and its application to amino acid analysis. Int J Pept Protein Res 30:572–576
13. Gilman LB, Woodward C (1990) An evaluation of microwave heating for vapour phase hydrolysis of proteins. Comparison to vapour phase hydrolysis for 24 hours. In Gilman LB (ed) Current research in protein Chemistry, Academic Press: New York, NY
14. Woodward C, Gilman LB, Engelhart WG (1990) An evaluation of microwave heating for vapour phase hydrolysis of proteins. Int Lab 20:40–46

15. Marconi E et al (1995) Comparative study on microwave and conventional methods for protein hydrolysis in food. Amino Acids 8:201–208
16. Spackman DH, Stein WH, Moore S (1958) Chromatography of amino acids on sulphonated resins. Anal Chem 30:1190–1196
17. Polta JA, Johnson DC (1983) The direct electrochemical detection of amino-acids at platinum electrode in an alkaline chromatographic effluent. J Liq Chromatogr 6:1727–1743
18. Marten DA, Frankenberg WT (1992) Pulsed amperometric detection of amino-acids separated by anion-exchange chromatography. J Liq Chromatogr 15:423–439
19. Welch LE et al (1989) Comparison of pulsed coulometric detection and potential-sweep pulsed coulometric detection for underivatized amino-acids in liquid chromatography. Anal Chem 61:555–559
20. Clarke AP et al (1999) An integrated amperometry waveform for the direct, sensitive detection of amino acids and amino sugars following anion-exchange chromatography. Anal Chem 71:2774–2781
21. Messia MC et al (2008) Rapid determination of collagen in meat based foods by microwave hydrolysis and HPAEC-PAD analysis of 4-hydroxyproline. Meat Sci 80:401–409
22. AOAC (2000). OFFICIAL METHOD 990.26, and 928.08, Official Methods of Analysis, 17th edn. Gaithersburg, MD; Association of Official Analytical Chemists.

Chapter 23

Multiple Reaction Monitoring for the Accurate Quantification of Amino Acids: Using Hydroxyproline to Estimate Collagen Content

Michelle L. Colgrave, Peter G. Allingham, Kerri Tyrrell, and Alun Jones

Abstract

Multiple reaction monitoring (MRM) mass spectrometry may be regarded as the gold standard methodology for quantitative mass spectrometry and has been adopted for the analysis of small molecules especially within the pharmaceutical industry. It can also be applied to the analysis of peptides and proteins and to the measurement of the basic building blocks of proteins, amino acids. Here, we describe the application of MRM mass spectrometry to the measurement of hydroxyproline after acid hydrolysis of various animal tissues. We show that the measurement of hydroxyproline provides an accurate and reliable estimate of the collagen content of such tissues and may be a useful indicator of meat tenderness.

Key words: Amino acid, Collagen, Hydroxyproline, Mass spectrometry, Multiple reaction monitoring, Quantitation, Amino acid analysis

1. Introduction

Collagen is the most abundant protein family in mammals making up almost a quarter of the total protein content (1) providing support and structural organization to cells in most tissue types. Collagen has great tensile strength and is a principle protein component of intramuscular connective tissue, the fascia, cartilage, ligaments, tendons, bone, teeth, and cornea (2, 3). Collagen, in conjunction with soft keratin, is responsible for skin strength and elasticity. Variation in the total collagen concentration in tissue as well as changes in its thermo-dependent solubility has been implicated in the variability in tenderness of meat from production species, such as sheep and cattle (4–6).

Michail A. Alterman and Peter Hunziker (eds.), *Amino Acid Analysis: Methods and Protocols*, Methods in Molecular Biology, vol. 828, DOI 10.1007/978-1-61779-445-2_23, © Springer Science+Business Media, LLC 2012

Collagen is a long, fibrous structural protein with the distinctive feature of having a regular arrangement of amino acids. The sequence consists of Gly–Pro–*X* or Gly–*X*–Hyp repeats, where *X* may be any of the various other amino acid residues with the Gly–Pro–Hyp motif occurring frequently (1). Collagen contains a high concentration of the imino acid 4-hydroxyproline (7) with up to 14% of the dry weight of collagen composed of Hyp, compared to ~1% Hyp found in elastin, the mammalian protein with the next highest proportion of Hyp (7). Owing to its abundance in collagen, the concentration of hydroxyproline has been historically used as an estimate of collagen content (8, 9).

To overcome difficulties encountered with the measurement of Hyp concentrations by conventional methodologies, such as colourimetric assays and HPLC which lack sensitivity, we recently demonstrated that multiple reaction monitoring (MRM) mass spectrometry could be successfully employed (10).

MRM mass spectrometry relies upon the inherent chemical and physical properties of the analyte, that is, its mass and its fragmentation pattern. MRM experiments may be conducted on any triple quadrupole instrument. As depicted in Fig. 1, the first quadrupole is used to select the mass-to-charge ratio (*m/z*) of the analyte, the so-called precursor ion. The precursor ion is then transmitted to the collision cell (the second quadrupole). Collision-induced dissociation (CID) (11) occurs resulting in the production of fragment ions that are transmitted to the third quadrupole. A second stage of mass selection occurs specifically targeting the *m/z* values of the known fragment ions. The two stages of mass selection are known as Q1 and Q3 referring to the quadrupole in which they occur. The Q1–Q3 transition is thus known as

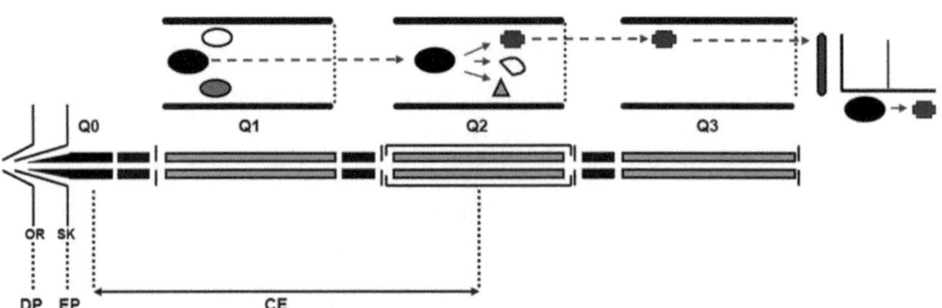

Fig. 1. Schematic diagram of the 4000 QTRAP mass spectrometer used for MRM analysis. The precursor ion (or Q1 mass) is selected in the first quadrupole (*Q1*). This ion is transmitted to the collision cell (*Q2*) where collision-induced dissociation results in the formation of product ions. The product ion of interest (or Q3 mass) is then selected in the third quadrupole (*Q3*). The collision energy (*CE*) is the voltage difference between the ion transmission quadrupole (*Q0*) in the source and the collision cell (*Q2*). The declustering potential (*DP*) is the voltage applied to the orifice plate (*OR*) and the entrance potential (*EP*) is the voltage applied to the skimmer. All of these parameters are optimized to increase the intensity of the analyte's MRM transition and hence the sensitivity of the analytical method.

the MRM transition and is highly specific and selective for the analyte of choice.

This chapter describes the development of the LC–MS/MS analytical method for the quantification of hydroxyproline and its application to the determination of the total collagen content of ovine *semimembranosus* muscle (SM). This procedure can be applied to the development of MRM assays for any other amino acid.

2. Materials

Prepare all solutions using ultrapure (UP) water (prepared by purifying deionized water to attain a resistance of 18 MΩ·cm at 25°C) and analytical grade reagents. Prepare and store all reagents at room temperature (unless indicated otherwise). Follow all waste disposal regulations when disposing of waste materials and check Material Safety Data Sheets (MSDS) for reagents prior to use.

2.1. Chemicals

1. The amino acids, L-proline (Pro), L-leucine (Leu), L-isoleucine (Ile) and *trans*-4-L-hydroxyproline (Hyp) (10 mg), can be purchased from Sigma-Aldrich (Castle Hill, Australia) (see Note 1). The internal standard, DL-Proline (DL-Pro, 2,3,3,4,4,5,5-D7-proline) can be purchased from Cambridge Isotope Labs (Andover, MA, USA).
2. Diluent for standards: 0.1% formic acid. Add 0.1 mL of concentrated formic acid to 99.9 mL of UP water (see Note 2).
3. Amino acid solutions used for MRM determination: 25 μM solutions of L-proline, L-leucine, L-isoleucine, and *trans*-4-L-hydroxyproline in 0.1% formic acid in 50% acetonitrile. From a 100 μM stock solution (see Note 3), add 250 μL of each amino acid solution to 500 μL of acetonitrile and 250 μL of 0.1% formic acid.

2.2. HPLC Components

1. HPLC buffer A: 0.1% formic acid in water. Add 1 mL of concentrated formic acid to 999 mL of UP water. Mix by inversion.
2. HPLC buffer B: 0.1% formic acid in 90% acetonitrile/9.9% water. Add 1 mL of concentrated formic acid to 900 mL of acetonitrile and 99 mL of UP water. Mix by inversion.
3. Filter HPLC buffers A and B through 0.22 μm Whatman filter paper (Millipore, North Ryde, Australia) (see Note 4).

2.3. Sample Components

Rat tail collagen (α-1(I) chain) is used for method validation (Vitrogen, Collagen Corp., Palo Alto, CA, USA).

3. Methods

Carry out all procedures at room temperature unless specified otherwise.

3.1. Tissue Extraction, Delipidation, and Hydrolysis

1. Excise 20 g of tissue from the belly of the *semimembranosus* muscle (away from tendons and fascia) of sheep, 24 h *post-mortem* and freeze at –20°C until required.
2. Mill frozen samples to a fine powder under liquid nitrogen. Freeze-dry the homogenate in a vacuum centrifuge.
3. Remove the lipid from the dried muscle homogenate by overnight extraction with di-ethyl ether (10 mL/100 mg tissue) in a fume hood under continuous agitation.
4. Pellet the sample by centrifugation at 3,000 × *g* at room temperature for 5 min, discard the supernatant, and re-dry the sample for 24 h at 65°C.
5. Hydrolyze approximately 3 mg of dry defatted muscle homogenate using 2 mL of 6 M HCl under nitrogen in sealed vials at 110°C overnight.
6. Following hydrolysis, cool the vials and centrifuge at 14,000 × *g* for 1 min at room temperature to remove particulate matter.
7. Take aliquots (100 μL) of the hydrolysates and lyophilize. Reconstitute the lyophilized material in 1 mL of 0.1% formic acid.

3.2. Rat Tail Collagen for Method Validation

1. Dry 1 mL of soluble collagen derived from rat tail (Vitrogen 100, supplied as 3.1 mg/mL solution in 0.01 M HCl) in a vacuum drier.
2. Hydrolyze the dried collagen in 2 mL of 6 M HCl under nitrogen in sealed vials at 110°C overnight.
3. Thereafter, treat the sample the same as for the muscle hydrolysates.

3.3. Standard Preparation

1. Stock solutions: 50,000, 10,000, 5,000, 1,000, 500, and 100 nM hydroxyproline in 0.1% formic acid. Prepare a 50,000 nM stock solution by adding 500 μL of the 100 μM stock solution of Hyp prepared previously (see Note 3) to 500 μL of 0.1% formic acid. Prepare tenfold serial dilutions by dissolving 100 μL in 900 μL of 0.1% formic acid to yield the 5,000 and 500 nM stock solutions. Likewise, prepare a 10,000 nM stock solution by adding 100 μL of the 100 μM stock solution of Hyp to 900 μL of 0.1% formic acid. Prepare tenfold serial dilutions by dissolving 100 μL in 900 μL of 0.1% formic acid to yield the 1,000 and 100 nM stock solutions.

Table 1
Preparation of spiked standards

Std #	Volume	Hyp concentration (nM)	Hyp vol. (μL)	Hyp stock concentration (nM)	Sample vol. (μL)	IS vol. (μL)[a]	0.1% FA vol. (μL)
0	200	0.0	0.0	0	100	50	50.0
1	200	19.5	39.1	100	100	50	10.9
2	200	39.1	15.6	500	100	50	34.4
3	200	78.1	15.6	1,000	100	50	34.4
4	200	156.3	31.3	1,000	100	50	18.8
5	200	312.5	12.5	5,000	100	50	37.5
6	200	625.0	12.5	10,000	100	50	37.5
7	200	1,250.0	25.0	10,000	100	50	25.0
8	200	2,500.0	10.0	50,000	100	50	40.0
9	200	5,000.0	20.0	50,000	100	50	30.0

IS internal standard DL-Pro

2. Internal standard (IS) stock solution: 200 nM DL-Pro in 0.1% formic acid. Dissolve 10 mg of DL-Pro in 1 mL of UP water (10 mg/mL = 81.2 mM). Dilute 123.1 μL of this solution in 876.9 μL of UP water and mix well (10 mM). Prepare tenfold dilutions of the 10 mM DL-Pro solution by dissolving 100 μL in 900 μL of 0.1% formic acid to yield 1 mM, 100, 10, and 1 μM solutions. For 200 nM DL-Pro stock solution, dissolve 200 μL of 1 μM in 800 μL of 1% formic acid.
3. Spiked standard preparation: standards ranging in concentration from 19.5 to 5,000 nM Hyp prepared by addition of Hyp stock solutions to the hydrolysate of a freeze-dried ovine muscle sample (i.e. that contains endogenous hydroxyproline). Prepare a series of standards by addition of Hyp stock solution and 50 μL of 200 nM DL-Pro internal standard solution to 100 μL of hydrolysate as shown in Table 1. Prepare all standards in triplicate.

3.4. Quality Control Preparation

Quality control (QC) samples should be prepared from a separate weighing of Hyp and subsequent dilution to yield final concentrations for QC1 (low range) of 78.1 nM, QC2 (medium range) of 625 nM and QC3 (high range) of 2,500 nM. The QC samples are prepared in the ovine muscle hydrolysate to account for matrix effects (Table 2).

Table 2
Preparation of quality controls (QCs)

Std #	Volume	Hyp concentration (nM)	Hyp vol. (μL)	Hyp stock concentration (nM)	Sample vol. (μL)	IS vol. (μL)	0.1% FA vol. (μL)
QC-L	200	78.1	15.6	1,000	100	50	34.4
QC-M	200	625.0	12.5	10,000	100	50	37.5
QC-H	200	2,500.0	10.0	50,000	100	50	40.0

IS internal standard DL-Pro

Table 3
Preparation of samples

Std #	Volume	Hyp concentration (nM)	Hyp vol. (μL)	Hyp stock concentration (nM)	Sample vol. (μL)	IS vol. (μL)[a]	0.1% FA vol. (μL)
Any	200	N/A	N/A	N/A	100	50	50

IS internal standard DL-Pro

3.5. Sample Preparation

To each of the reconstituted samples (100 μL), add 50 μL of 200 nM internal standard (DL-Pro) and 50 μL of 0.1% formic acid and aliquot in a 384-well plate (Table 3).

3.6. MRM Determination

To optimize the MRM transitions for each of the analytes, solutions of hydroxyproline, proline, leucine, isoleucine, and DL-proline at a concentration of 25 μM (in 0.1% formic acid, 50% acetonitrile) should be used.

A 4000 QTRAP mass spectrometer (Applied Biosystems, Foster City, CA, USA) should be employed for MRM determination and subsequent LC–MRM-MS assays. Configure the 4000 QTRAP mass spectrometer with the turboionspray source and set the source parameters as follows: ionspray voltage 5,500 V, temperature 450°C, curtain gas flow 20, nebulizer gas 1 (GS1) 35, nebulizer gas 2 (GS2) 20, and collision cell (CAD) gas pressure set to high (see Note 5). Data acquisition and processing should be performed using Analyst v1.5 software.

1. Directly infuse the individual amino acids into the mass spectrometer using a Harvard syringe pump at a flow rate of 10 μL/min.

2. The intact protonated molecular ion (Q1 mass) can be observed in a Q1 MS scan (see Notes 6 and 7).

3. While observing the Q1 mass, the declustering potential (DP) should be optimized. Vary the DP in 5 V increments monitoring the intensity of the Q1 mass and the total ion chromatogram (TIC). The aim is to obtain maximum signal for the Q1 mass while decreasing the background (chemical noise) (see Note 8).

4. Determine the fragment ions (Q3 masses) produced by CID using a product ion (MS2) scan for each amino acid. Figure 2a shows an example product ion scan for hydroxyproline (precursor

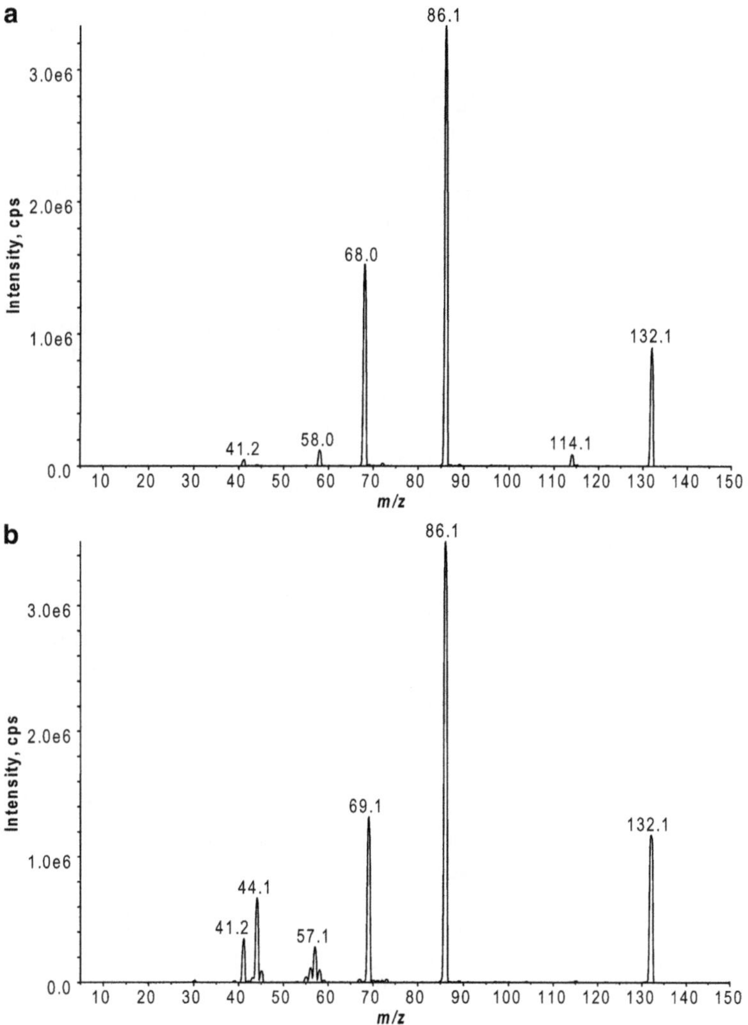

Fig. 2. Product ion (MS2) scan for hydroxyproline (**a**) and isoleucine (**b**). The precursor ion of both amino acids is *m/z* 132.1, but these isobaric amino acids can be differentiated by comparing their fragment ions. Hydroxyproline yields characteristic ions at *m/z* 114.1, 68.0 and 58.0, whereas isoleucine and leucine (data not shown) yield characteristic ions at 69.1 and 57.1.

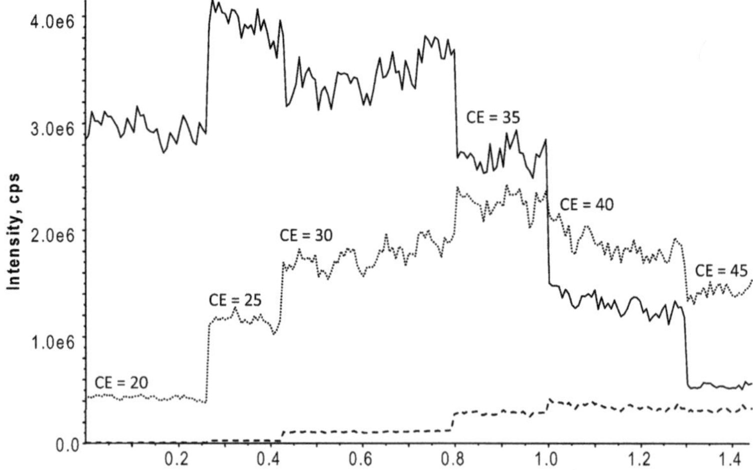

Fig. 3. Optimization of collision energy for MRM transitions. Three MRM transitions of hydroxyproline are shown: 132.1–86.1 (*solid line*, common to Leu and Ile), 132.1–68.0 (*dotted line*), and 132.1–58.0 (*dashed line*). The collision energy was increased in 5 V increments as indicated. The optimum CE was determined to be 25 V for 132.1–86.1, 32 V for 132.1–68.0, and 45 V for 132.1–58.0.

ion *m/z* 132.1). Five major fragment ions may be observed and Q3 masses should be selected based on their intensity and uniqueness (see Note 9). Figure 2b shows an example product ion for isoleucine (precursor ion *m/z* 132.1), but it should be noted that Ile and Leu may be differentiated from Hyp based on the presence of characteristic fragment ions.

5. Optimization of collision energy for each MRM transition (see Note 10). Using an MRM scan, increase the collision energy (from 20 V) in 5 V increments noting the optimum collision energy range for each MRM transition (Fig. 3) (see Note 11). The collision energy can be further refined by varying the CE range in 1–2 V increments.

3.7. LC Optimization

An Ultimate 3000 capillary HPLC system (Dionex, Sunnyvale, USA) and a Zorbax SB-Aq column (Agilent, Forest Hill, NSW, Australia; 250 mm × 4.6 mm ID) are recommended for use (see Note 12).

1. Set the flow rate to 700 μL/min using an isocratic flow (consisting of 80% HPLC buffer A and 20% HPLC buffer B) over 5 min.
2. Inject 20 μL of each standard, QC and/or sample (see Note 13). An example LC–MRM-MS chromatogram is given in Fig. 4.
3. Clean the column by washing with 90% solvent B for 20 min at the end of each batch. Blank samples should be strategically placed within each batch to ensure no sample carryover is occurring.

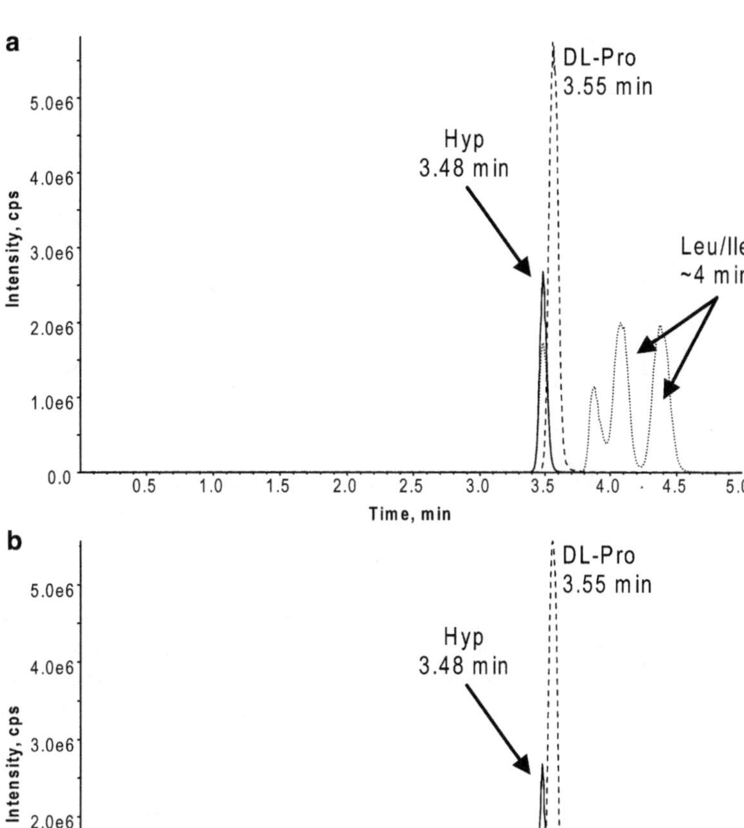

Fig. 4. LC–MRM-MS analysis of amino acids following acid hydrolysis. (**a**) The MRM transition common to Hyp, Leu, and Ile is shown (*dotted line*). Performing an extracted ion chromatogram (**b**) on the unique MRM transitions for Hyp (132.1–68.0, *black solid line*) and DL-Pro (123.1–77.1, *grey dashed line*) allows quantification of these analytes without interference.

3.8. Data Analysis and Interpretation

Calibration curves can be constructed from the peak area ratios of the analyte to the internal standard versus the theoretical concentrations (see Note 14). An example of a calibration curve generated for Hyp is given in Fig. 5.

1. Linear regression analysis of the standard curves and calculation of hydroxyproline concentration can be undertaken using the Analyst Quantitation toolbox, GraphPad Prism v4 or a similar graphical software package.
2. Unknown sample concentrations can be calculated using linear equations ($y = mx + b$) fitted for the calibration curves, where y

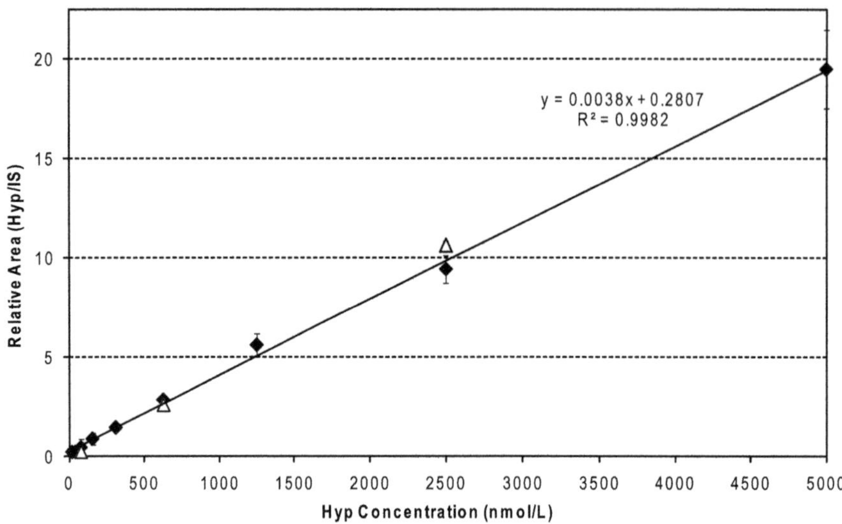

Fig. 5. Calibration curve for quantification of hydroxyproline in ovine tissue. Known amounts of Hyp were spiked into tissue hydrolysates (containing endogenous Hyp). The relative area of Hyp/IS was calculated for each standard and the endogenous Hyp level was subtracted to yield the normalized values. Spiked standards were prepared and analyzed in triplicate (represented by *black diamonds*). Quality control (QC) standards were prepared from a separate preparation and are plotted on the curve at three concentrations (78.1, 625, and 2,500 nM; represented by *white triangles*).

is the relative area (analyte/IS), x is the concentration of analyte, m is the slope of the line and b is the intercept.

3. The precision of the method can be determined by replicate analyses ($n \geq 3$) of hydrolysates at different concentrations within the calibration range and presented as the standard deviation (SD).
4. The accuracy of the method can be expressed as (mean calculated concentration)/(spiked concentration) $\times$ 100.
5. The limit of detection (LOD) is defined as the lowest sample concentration that can be detected with a signal-to-noise ratio (S/N) >5.
6. The limit of quantification (LOQ) is defined as the lowest sample concentration that can be measured (quantified) (12).

4. Notes

1. It is necessary to determine MRM transitions for the isobaric amino acids L-leucine, L-isoleucine, and L-hydroxyproline. As all three amino acids share the same molecular weight, it is possible that they could also share a similar fragmentation pattern. The determination of the fragmentation spectra for L-leucine and L-isoleucine enabled determination of unique MRM

transitions for L-hydroxyproline ensuring specificity in this assay for this amino acid alone.

2. Formic acid is a strong acid, is corrosive and may cause burns by inhalation, ingestion, or contact. Use in a fumehood with appropriate personal protective equipment.
3. Prepare a 10 mg/mL stock solution of each of L-leucine, L-isoleucine, *trans*-4-L-hydroxyproline (76.3 mM), and L-proline (86.8 mM). Dissolve 10 mg of each amino acid in 1 mL of UP water. Prepare a 10 mM solution of each amino acid by adding 131.1 μL of 10 mg/mL Leu, Ile, or Hyp solution to 868.8 μL of UP water. For Pro, add 115.1 μL of 10 mg/mL stock solution to 884.9 μL of UP water. Prepare tenfold serial dilutions (to 1 mM and subsequently to 100 μM) of these solutions by dissolving 100 μL in 900 μL of 0.1% formic acid.
4. The HPLC buffers should be degassed if the HPLC to be used is not fitted with a degassing module. This can be achieved by sonication of the solutions in an ultrasonic bath for 10 min.
5. The source conditions will vary depending on the instrument, the flow rate and the solvents used. The parameters detailed here act as a starting point, but can be optimized for improved performance. In positive ion mode, typical ionspray voltages range from 4,000 to 5,500 V. The curtain gas is used to prevent air or solvent entering the analyzer region of the MS while permitting ions to enter. The curtain gas should be set as high as possible without reducing the signal significantly. The position of the probe may also be adjusted in both the vertical and horizontal directions and should be set to maximize the signal. Nebulizer gas 1 (GS1) should be optimized for stability and sensitivity, typically a value between 20 and 40 is suitable. Nebulizer gas 2 (GS2) aids in the evaporation of solvent and thus increases the ionization of the analyte and should be optimized for stability and sensitivity. Likewise, the source temperature should be optimized. As a general rule, the higher the flow rate, the higher the temperature that is required.
6. The theoretical m/z values for each amino acid can be calculated using the mass property calculator in Analyst v1.5 software (Applied Biosystems). For analytes of molecular weight less than 500 Da, it is expected that the resultant ions will be of single charge as the result of electrospray ionization.
7. An enhanced mass spectrum (EMS) scan or Q3 scan may also be used to determine the experimental m/z of the analyte, however, we chose to use the Q1 scan as quadrupole1 is used for precursor ion selection in MRM analyses in contrast to the linear ion trap ion selector that would be used in either EMS scans or Q3 scans.

8. Declustering potential should be set high enough to reduce chemical noise, but while avoiding fragmentation of the analyte. The higher the DP, the more energy is transferred to the molecules entering the MS and this energy transfer aids in declustering the ions. However, if the energy is too high, the ions may fragment in-source causing a decrease in sensitivity.
9. The most intense ions are usually selected for Q3 transitions, however, in the case of hydroxyproline several of the Q3 fragment ions are identical to those produced by L-leucine and L-isoleucine. For example, the most obvious Q1–Q3 transition would be 132.1–86.1, but this transition is common to Hyp, Leu, and Ile. In this case, the Q3 masses 114.1, 68.0, and 58.1 were selected for Hyp and 69.1 and 58.0 were selected for Leu and Ile.
10. After determination of each Q1–Q3 MRM transition, it is necessary to optimize the collision energy as increasing the collision energy will decrease the intensity of high mass fragment ions and increase the intensity of low mass ions.
11. The optimum collision energy is the energy in V at which the intensity of the extracted ion chromatogram (XIC) of the individual MRM trace is at its maximum.
12. A comparison of a range of HPLC columns was undertaken in a previous study (10). The optimum results were achieved using the Zorbax SB-Aq column (Agilent; 250 mm × 4.6 mm ID) and as such it was selected for use in the analytical method.
13. As the concentration of the standards span several orders of magnitude, standards may be run in the order of increasing concentration to avoid any effects from sample carryover. After each set of standards, 1–2 blank samples should be run. Samples may alternatively be run in random order. This procedure ensures that no sample is unfairly biased.
14. Analyst v1.5 software contains a quantitation module that enables peak integration (Contact AB Sciex for more details).

Acknowledgements

This work was supported by the Cooperative Research Centre for Sheep Innovation. The authors thank Dr. David Hopkins of NSW Agriculture, Cowra, NSW, Australia and Mr. Malcolm Boyce of Murdoch University, WA, Australia for the supply of the freeze-dried muscle tissue samples used in this study. The authors also thank the Molecular and Cellular Proteomics Facility at The University of Queensland Institute for Molecular Bioscience for access to equipment.

References

1. Gelse K, Poschl E, and Aigner T (2003) Collagens – structure, function, and biosynthesis. Adv Drug Deliv Rev 55:1531–1546
2. Fratzl P (2008) Collagen: Structure and Mechanics, Springer, New York
3. Buehler MJ (2006) Nature designs tough collagen: Explaining the nanostructure of collagen fibrils. Proc Natl Acad Sci USA 103:12285–12290
4. Bailey AJ (1972) The basis of meat texture. J Sci Food Agric 23:995–1007
5. Light N et al (1985) The role of epimysial, perimysial and endomysial collagen in determining texture in six bovine muscles. Meat Sci 13:137–149
6. Purslow PP (2002) The structure and functional significance of variations in the connective tissue within muscle. Comp Biochem Physiol A Mol Integr Physiol 133:947–966
7. Etherington DJ (1987) In: Pearson AM Dutson TR and Bailey AJ (eds.) Advances in Meat Research, Van Nostrand Reinhold Co., New York
8. Flores J (1980) Control de calidad de los productos carnicos. Rev Agroquim Tecnol Aliment 20:180
9. Vazquez-Ortiz FA and Gonzalez-Mendez NF (1996) Determination of collagen as a quality index in Bologna from Northwestern Mexico. J Food Comp Anal 9: 269–276
10. Colgrave ML, Allingham PG, and Jones A (2008) Hydroxyproline quantification for the estimation of collagen in tissue using multiple reaction monitoring mass spectrometry. J Chrom A 1212:150–153
11. Mitchell Wells J, and McLuckey SA (2005) Collision-induced dissociation of peptides and proteins. In: Burlingame AL (ed.) Methods in Enzymology, Academic Press
12. MacDougall D, and Crummett WB (1980) Guidelines for data acquisition and data quality evaluation in environmental chemistry. Anal Chem 52:2242–2249

Chapter 24

Sequential Injection Chromatography for Fluorimetric Determination of Intracellular Amino Acids in Marine Microalgae

Marilda Rigobello-Masini and Jorge C. Masini

Abstract

This chapter describes a sequential injection chromatography method to automate the fluorimetric determination of amino acids after precolumn derivatization with *o*-phthaldialdehyde in presence of 2-mercaptoethanol using reverse-phase liquid chromatography in C_{18} silica-based monolithic column. The method is low-priced and based on six steps of isocratic elutions. At a flow rate of 30 μl/s, a 25 mm long-column coupled to 5-mm guard column is capable to separate aspartic acid (Asp), glutamic acid (Glu), asparagine (Asn), serine (Ser), glycine (Gly), threonine (Thr), citrulline (Ctr), arginine (Arg), alanine (Ala), tyrosine (Tyr), phenylalanine (Phe), ornithine (Orn), and lysine (Lys). Under these conditions, histidine (His) and glutamine (Gln), methionine (Met) and valine (Val), and isoleucine (Ile) and leucine (Leu) coelute. The entire cycle of amino acids derivatization, chromatographic separation, and column conditioning at the end of separation lasts 16 min. The method was successfully applied to the determination of the major intracellular free amino acids in the marine green alga *Tetraselmis gracilis*.

Key words: Intracellular amino acids, Sequential injection chromatography, Precolumn derivatization, Fluorimetry, Marine microalgae, Automation, Amino acid analysis

1. Introduction

Amino acids play a key role in plant metabolism. Photosynthetic process and nitrogen assimilation are the main modulators of the types and total concentration of dissolved free amino acids in the intracellular environments of phytoplankton (1). Amino acids excreted by the biota in marine media are responsible for 13–30% of the dissolved nitrogen, which contribute to feed the heterotrophic consumption (2). Thus, these carbon–nitrogen compounds

Michail A. Alterman and Peter Hunziker (eds.), *Amino Acid Analysis: Methods and Protocols*, Methods in Molecular Biology, vol. 828, DOI 10.1007/978-1-61779-445-2_24, © Springer Science+Business Media, LLC 2012

can be used as markers to characterize distinct physiological conditions of the phytoplankton and to help understand oceanic food webs.

Determination of dissolved free amino acids (DFAA) in intra or extracellular fluids can be made by reverse-phase HPLC with fluorescence detection and precolumn derivatization using the *o*-phthaldialdehyde in presence of 2-mercaptoethanol (OPA-2MCE) reagent in borate buffer (pH >9.0) (3). The chromatographic runs can last about 50–60 min (4), consuming large amounts of organic solvents. Modern instrumental developments have allowed amino acid analyses to be made in less than 8 min by Ultra Performance Liquid Chromatography (UPLC™) using either UV (5) or tandem mass spectrometry (6) detection. Anion exchange high-performance chromatography with integrated amperometry waveform direct detection of amino acids is another relatively new but well-established technique widely used for amino acids analyses (7). These modern methods, however, employ expensive and dedicated instrumentation.

Reversed-phase liquid chromatography in monolithic columns has been proposed as an approach to shorten the length of chromatographic analyses using conventional HPLC instrumentation (8–10). Monolithic columns are separation devices that provide low-pressure drops and high rates of mass transfer, being constituted by a continuous rod of solid silica instead of packed particles. These columns have bimodal pore structure of macro and mesopores. Macropores act as a low-pressure flow-through pathway, having about 2 μm in diameter. Mesoporous have diameters of approximately 13 nm, creating a large and uniform surface area on which adsorption takes place, allowing high-performance chromatographic separation to be accomplished at flow rates much higher than those used with packed particles columns (11).

The low backpressure provided by the monolithic columns enabled Satinsky et al. (12) to propose a new approach to perform reverse-phase liquid chromatography coupling a monolithic column to the simple instrumentation of a Sequential Injection Analysis (SIA) system, which is consisted basically by a multi-position selection valve and a syringe pump (13). This new technique was named sequential injection chromatography (SIC), being mostly exploited for determination of pharmaceutical compounds in medicines, although more recent applications have described its application to environmental samples (14). Flow programming and precise time control are remarkable features of sequential injection systems that were explored to develop a protocol for amino acid analysis based on precolumn derivatization. Separation was achieved by five steps of quasi-isocratic elutions followed by fluorimetric detection (15). The method was successfully employed for determination of dissolved free amino acids (DFAA) in intracellular media of a marine microalgae cultivated in batch conditions.

2. Materials

Prepare all solutions using deionized water (18.2 Ω cm resistivity at 25°C).

Salts, acids, and bases must be of analytical grade.

Organic solvents used in mobile phases must be of HPLC grade.

Filter mobile phases through 0.45-μm regenerated cellulose membranes and sonicate them by ultrasound for about 20 min, followed by 30 min degassing with high-purity He prior to the use.

2.1. Buffers

1. Phosphate buffer: 0.1 M KH_2PO_4, pH 7.2: weigh 13.609 g of KH_2PO_4 and dissolve in 800 ml of deionized water. Add 1.0 M KOH solution dropwise to adjust the solution pH to 7.20, monitoring the pH with a combination glass electrode previously calibrated with pH 4.0 and 7.0 buffers (25°C). Complete the volume to 1.0 L in a volumetric flask with deionized water.
2. Borate buffer: 0.40 M borate buffer, pH 9.5: weigh 12.366 g of boric acid and dissolve in 400 ml of deionized water. Add 1.0 M KOH solution dropwise to adjust the solution pH to 9.50, monitoring the pH with a combination glass electrode previously calibrated with pH 7.0 and 9.0 buffers (25°C). Complete the volume to 500 ml in a volumetric flask with deionized water.

2.2. Fluorescence Reagent (OPA-2MCE)

1. OPA-2MCE reagent: dissolve 27 mg of *o*-phthaldialdehyde (OPA) in 0.50 ml of absolute ethanol, add 20 μl of 2-mercaptoethanol (2MCE), and dilute the mixture to a final volume of 5.0 ml with 0.40 M borate buffer at pH 9.5. Protect from light and store at 4°C (see Note 1).

2.3. Mobile Phases

1. Mobile phase 1 (MP_1): Methanol (MeOH):Tetrahydrofuran (THF):10 mM Phosphate buffer (8:1:91): in a 1.0 L volumetric flask, add 100 ml of the 0.10 M phosphate buffer, pH 7.20, 80 ml of MeOH, and 10 ml of THF. Complete to the mark with deionized water.
2. Mobile phase 2 (MP_2): MeOH: 10 mM Phosphate buffer (20:80): in a 100 ml volumetric flask, add 20.0 ml of MeOH (see Note 2), 10.0 ml of the 0.10 M phosphate buffer and complete the volume to the mark with deionized water.
3. Mobile phase 3 (MP_3): MeOH: 10 mM Phosphate buffer (35:65): in a 100 ml volumetric flask, add 35.0 ml of methanol (see Note 2), 10.0 ml of the 0.10 M phosphate buffer and complete the volume to the mark with deionized water.
4. Mobile phase 4 (MP_4): MeOH: 10 mM Phosphate buffer (50:50): In a 100 ml volumetric flask, add 50.0 ml of methanol

(see Note 2), 10.0 ml of the 0.10 M phosphate buffer and complete the volume to the mark with deionized water.

5. Mobile phase 5 (MP_5): MeOH: 10 mM Phosphate buffer (65:35): in a 100 ml volumetric flask add 65.0 ml of methanol (see Note 2), 10.0 ml of the 0.10 M phosphate buffer and complete the volume to the mark with deionized water.

2.4. Apparatus

1. The SIChrom – accelerated liquid chromatography system is provided by FIAlab Instruments (Bellevue, WA, USA). The apparatus scheme is shown on Fig. 1, where SV is a 10-port high pressure (5,000 psi) selection valve constructed in stainless steel, provided by Valco Instruments (Houston, TX, USA).
2. PP is a medium pressure (1,000 psi) piston pump model S17 PDP from Sapphire Engineering™ (Pocasset, MA, USA) with capacity of 4,000 μl (volume of the MP_1 compartment), built in ULTEM, having a ceramic piston (P) for solution propelling and aspiration. A frontal port in PP connects the solvent compartment to the central port of SV by the holding coil (HC),

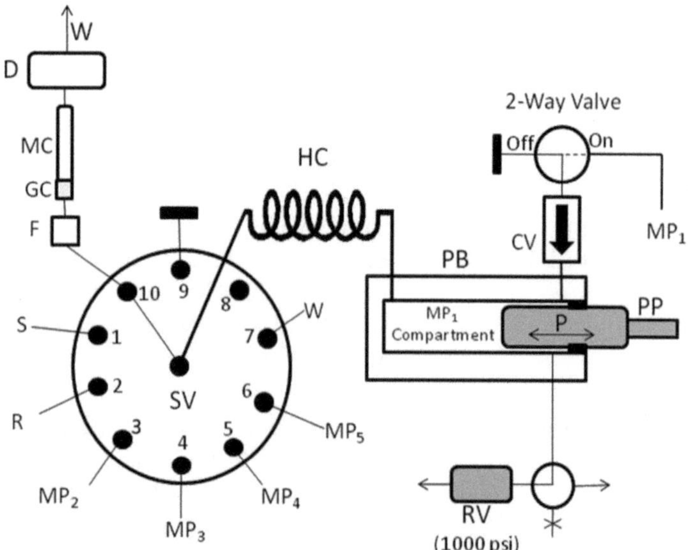

Fig. 1. SIC manifold to perform reverse-phase liquid chromatography separation of amino acids after precolumn derivatization with OPA-2MCE reagent and fluorescence detection. *PP* piston pump, *P* piston, *PB* pump body, *CV* check-valve, *RV* relief valve (open at pressure >1,000 psi), *HC* holding coil (4 m of 0.8 mm I.D. PTFE tubing), *W* waste, *SV* ten port selection valve, *F* in-line 10 μm porosity filter, *GC* C_{18} monolithic guard column, *MC* C_{18} monolithic column, *D* fluorescence detector ($\lambda_{ex.}$ = 340 nm, λ_{em} = 450 nm), *S* sample/standard solution, *R* reagent [0.040 M *o*-phthaldialdehyde in 0.0572 M 2-mercaptoethanol, 10% (*v*/*v*) ethanol and 0.40 M borate buffer at pH 9.5], *MP1* methanol, tetrahydrofuran and 10 mM phosphate buffer at pH 7.2 (8:1:91), *MP2* methanol and phosphate buffer: (20:80), *MP3* methanol and phosphate buffer (35:65), *MP4* methanol and phosphate buffer (50:50), *MP5* methanol and phosphate buffer (65:35). *Black small rectangles* mean that the port is closed with solid Teflon tubing.

which is made of 4 m of 0.8 mm internal diameter Teflon tubing (capacity of 2,000 μl). The rear port of PP is connected to the main solvent reservoir (MP_1) through a check valve (CV) and a two-way solenoid valve (Fig. 1). An additional port in the bottom of the pump body is connected through a four-way valve to a Relief Valve (RV) from Up-Church Scientific (Oak Harbor, WA, USA) that opens at pressure >1,000 psi. Port 10 of SV is connected (see Note 3) to a 10 μm in-line filter and to a 5×4.3 mm long guard column coupled to a 25×4.3 mm reverse-phase C_{18} monolithic column, both from Phenomenex (Torrance, CA, USA) (see Note 4). Pump and valve movements and synchronization, as well as data acquisition from detector are controlled by the FIAlab 5.0 software.

3. Fluorescence detection is performed using a PMT-FL flow-through detector from FIAlab Instruments. Excitation is achieved at 340 nm using a LED as energy source. Emission is measured at 450 nm using an interference filter for wavelength selection. A 1-cm path-length flow-through (10 μl) cuvette can be used for all measurements.

3. Methods

3.1. Cultivation and Extraction

1. Cultivate microalgae without forced aeration at 20°C in 6-L flasks inside a thermostatized incubation chamber using sterilized seawater with addition of Guillard f/2 medium (see Note 5).
2. Count the algal cells daily using a hemocytometer-type chamber in an optic microscope to allow the DFAA concentration to be expressed as μmol/cell. At the fifth cultivation day, during the exponential growth phase, collect sample aliquots of 100 ml and filter 20 ml portions using 0.22-μm cellulose acetate membranes (see Note 6).
3. Extract the intracellular DFAA with 1.0 ml of deionized water after grinding the filter membranes in a porcelain mortar. Centrifuge for 5 min at 1,700×g and store the supernatant extracts in freezer at −20°C until the moment of analysis.

3.2. Analysis

1. The automated procedure of analysis is based on the scheme of the instrument shown in Fig. 1.
   (a) A first nonautomated step is filling the system (MP_1 compartment, holding coil, column and detector) with the mobile phase MP_1.
   (b) Next, the tubing connecting the selection valve (SV) ports with sample (port 1), reagent (port 2), MP_2 (port 3), MP_3 (port 4), MP_4 (port 5), and MP_5 (port 6) are filled with

the respective solution. This is made by aspirating 300 μl of each solution to the holding coil, discarding the excess through the auxiliary waste port (port 7).

(c) For column conditioning, there is a specific program in the library of programs available in the software. For this, the pump aspirates 4,000 μl of MP_1 into the MP_1 compartment and delivers this volume through the column at 30 μl/s. This procedure is repeated at least three times (see Note 7).

2. From this point on, the analysis is completely automated according the following steps:

   (a) Fill the MP_1 compartment: Select port 9 of SV and turn on the two-way solenoid valve. Aspirates 3,900 μl of MP_1 into the MP_1 compartment at 150 μl/s. At the end of the step (see Note 8), turn off the two-way valve (see Note 9).

   (b) Clean the sampling line with representative solution to be analyzed: Select port 1 of SV and aspirate 100 μl of sample/standard into the holding coil (HC) at 100 μl/s. Discard the excess by injecting 200 μl (sample/standard plus MP_1) through port 7 of SV.

   (c) Precolumn derivatization: Set SV to port 1 and aspirate 20 μl of sample into HC at 10 μl/s. Wait the operation finishes. Set SV to port 2 and aspirate 10 μl of OPA-2MCE reagent into HC at 10 μl/s. Wait the operation finishes and repeat all the operations described in this item three more times (see Note 10).

   (d) Complete the capacity of the pump MP_1 reservoir: set SV to port 9, turn on the two-way solenoid valve and aspirate 80 μl (see Note 11) of MP_1 at flow rate of 30 μl/s, filling the pump reservoir with 4,000 μl of MP_1. At the end of the step, turn off the two-way valve.

   (e) First step of elution: Set SV to port 10, connecting HC to the column and detector through the central port in SV. Turn on the acquisition data system at frequency of 8 Hz and integration time of 25 ms. Dispense 4,000 μl of MP_1 through the column at 30 μl/s (see Note 12), eluting Asp, Glu, Asn, and Ser (Fig. 2) (see Note 13).

   (f) Refilling the solvent compartment: Set SV to port 9, turn on the two-way solenoid valve, and aspirate 2,500 μl of MP_1 to the MP_1 compartment at 150 μl/s. After operation finishes, turn off the two-way valve.

   (g) Aspirate MP_2 into the holding coil: Set SV at port 3 and aspirate 1,500 μl of MP_2 into HC at 100 μl/s (see Note 14).

   (h) Second step of elution: Set SV to port 10 and dispense 1,200 μl of solvent (predominantly MP_2) through the column and detector at 30 μl/s, eluting Gln and His (as a single peak).

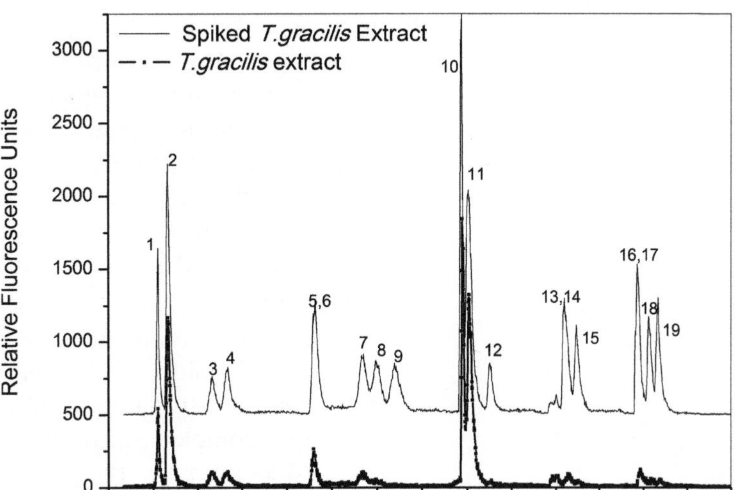

Fig. 2. Chromatogram of an intracellular extract of the algae *Tetraselmis gracilis* superimposed by the chromatogram of the same extract spiked with a mixture of 19 amino acids fluorescent derivatives obtained by reaction with the OPA-2MCE reagent. The chromatogram was obtained at 30 μl/s, using sample and reagent volumes of 80 and 40 μl, respectively. Excitation and emission wavelengths were 340 and 450 nm, respectively. Peak assignment and amino acids concentrations (μM) are as follows: (1) Asp, 5.08; (2) Glu, 6.80; (3) Asn, 3.78; (4) Ser, 5.14; (5,6) His, 10.8 and Gln, 5.36; (7) Gly, 8.80; (8) Thr, 10.2; (9) Ctr, 10.6; (10) Arg, 10.5; (11) Ala, 6.86; (12) Tyr, 5.58; (13,14) Met, 6.80 and Val, 5.36; (15) Phe, 10.0; (16,17) Ile, 4.82 and Leu, 8.54; (18) Orn, 26.7; and (19) Lys, 25.0.

(i) Third step of elution: Set SV at port 3 and aspirate 1,200 μl of MP_2 into HC at 100 μl/s. After finishing, set SV to port 10 and dispense 2,000 μl of solvent through column and detector (predominantly MP_2) at 30 μl/s, eluting Gly, Thr, and Ctr (see Note 15).

(j) Fourth step of elution: Set SV to port 4 and aspirate 2,000 μl MP_3 into HC at 100 μl/s, filling the MP_1 compartment with the remaining MP_1 that was inside HC (see Note 16). After finishing the aspiration step, turn SV to port 10 and dispense 2,000 μl of solvent through column and detector at 30 μl/s (predominantly MP_3), eluting Arg, Ala, and Tyr (see Note 17).

(k) Fifth step of elution: Set SV to port 5 and aspirate 2,000 μl of MP_4 into HC (see Note 18) at 100 μl/s, filling the pump compartment with the remaining MP_1 that was inside HC. After finishing the aspiration step, turn SV to port 10 and dispense 2,000 μl of solvent through column and detector at 30 μl/s (predominantly MP_4), eluting Met and Val as a single peak and Phe (see Note 19).

(l) Sixth step of elution: Set SV to port 6 and aspirate 2,000 μl of MP_5 (see Note 20) into HC at 100 μl/s, filling the MP_1 compartment with the remaining MP_1 that was inside HC. After finishing the aspiration step, turn SV to port 10 and dispense 4,000 μl (see Note 21) of solvent through column and detector at 30 μl/s, eluting Ile and Leu as a single peak, Orn and Lys (see Note 22), starting the column reconditioning. At the end of this step, turn off the acquisition data.

(m) Column reconditioning: Set SV to port 9, turn on the two-way solenoid valve and fill the solvent compartment of the pump with 4,000 μl of MP_1 at 100 μl/s. After the refilling is complete, turn off the two-way valve, set SV at port 10 and empties the pump through the column and detector at 30 μl/s (see Note 23).

4. Notes

1. The OPA-2MCE reagent solution must be aged for 24 h prior to use, and used in a time interval between 1 and 9 days.
2. To assure high reproducibility of retention times from lot-to-lot of mobile phases (MP_2, MP_3, MP_4, and MP_5), these solutions are prepared using burette for transferring methanol and volumetric pipette to transfer the phosphate buffer into the volumetric flasks.
3. Connections of port 10 of SV to the in-line filter, to the precolumn and from the column outlet to flow cell are made with 0.25 mm I.D. PEEK (polyetheretherketone polymer) tubing. Connection of the MP_1 reservoir to the solvent compartment in the pump body is made with 1/8 O.D. Teflon tubing. Port 1 of SV is connected to the sample reservoir (S), and to minimize the sample consumption, this connection is made with 10 cm of 0.25 mm I.D. PEEK tubing (5 μl). Port 2 is connected to the reagent (R) and ports 3–6 are connected to solvent reservoir of increasing methanol concentrations (MP_2–MP_5). These connections are made with 0.5 mm I.D. Teflon tubing.
4. The present configuration of pump and selection valve can handle with 50 mm long column, improving separation, but with limitation in the flow rate, which has to be handled with care to avoid pressure increase that may cause leaking of mobile phase through the relief valve (RV).
5. The Guillard medium is given in ref. 16.

6. To avoid cell rupture during filtration, filter aliquots of up to 20 ml of cultivation media (17).
7. To fill the solvent compartment with the mobile phase MP_1, the piston of PP is pushed forward (via software) and SV is set to port 9, which is closed with solid Teflon tubing; the two-way solenoid valve is turned on, connecting the solvent compartment of the pump and the solvent reservoir bottle. Then, the piston retracts aspirating MP_1 into the MP_1 compartment; no solutions connected to SV are aspirated because port 9 is selected. When the pump is delivering mobile phase to the column (port 10), the check valve (CV) prevents the returning of MP_1 to the reservoir bottle.
8. If the pump is aspirating or dispensing solution, a "delay until done" command is sent to the system, avoiding change of valve positions during pump operation. This command avoids strong variations of pressure, which can lead to overpressure in case of dispensing solution to the column, or surge of bubbles in case of aspiration.
9. If the two-way valve is off, the pump can aspirate solutions connected to the selection valve without entrance of MP_1 (from the solvent reservoir) into the MP_1 compartment.
10. In the present case, total sample and reagent volumes are 80 and 40 μl, respectively. These volumes are fractionated four times to facilitate interpenetration of sample and reagent zones. The flow rate is decreased in this step to increase the residence time of the reaction zone in the system. This approach allows the reaction to proceed almost to completion. The high precision of the pump provides high reproducibility of volumes and reaction time, leading to highly reproducible results. Additionally, the sample volume is a parameter directly related to sensitivity: increase in the sample volume increases the sensitivity if reagent is in stoichiometric excess along the sample zone; decreasing the sensitivity is achieved by decreasing the sample volume, an approach that can be used in case of concentrated samples.
11. The volume to complete the 4,000 μl depends on the sample and reagent volumes used.
12. Amino acid analyses have been made at flow rates between 10 and 40 μl/s (15). At 10 μl/s an improvement in the separation is achieved in comparison with the higher flow rates, but the chromatographic run lasts more than 25 min. Flow rate of 40 μl/s make the separation worse and leaking of mobile phase through the relief valve (RV, Fig. 1) is likely to occur, so that 30 μl/s is a compromise option.
13. Linear response between relative fluorescence units (RFU) read at peak maximum and amino acid concentration is observed

for concentrations between 0.50 and 12.5 μM. The limits of detection (LOD) are 0.25 μM for the four amino acids eluted in this step.

14. At the end of this step, MP_2 will fill the portion of HC closer to the central port of SV and column. MP_1 that was inside HC goes back to MP_1 compartment. Reproducible interdispersion between MP_1 and MP_2 occurs at the interfaces of these solutions inside the tubing. Upon reversing the flow direction toward the column, the frontal portion of MP_2 will suffer some dispersion in MP_1 which was filling the column, but MP_1 is quickly changed by MP_2 increasing the solvent strength, facilitating the elution of the compounds not eluted by MP_1. The reproducible dispersion of one mobile phase in another can be explored to create limited solvent gradients in the holding coil (18).
15. Linear response is observed in the concentration ranges: 0.50–12.5 μM for Gly, and 1.25–25.0 μM for Thr and Ctr. The LOD values are 0.31 μM for Gly, 0.42 μM for Thr, and 0.42 μM for Ctr.
16. Similar to Note 14, but with MP_3 taking place in HC.
17. Linear response is observed for concentration ranges 1.30–26.0 μM of Arg, 0.80–17 μM of Ala, and 0.70–14 μM of Tyr. The LOD values are 0.46 μM for Arg, 0.21 μM for Ala, and 0.50 M for Tyr.
18. Similar to Note 14, but with MP_4 taking place in HC.
19. Linear response is observed for Phe concentrations between 0.70 and 14 μM, with LOD of 0.50 μM.
20. Similar to Note 14, but with MP_5 taking place in HC.
21. This volume corresponds to 2,000 μl of MP_5 plus MP_1 that was filling the MP_1 compartment.
22. Linear response for Orn and Lys is observed for the concentration range 3.0–75 μM; LOD values are 1.5 and 0.70 μM for Orn and Lys, respectively.
23. At end of this step, the column is reconditioned with MP_1, leading to reproducible retention times among sequential chromatographic analysis.

Acknowledgments

This work was supported by FAPESP (grant 2006/06410-2) and CNPq (grant 4769921/2007-9. Fellowships from CNPq (processes 307761/2006-1 and 151521/2007-7) are acknowledged.

References

1. Noctor G, de Paepe R, Foyer CH (2007) Mitochondrial redox biology and homeostasis in plants. Trends Plant Sci 12:125–134
2. Braven J, Evens R, Butler EI (1984) Amino acids in sea water. Chem Ecol 2:11–21
3. Lindroth P, Mopper K (1979) High-performance liquid-chromatographic determination of subpicomole amounts of amino-acids by pre-column fluorescence derivatization with ortho-phthaldialdehyde. Anal Chem 51:1667–1674
4. Pereira V et al (2008) Simultaneous analysis of free amino acids and biogenic amines in honey and wine samples using in loop orthophthala-ldeyde derivatization procedure. J. Chromatogr A 1189:435–443
5. Boogers I et al (2008) Ultra-performance liquid chromatographic analysis of amino acids in protein hydrolysates using an automated pre-column derivatisation method. J Chromatogr A 1189:406–409
6. Freeto S et al (2007) A rapid ultra performance liquid chromatography tandem mass spectrometric method for measuring amino acids associated with maple syrup urine disease, tyrosinaemia and phenylketonuria. Ann Clin Biochem 44:474–481
7. Clarke AP et al (1999) An integrated amperometry waveform for the direct, sensitive detection of amino acids and amino sugars following anion-exchange chromatography. Anal Chem 71:2774–2781
8. Dawson LA et al (2004) Rapid high-throughput assay for the measurement of amino acids from microdialysates and brain tissue using monolithic C-18-bonded reversed-phase columns. J Chromatogr B 807:235–241
9. Devall AJ et al (2007) Monolithic column-based reversed-phase liquid chromatography separation for amino acid assay in microdialysates and cerebral spinal fluid. J Chromatogr B 848:323–328
10. Penteado JCP et al (2009) Fluorimetric determination of intra- and extracellular free amino acids in the microalgae *Tetraselmis gracilis* (Prasinophyceae) using monolithic column in reversed phase mode. J Sep Sci 32:2827–2834
11. Cabrera K (2004) Applications of silica-based monolithic HPLC columns. J Sep Sci 27:843–852
12. Satinsky D et al (2003) Monolithic columns – a new concept of separation in the sequential injection technique. Anal Chim Acta 499, 205–214
13. Ruzicka J, Marshall GD (1990) Sequential injection – A new concept for chemical sensors, process analysis and laboratory assays. Anal Chim Acta 237:329–343
14. Chocholous P, Solich P, Satinsky D (2007) An overview of sequential injection chromatography. Anal Chim Acta 600:129–135
15. Rigobello-Masini M et al (2008) Implementing stepwise solvent elution in sequential injection chromatography for fluorimetric determination of intracellular free amino acids in the microalgae *Tetraselmis gracilis*. Anal Chim Acta 628:123–132
16. Guillard RRL, Ryther JH (1962) Studies of marine planktonic diatoms. I. *Cyclotella nana* Hustedt, and *Detonula confervacea* (Cleve). Can J Microbiol 8:229–239
17. Fuhrman JA, Bell TM (1985) Biological considerations in the measurement of dissolved free amino acids in seawater and implications for chemical and microbiological studies. Mar Ecol Prog Ser 25:13–21
18. Lähdesmäki I et al (2008) Sequential injection chromatography: automated sample preparation, derivatization and separation of amino acids. FIAlab Instruments. http://www.flowinjection.com/Brochures/sichrom.pdf. Accessed 25 November 2010

Chapter 25

Direct Analysis of Underivatized Amino Acids in Plant Extracts by LC-MS-MS

Björn Thiele, Nadine Stein, Marco Oldiges, and Diana Hofmann

Abstract

In this chapter, we describe a method for quantification of 20 proteinogenic amino acids as well as 13 ^{15}N-labeled amino acids by liquid chromatography–mass spectrometry without the need for derivatization and use of organic solvents. Analysis of the underivatized amino acids is performed by liquid chromatography–electrospray ionization-tandem mass spectrometry (LC-ESI-MS-MS) in the positive ESI mode. Separation is achieved on a strong cation exchange (SCX) column (Luna 5 μm SCX 100 Å) with 30 mM ammonium acetate in water (A) and 5% acetic acid in water (B). Quantification is accomplished by use of d_5-phenylalanine as internal standard achieving limits of detection of 0.1–3.0 μM. The method was successfully applied for the determination of proteinogenic and ^{15}N-labeled amino acids in plant extracts.

Key words: Liquid chromatography–electrospray ionization-tandem mass spectrometry, Amino acids, Strong cation exchange column, Plant metabolism, Amino acid analysis

1. Introduction

Amino acids as one main class of primary plant metabolites are not only playing a key role in the fixation of nitrogen from nitrate but also in intracellular and intercellular nitrogen transport and in the biosynthesis of numerous other metabolites (1–4). Plant reactions to biotic and abiotic factors cause changes in amino acid levels (5). Therefore, plant physiological investigations to the stress-induced regulation of amino acid levels demand powerful analytical methods. Analysis of amino acids has been commonly performed by precolumn derivatization with *ortho*-phthalaldehyde (OPA) and/or 9-fluorenylmethyl chloroformate (FMOC) followed by reversed-phase HPLC and fluorescence detection (6–8).

Michail A. Alterman and Peter Hunziker (eds.), *Amino Acid Analysis: Methods and Protocols*, Methods in Molecular Biology, vol. 828, DOI 10.1007/978-1-61779-445-2_25, © Springer Science+Business Media, LLC 2012

Spectrophotometric detection systems, however, are unable for the simultaneous determination of ^{14}N- and ^{15}N-labeled amino acids, which is essential for ^{15}N-labeling experiments. For that reason only mass spectrometry can be considered as a detection system. Numerous methods had been published for the determination of amino acids using gas chromatography–mass spectrometry (GC-MS) (9–12), capillary electrophoresis–mass spectrometry (CE-MS) (13–15) or liquid chromatography–mass spectrometry (LC-MS) (16–20). GC requires derivatization of the amino acids which has been effectively performed with, e.g., chloroformates (21, 22). Although this GC method shows high sensitivity and excellent resolution, it has the drawback that a laborious manual derivatization procedure is necessary and the method fails in the determination of Arg. The advantage of CE-MS and LC-MS is that all nonvolatile polar amino acids can be analyzed without derivatization. A CE-ESI-MS method has been developed for the determination of free amino acids using an electrolyte at low pH (14, 15). Short analysis times, selectivity, and almost no matrix interferences have been reported for this CE method. Reversed-phase liquid chromatography (RPLC) is commonly used to analyze nonvolatile components in biological matrices. However, polar low molecular weight compounds such as hydrophilic amino acids are not sufficiently retained and consequently alternatives are required. Ion pairing reagents such as perfluorinated carboxylic acids had been presented to improve the separation of amino acids on C18 columns and to be suitable for coupling of LC with ESI-MS (16–19). However, the use of ion pairing reagents in combination with ESI-MS might lead to difficulties such as ion suppression, memory effects, and contamination of the MS source (23, 24). Hydrophilic LC columns in combination with hydrophobic mobile phases offer an opportunity for separation of polar analytes. This variant of liquid chromatography was termed hydrophilic interaction liquid chromatography (HILIC) (25). HILIC-ESI-MS/MS was successfully employed to separate and quantify amino acids in different matrices (26, 27). A further alternative is the separation on ion exchange columns which had been carried out only with electrochemical detectors so far (28, 29).

In this chapter, we report a protocol for extraction and determination of amino acids from plant samples. Chromatographic separation on a strong cation exchange (SCX) column in combination with ESI-MS-MS has enabled the simultaneous determination of 20 ^{14}N- and 13 ^{15}N-amino acids without derivatization. Organic solvents as eluents in HPLC are not needed (30).

2 Materials

Prepare all aqueous solutions with ultrapure water (18 MΩ cm at 25°C) and analytical grade reagents.

2.1. Plant Sample Extraction

1. Scalpel.
2. Tweezer.
3. Crucible tongs.
4. Porcelain mortar (∅ 63 mm) and pestle.
5. Polystyrene box (see Note 1).
6. 1 l Dewar vessel.
7. Liquid nitrogen.
8. Latex and cotton gloves (see Note 2).
9. 1.5 ml Eppendorf vials (P/N 0030 120.086).
10. Extraction solvent: 0.05 M HCl–ethanol (1:1, v/v). Add about 50 ml water to a 100-ml volumetric flask and transfer 0.41 ml concentrated HCl (12 M) to it. Make up to 100 ml with water. Mix 50 ml 0.05 M HCl and 50 ml ethanol in a 125-ml polypropylene narrow mouth bottle.
11. Internal standard solution (1.0 mM d_5-Phe): Add about 5 ml water to a 10-ml volumetric flask. Weigh 17.02 mg d_5-Phe and transfer to the flask. By making up to 10 ml with water a 10.0-mM stock solution is obtained. Dilute this solution to 1 mM. Store the 10.0 mM stock solution at –20°C and the 1 mM internal standard at 4°C.
12. Centrifuge MiniSpin (Eppendorf, Hamburg, Germany).

2.2. LC-ESI-MS-MS

1. Liquid chromatography–electrospray ionization-tandem mass spectrometry (LC-ESI-MS-MS) system: Agilent 1100 series HPLC (Santa Clara, CA, USA) equipped with a binary pump, a vacuum degasser, and a thermostated column oven.
2. HTC Pal autosampler (CTC Analytics, Zwingen, Switzerland) equipped with a 20-μl sample loop.
3. Thermo Electron (Waltham, MA, USA) TSQ Quantum triple quadrupole mass spectrometer with a electrospray ionization source.
4. Luna 5 μm SCX 100 Å column (150 × 2.0 mm internal diameter) with a security cartridge (4 × 2.0 mm) (Phenomenex, Torrance, CA, USA), screwed on top (see Note 3).
5. Elution solvents: A: 5% acetic acid: Add 50 ml concentrated acetic acid to a 1-l volumetric flask and make up to 1 l with water. B: 30 mM ammonium acetate (pH 6.0): Prepare a 150-mM

NH_4AC stock solution by adding 5.781 g NH_4AC in a 500-ml volumetric flask and making up to 500 ml with water. Store it at 4°C in a brown glass bottle. Add 100 ml 150 mM NH_4AC stock solution to a 500-ml volumetric flask in order to obtain a 30-mM NH_4AC solution. Adjust the pH to 6.0 with concentrated acetic acid (see Note 4) and filter through a 0.2-µm nylon filter. Store at 4°C in a brown glass bottle (see Note 5).

6. 10 mM (^{15}N-)amino acid stock solutions/200 µM mixed (^{15}N-)amino acids standard: According to Tables 1 and 2 10 mM stock solutions of each amino acid and ^{15}N-amino acid, respectively,

Table 1
Amounts of 20 amino acids and d_5-Phe (IStd) for preparation of standard stock solutions

No.	Amino acid	Molecular weight (g/mol)	Mass (mg)[a]
1	Ala	89.09	8.9
2	Gly	75.07	7.5
3	Val	117.15	11.7
4	Leu	131.17	13.1
5	Ile	131.17	13.1
6	Thr	119.12	11.9
7	Ser	105.09	10.5
8	Pro	115.13	11.5
9	Asp	133.10	13.3
10	Met	149.21	14.9
11	Glu	147.13	14.7
12	Phe	165.19	16.5
13	Lys	146.19	14.6
14	His	155.16	15.5
15	Tyr	181.19	18.1
16	C-C	240.30	24.0
17	Arg	174.20	17.4
18	Trp	204.23	20.4
19	Asn	132.12	13.2
20	Gln	146.15	14.6
21	d_5-Phe (IStd)	170.22	17.0

[a] For preparation of 10 ml 10 mM solution

Table 2
Amounts of 13 ^{15}N-amino acids for preparation of standard stock solutions

No.	^{15}N-amino acid	Molecular weight (g/mol)	Mass (mg)[a]
1	^{15}N-Ala	90.09	9.0
2	^{15}N-Gly	76.06	7.6
3	^{15}N-Val	118.14	11.8
4	^{15}N-Leu	132.17	13.2
5	^{15}N-Ile	132.17	13.2
6	^{15}N-Ser	106.09	10.6
7	^{15}N-Met	150.20	15.0
8	^{15}N-Glu	148.12	14.8
9	^{15}N-Phe	166.18	16.6
10	^{15}N-Lys	147.18	14.7
11	^{15}N-Tyr	182.18	18.2
12	$^{15}N_2$-Arg	176.19	17.6
13	^{15}N-Gln	147.14	14.7

[a] For preparation of 10 ml 10 mM solution

are prepared in water and stored at −20°C. Prepare a 200-μM mixed amino acids standard and a 200 μM mixed ^{15}N-amino acids standard by combining 200 μl of each (^{15}N-)amino acid stock solution in 10-ml volumetric flasks and making up to 10 ml with water.

3. Methods

3.1. Plant Sample Extraction

1. All steps of leaf grinding are carried out with tools cooled with liquid nitrogen prior to use.
2. For handling the cooled tools wear the cotton/latex gloves (see Note 2).
3. After harvesting the leaves they are immediately frozen in liquid nitrogen.
4. Weigh approximately 100 mg in a mortar and grind it to a fine powder in liquid nitrogen by use of a pestle.
5. Add 0.5 ml of the extraction solvent [0.05 M aqueous HCl–ethanol (1:1, v/v)] and 10 μl of 1 mM d_5-Phe as an internal standard to the leaf powder and continue grinding.

6. Transfer the homogenate to an Eppendorf vial with additional 0.5 ml extraction solvent.
7. Centrifuge the sample solution at 12,100×*g* for 15 min at 4°C. The supernatant is directly used for analysis.

3.2. LC-ESI-MS-MS Analysis

1. Operate the mass spectrometer in the ESI(+) mode. The ion spray voltage is set at 4 kV. Nitrogen is used as sheath gas (25 psi) and auxiliary gas (1.0 arbitrary unit). The ion transfer capillary is heated to 300°C.
2. Determine the MS-MS parameters for each (^{15}N-)amino acid by flow injection analysis (FIA) of amino acid standard solutions in order to run amino acid analysis in the multiple reaction monitoring (MRM) mode. Argon is used as a collision gas at a pressure of 1.0 mTorr. The scan time of the mass spectrometer is set at 0.4 s. Let the amino acid standard solutions infuse in the flow of the HPLC via a tee piece (see Note 6). Determine the tube lens voltages and collision energies for all amino acids (see Note 7).
3. Flush a new SCX column with 150 mM ammonium acetate solution at 0.2 ml/min overnight followed by flushing with 30 mM ammonium acetate (pH 6, solvent A) – 5% acetic acid (solvent B) (12.5:87.5) for 4 h prior to use (see Note 8).
4. Set the column oven temperature at 40°C. Perform the analysis of amino acids with two consecutive isocratic elution steps at a flow rate of 0.2 ml/min: Start with A:B (12.5:87.5) for 42 min; change the composition to 100% A within 2 min and hold it for 19 min; change it to the initial condition within 2 min and keep it constant for 10 min.
5. Calibration standards: Prepare a series of three calibration standards using the 200 µM mixed amino acids standard and the 1 mM d_5-Phe standard:
   (a) 5 µM: Add 25 µl 200 µM mixed amino acids standard, 10 µl 1 mM d_5-Phe standard, and 965 µl water in a 1.5-ml sample vial.
   (b) 10 µM: Add 50 µl 200 µM mixed amino acids standard, 10 µl 1 mM d_5-Phe standard, and 940 µl water in a 1.5-ml sample vial.
   (c) 20 µM: Add 100 µl 200 µM mixed amino acids standard, 10 µl 1 mM d_5-Phe standard, and 890 µl water in a 1.5-ml sample vial. Do exactly the same for preparation of ^{15}N-amino acids calibration standards.
6. Run the LC-MS-MS analysis of calibration standards and plant samples acquiring the chromatograms in the MRM mode. Control of the LC-MS system, as well as data acquisition is performed by the XCalibur Version 1.3 software.

3.3. Method Validation and Quantification

1. The analytical method has been evaluated with regard to linearity, accuracy, precision, and limits of detection (30) (see Note 9).
2. Peak detection and integration are automatically carried out by XCalibur Version 1.3 software. The areas of those peaks with the same retention times as the calibration standard as well as concurrent MS-MS spectra are integrated (Fig. 1). Nevertheless, check the chromatograms for correctness of peak integration manually (see Note 10).
3. Export the quantification results to Excel and perform the calculation of each individual amino acid concentration applying the internal standard method. Carry out statistical analysis like average and relative standard deviation (RSD).
4. The concentrations of ^{15}N-amino acids are raised by the amounts of the ^{13}C-isotopes (1.08% relative abundance) of ^{14}N-amino acids. Therefore, correct ^{15}N-amino acid concentrations by the equation:

$c_{15N/corr} = c_{15N} - c_{14N} \times 0.0108 \times n_C$	$c_{15N/corr}$	Corrected ^{15}N-amino acid concentration
	c_{15N}	Measured ^{15}N-amino acid concentration
	c_{14N}	Measured ^{14}N-amino acid concentration
	n_C	Number of C atoms of amino acid

4. Notes

1. Polystyrene boxes which are used for transport of solvents are well suited.
2. Put on the cotton gloves first then the latex gloves. This combination allows a still comfortable handling of deep-frozen tools and protects from freezing a while.
3. Connection of injector port to the security cartridge is made with 0.13 mm I.D. steel capillary while the analytical column is connected to the ESI source via 0.13 mm PEEK (polyetheretherketone) capillary.
4. pH adjustment is easily done by the addition of four drops of concentrated acetic acid from a Pasteur pipette.
5. A common problem with ammonium acetate solutions is the rapid growth of microalgae especially at room temperature. Besides storage of the solution in brown glass bottles, it is recommended to filter it through 0.2 μm nylon filter daily. Due to its application on top of the HPLC system at daylight and room

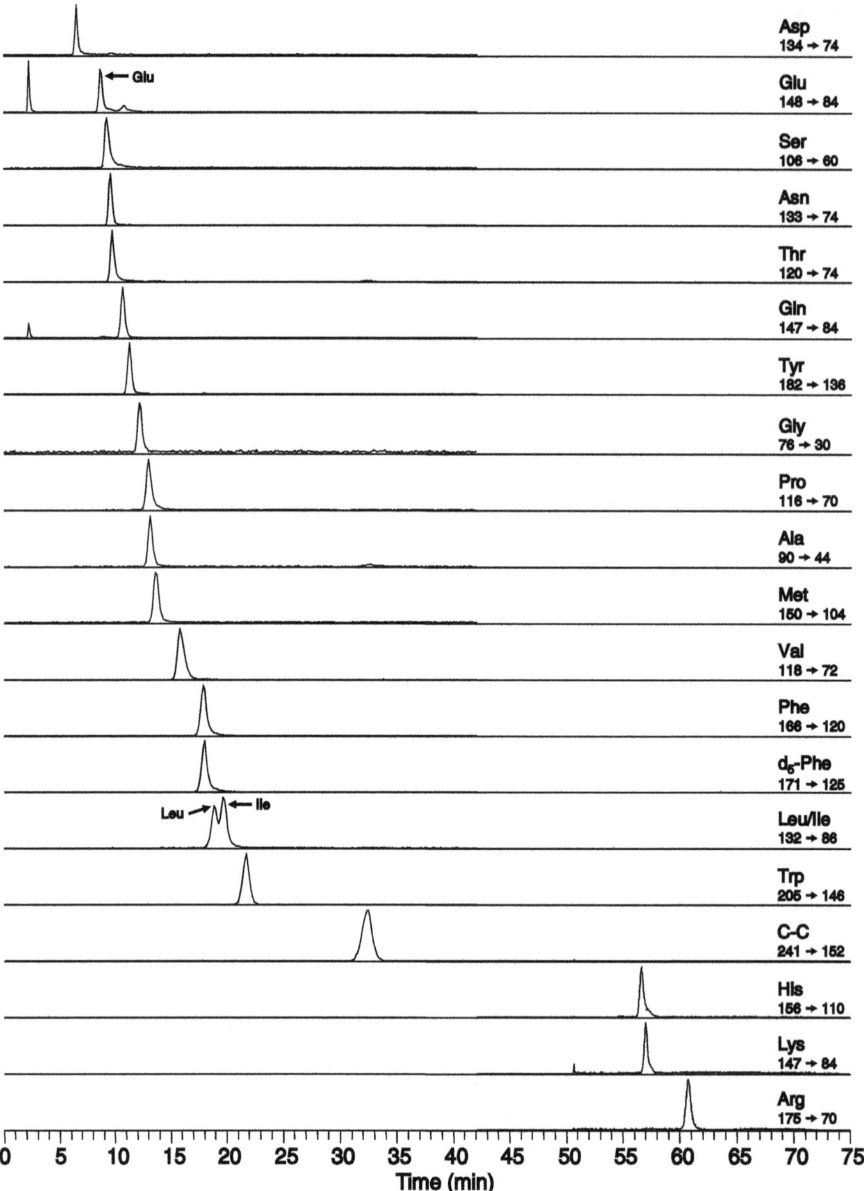

Fig. 1. Selected ion chromatograms of 20 amino acids (20 μM each) and d_5-Phe (10 μM) obtained in MRM mode. Separation was achieved isocratically at pH 2.5 during the first 42 min and in a second isocratic step at pH 6.0 for 19 min using a SCX column. The numbers in the *right corner* are *m/z* ions of Q1 (protonated precursor ion) and Q3 (product ion) in MRM for each analyte.

temperature the solution should not be used later than 1 week and better prepared fresh.

6. Conditions of the FIA experiments: flow rate of the syringe pump, 5 μl/min; flow rate of the HPLC, 0.2 ml/min (30 mM ammoniumacetate–5% acetic acid, 12.5:87.5).
7. Tables 3 and 4 show the collision-induced dissociation (CID) product ions of 20 amino acids and 13 ^{15}N-amino acids together

Table 3
Precursor and product ions for LC-ESI-MS-MS analysis of 20 underivatized amino acids and d_5-Phe (internal standard) with their optimized values for collision energy and tube lens

Compound	Precursor ion $[M+H]^+$ (*m/z*)	Product ion (*m/z*)	Collision energy (eV)	Tube lens (*V*)
Asp	134	74 $[HO_2C–CH=NH_2]^+$	22	151
Glu	148	84 $[C_4H_6NO]^{+a}$	22	158
Ser	106	60 $[R–CH=NH_2]^+$	20	131
Asn	133	74 $[HO_2C–CH=NH_2]^+$	22	158
Thr	120	74 $[R–CH=NH_2]^+$	16	160
Gln	147	84 $[C_4H_6NO]^{+a}$	24	151
Tyr	182	136 $[R–CH=NH_2]^+$	20	157
Gly	76	30 $[R–CH=NH_2]^+$	16	140
Pro	116	70 $[C_4H_8N]^{+b}$	20	165
Ala	90	44 $[R–CH=NH_2]^+$	14	168
Met	150	104 $[R–CH=NH_2]^+$	16	156
Val	118	72 $[R–CH=NH_2]^+$	16	151
Phe	166	120 $[R–CH=NH_2]^+$	20	180
d_5-Phe	171	125 $[R–CH=NH_2]^+$	18	152
Leu	132	86 $[R–CH=NH_2]^+$	14	170
Ile	132	86 $[R–CH=NH_2]^+$	14	170
Trp	205	146 $[M+H–59]^+$	24	180
C–C	241	152 $[M+H–89]^+$	20	155
His	156	110 $[R–CH=NH_2]^+$	20	145
Lys	147	84 $[C_5H_{10}N]^{+c}$	24	166
Arg	175	70 $[C_4H_8N]^{+b}$	28	156

a
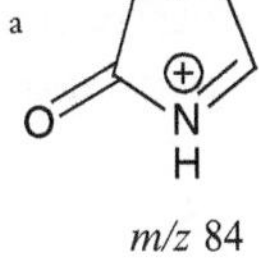

m/z 84

b
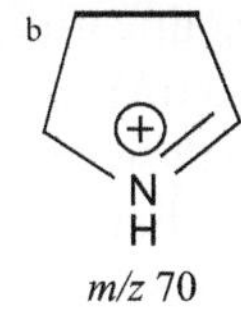

m/z 70

c
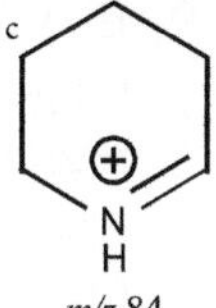

m/z 84

Table 4
Precursor and product ions for LC-ESI-MS-MS analysis of 13 underivatized ^{15}N-labeled amino acids with their optimized values for collision energy and tube lens

Compound	Parent ion $[M+H]^+$ (*m/z*)	Product ion (*m/z*)	Collision energy (eV)	Tube lens (*V*)
^{15}N-Glu	149	85 $[C_4H_6{}^{15}NO]^{+a}$	20	138
^{15}N-Ser	107	61 $[R–CH={}^{15}NH_2]^+$	16	113
^{15}N-Gln	148	84 $[C_4H_6NO]^{+a}$	24	150
^{15}N-Tyr	183	137 $[R–CH={}^{15}NH_2]^+$	10	169
^{15}N-Gly	77	60 $[M+H–{}^{15}NH_2]^+$	10	121
^{15}N-Ala	91	45 $[R–CH={}^{15}NH_2]^+$	18	193
^{15}N-Met	151	92 $[M+H–59]^+$	18	117
^{15}N-Val	119	73 $[R–CH={}^{15}NH_2]^+$	16	142
^{15}N-Phe	167	121 $[R–CH={}^{15}NH_2]^+$	18	143
^{15}N-Leu	133	87 $[R–CH={}^{15}NH_2]^+$	16	156
^{15}N-Ile	133	87 $[R–CH={}^{15}NH_2]^+$	16	156
^{15}N-Lys	148	131 $[M+H–{}^{15}NH_2]^+$	10	139
$^{15}N_2$-Arg	177	70 $[C_4H_8N]^{+b}$	36	150

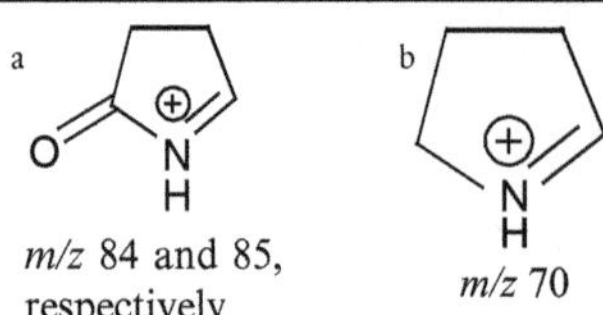

with their tube lens voltages and collision energies obtained with the TSQ Quantum mass spectrometer in ESI(+) mode.

8. With decline of the column performance due to matrix contamination column flushing should be performed again.
9. In summary, linearity was determined in the concentration ranges 0.5–50 μM for amino acids and ^{15}N-amino acids. Linear regression analysis from calibration curves showed correlation coefficients R^2 between 0.9943 and 0.9999. Method of detection limits corresponding to concentrations with signal-to-noise ratios of three were in the range 0.1–3.0 μM. The interday precision of all amino acids for unspiked leaf samples was less than 10% RSD and for spiked samples less than 8% RSD. Spike recoveries for most amino acids were within 80–120%, with the exception of cystine (C–C) (75%), His (70%), and Lys (64%).
10. The peaks of closely eluted Leu/Ile have to be particularly checked with regard to correct attribution and integration.

Acknowledgments

The authors are grateful to Matthias Gehre (Helmholtz Centre for Environmental Research – UFZ Leipzig, Germany) for providing the ^{15}N-labeled amino acids. We also like to thank Bernd Kastenholz for his assistance in the optimization of the leaf extraction procedure.

References

1. Atkins C and Beevers L (1990) Synthesis, transport and utilization of translocated solutes of nitrogen. In: Abroyl, YP (ed) Nitrogen in higher plants, John Wiley and Sons, New York
2. Guerrero MG, Vega JM and Losada M (1981) The assimilatory nitrate-reducing system and its regulation. Annu Rev Plant Physiol 32: 169–204
3. Miflin BJ and Lea PJ (1977) Amino acid metabolism. Annu Rev Plant Physiol 28:299–329
4. Hewitt EJ (1975) Assimilatory nitrate-nitrite reduction. Annu Rev Plant Physiol 26:73–100
5. Fritz C et al (2006) Impact of the C-N status on the amino acid profile in tobacco source leaves. Plant Cell Environ 29:2055–2076
6. Roth M (1971) Fluorescence reaction for amino acids. Anal Chem 43:880–882
7. Einarsson S, Josefsson B and Lagerkvist S (1983) Determination of amino acids with 9-fluorenylmethyl chloroformate and reversed-phase high-performance liquid chromatography. J Chromatogr 282:609–618
8. Blankenship DT et al (1989) High-sensitivity amino acid analysis by derivatization with o-phthalaldehyde and 9-fluorenylmethyl chloroformate using fluorescence detection: Applications in protein structure determination. Anal Biochem 178:227–232
9. Calder AG et al (1999) Quantitation of blood and plasma amino acids using isotope dilution electron impact gas chromatography/mass spectrometry with U-^{13}C amino acids as internal standards. Rapid Commun Mass Spectrom 13:2080–2083
10. Wood PL, Khan MA and Moskal JR (2006) Neurochemical analysis of amino acids, polyamines and carboxylic acids: GC-MS quantitation of tBDMS derivatives using ammonia positive chemical ionization. J Chromatogr B 831: 313–319
11. Chace DH (2001) Mass spectrometry in the clinical laboratory. Chem Rev 101:445–478
12. Husek P and Simek P (2006) Alkyl chloroformates in sample derivatization strategies for GC analysis. Review on a decade use of the reagents as esterifying agents. Curr Pharm Anal 2:23–43
13. Monton MRN and Soga T (2007) Metabolome analysis by capillary electrophoresis-mass spectrometry. J Chromatogr A 1168:237–246
14. Soga T et al (2004) Qualitative and quantitative analysis of amino acids by capillary electrophoresis-electrospray ionization-tandem mass spectrometry. Electrophoresis 25:1964–1972
15. Soga T and Heiger DN (2000) Amino acid analysis by capillary electrophoresis electrospray ionization mass spectrometry. Anal Chem 72:1236–1241
16. Chaimbault P et al (1999) Determination of 20 underivatized proteinic amino acids by ion-pairing chromatography and pneumatically assisted electrospray mass spectrometry. J Chromatogr A 855:191–202
17. Petritis K et al (2000) Parameter optimization for the analysis of underivatized protein amino acids by liquid chromatography and ionspray tandem mass spectrometry. J Chromatogr A 896:253–263
18. Qu J et al (2002) Rapid determination of underivatized pyroglutamic acid, glutamic acid, glutamine and other relevant amino acids in fermentation media by LC-MS-MS. Analyst 127:66–69
19. Piraud M et al (2005) Ion-pairing reversed-phase liquid chromatography/electrospray ionization mass spectrometric analysis of 76 underivatized amino acids of biological interest: a new tool for the diagnosis of inherited disorders of amino acid metabolism. Rapid Commun Mass Spectrom 19:1587–1602
20. Armstrong M, Jonscher K and Reisdorph NA (2007) Analysis of 25 underivatized amino acids in human plasma using ion-pairing reversed-phase liquid chromatography/time-of-flight mass spectrometry. Rapid Commun Mass Spectrom 21: 2717–2726
21. Husek P. and Sweeley CC (1991) Gas chromatographic separation of protein amino acids

in 4 minutes. J High Resol Chromatogr 14: 751–753

22. Husek P (1991) Rapid derivatization and gas chromatographic determination of amino acids. J Chromatogr 552:289–299

23. Bonfiglio R et al (1999) The effects of sample preparation methods on the variability of the electrospray ionization response for model drug compounds. Rapid Commun Mass Spectrom 13:1175–1185

24. Hsieh Y et al (2001) Quantitative screening and matrix effect studies of drug discovery compounds in monkey plasma using fast-gradient liquid chromatography/tandem mass spectrometry. Rapid Commun Mass Spectrom 15:2481–2487

25. Alpert AJ (1990) Hydrophilic-interaction chromatography for the separation of peptides, nucleic acids and other polar compounds. J Chromatogr A 499:177–196

26. Langrock T, Czihal P and Hoffmann R (2006) Amino acid analysis by hydrophilic interaction chromatography coupled on-line to electrospray ionization mass spectrometry. Amino Acids 30:291–297

27. Schlichtherle-Cerny H, Affolter M and Cerny C (2003) Hydrophilic interaction liquid chromatography coupled to electrospray mass spectrometry of small polar compounds in food analysis. Anal Chem 75:2349–2354

28. Welch LE et al (1989) Comparison of pulsed coulometric detection and potential-sweep-pulsed coulometric detection for underivatized amino acids in liquid chromatography. Anal Chem 61:555–559

29. Luo P, Zhang F and Baldwin RP (1991) Constant-potential amperometric detection of underivatized amino acids and peptides at a copper electrode. Anal Chem 63: 1702–1707

30. Thiele B et al (2008) Analysis of amino acids without derivatization in barley extracts by LC-MS-MS. Anal Bioanal Chem 391:2663–2672

Chapter 26

Wheat Gluten Amino Acid Analysis by High-Performance Anion-Exchange Chromatography with Integrated Pulsed Amperometric Detection

Ine Rombouts, Bert Lagrain, Lieve Lamberts, Inge Celus, Kristof Brijs, and Jan A. Delcour

Abstract

This chapter describes an accurate and user-friendly method for determining amino acid composition of wheat gluten proteins and their gliadin and glutenin fractions. The method consists of hydrolysis of the peptide bonds in 6.0 M hydrochloric acid solution at 110°C for 24 h, followed by evaporation of the acid and separation of the free amino acids by high-performance anion-exchange chromatography with integrated pulsed amperometric detection. In contrast to conventional methods, the analysis requires neither pre- or postcolumn derivatization, nor a time-consuming oxidation or derivatization step prior to hydrolysis. Correction factors account for incomplete release of Val and Ile even after hydrolysis for 24 h, and for losses of Ser during evaporation. Gradient conditions including an extra eluent allow multiple sequential sample analyses without risk of Glu accumulation on the anion-exchange column which otherwise would result from high Gln levels in gluten proteins.

Key words: Amino acid analysis, Wheat, Gluten, HPAEC-IPAD, Acid hydrolysis

1. Introduction

The storage proteins of wheat represent an important fraction of the daily human protein intake. They form a cohesive viscoelastic dough mass when mixed with water, which is referred to as gluten, and their ability to form a strong protein network during heating impacts many wheat-based products (1, 2). Wheat gluten has high Gln (32–35) and Pro (12–17 mol%) levels. Other major amino acids are Leu (6–8), Ser (5–7), Gly (5–6), and Val (4–5 mol%).

Michail A. Alterman and Peter Hunziker (eds.), *Amino Acid Analysis: Methods and Protocols*, Methods in Molecular Biology, vol. 828, DOI 10.1007/978-1-61779-445-2_26, © Springer Science+Business Media, LLC 2012

While Cys is a minor amino acid (1–2 mol%), it is of paramount importance for gluten properties because 90–95% of it occurs as cystine, which forms intra- or intermolecular disulfide bonds (3). High temperature and/or pressure lead to additional intermolecular disulfide bond formation and thus gluten polymerization. Besides disulfide bonds, also cross-links derived from dehydroproteins, such as lysinoalanine and lanthionine, contribute to gluten polymerization (4). Amino acid analysis (AAA) of gluten is useful for wheat cultivar mapping and breeding purposes (5, 6) (7), for evaluating impact of wheat processing on nutritional value (8, 9), and for understanding the impact of high temperature and/or pressure on polymerization reactions of gluten. The present method allows quantification of all amino acids but Trp. Other compounds, including the unusual amino acids lysinoalanine and lanthionine can be quantified. Asn and Gln are deamidated during acid hydrolysis and are detected as Asp and Glu, respectively.

Conventional hydrolysis in 6.0 M hydrochloric acid (HCl) solution (24 h, 110°C) has been optimized for the specific case of gluten proteins. Due to the effect of hydrolysis time on peptide bond cleavage and amino acid degradation, acid hydrolysis during a single time interval leads to inaccurate quantification of acid-labile and/or acid-resistant amino acids (10). To compensate for these errors, correction factors, which depend on amino acid sequence and tertiary structure, were determined for 24 h hydrolysis of gluten. Also, we investigated which protective measures minimize amino acid degradation during hydrolysis of gluten proteins under the applied conditions. The presence of phenol in the hydrolysis mixture prevented Tyr degradation and samples were flushed with nitrogen prior to hydrolysis. However, Cys and Met were accurately determined in gluten proteins without additional oxidation or derivatization steps. Finally, correction factors were derived to account for losses occurring during evaporation of HCl at 110°C prior to chromatographic separation.

After hydrolysis, the liberated amino acids are quantified by high-performance anion-exchange chromatography with integrated pulsed amperometric detection (HPAEC-IPAD) (11–14). This method uses single-component mobile phases at low flow rates and requires no pre- or postcolumn derivatization. Another advantage of this technique is the low limits of detection (15). High Gln levels in gluten proteins complicate the determination method. Indeed, Gln is deamidated during acid hydrolysis to Glu, which is strongly retained by the AminoPac PA10 column. Thus, to allow multiple analyses of gluten hydrolyzates without column contamination, an extra eluent was included in the gradient conditions.

2. Materials

Prepare all solutions using deionized water unless indicated otherwise, and use analytical grade reagents.

2.1. Sample Material

As described, this procedure is optimal for gluten samples containing minimum 70% protein on dry matter basis. Amino acid analyses of other proteins or samples with lower protein content (e.g., wheat flour) require other protective measures and/or other correction factors (see Note 1).

2.2. Solutions Required for Hydrolysis

1. HCl solution: 12.0 M HCl, 2.0 g/l phenol. Cautiously weigh 0.20 g phenol directly in a 100-ml volumetric flask and make up to 100 ml with 37% (v/v) HCl solution (see Note 2). Prepare the solution 1 day in advance and stir overnight. Store in a dark bottle at room temperature.
2. Internal standard (IS) stock solution: 30 mM L-norleucine, 0.02% (w/v) sodium azide. Accurately weigh about 197 mg L-norleucine (MW = 131.18 g/mol), and transfer it to a 50-ml volumetric flask. Make up to 50 ml with 0.02% (w/v) sodium azide solution and store at 6°C (see Note 3).

2.3. Standard Mixture

1. Prepare a mixture of all amino acids to be quantified, by diluting a commercially available amino acid standard solution to 10–20 μM.
2. Add the IS, also at a concentration of 10–20 μM.
3. Push the mixture through a syringe filter (0.22 μm, no cellulose) into a vial.
4. Standard mixes can be prepared in advance (maximum 6 months) when stored at −18°C.

2.4. Eluents for the Gradient Mobile Phase

1. Prepare all eluents: sodium hydroxide solution (250 mM), sodium acetate solution (1.0 M), and acetic acid solution (0.1 M) with ultrapure water (prepared by purifying deionized water to attain a resistivity of at least 18.2 MΩ at 25°C).
2. Degas all eluents and keep them under slight helium overpressure to prevent accumulation of atmospheric carbon dioxide.

2.5. HPAEC-IPAD Components

1. Chromatography system: Dionex (Sunnyvale, CA, USA) BioLC, equipped with a GS50 gradient pump with online degasser, an AS50 autosampler with a thermal compartment, and an ED50 electrochemical detector containing both a gold working electrode and a pH reference electrode.
2. Columns: AminoPac PA10 guard (50 × 2 mm) and analytical (250 mm × 2 mm) column (Dionex).

3. Injection volume: 25 μl.
4. Column temperature: 30°C.
5. Operating back pressure: <20.7 MPa.
6. Flow rate: 0.25 ml/min.
7. Gradient conditions: see Table 1.
8. Waveform: see Table 2.
9. Software: Chromeleon Version 6.70 (Dionex).

3. Methods

3.1. Hydrolysis

1. Accurately weigh the sample (about 15 mg protein) in a test tube with screw cap. Make sure the test tube can be sealed airtight.
2. Dilute one part of IS stock solution with four parts of water and add 0.5 ml of the diluted IS stock solution to the test tube.
3. Flush the HCl solution for 30 s with nitrogen gas. Add 0.5 ml of the HCl solution to the sample, flush the headspace of the test tube during 10 s with nitrogen gas and close the test tube immediately (see Note 2).
4. Heat the samples for 24 h at 110°C (see Note 4).
5. Unscrew the test tubes and let the volatile compounds in the samples evaporate at 110°C for 180 min (see Note 5).
6. Resuspend the samples in 5.0 ml water (vortex) and transfer the hydrolyzate through a filter paper (diameter 90 mm) into a test tube (see Note 6). Dilute one part of the hydrolyzate with 39 parts of water and push the solution through a syringe filter (0.22 μm, no cellulose) into a vial.
7. Inject a standard mix, followed by 1–10 samples, then run a standard mix, again followed by 1–10 samples, and repeat until all samples are analyzed.

3.2. Calculation

In the mathematical equations, the following abbreviations are used: *Conc* concentration, *AA* amino acid, *IS* internal standard, and *Std mix* standard mix.

1. RF: response factor.

   Calculate the RF for each compound, based on its concentration in the standard mix and the corresponding area, using the following equation:

$$\text{RF} = \frac{\text{Conc of AA/Conc of IS}}{\text{Area of AA/Area of IS}}.$$

Table 1
Gradient conditions for separation of gluten amino acids by HPAEC-IPAD15

Time (min)	Water (resistivity 18.2 MΩ) (%)	Sodium hydroxide 250 mM (%)	Sodium acetate 1.0 M (%)	Acetic acid 0.1 M (%)	Curve[a]
0.0	76	24			
2.0	76	24			
8.0	64	36			8
11.0	64	36			
27.0	40	20	40		8
47.0	40	20	40		
47.1				100	8
49.1				100	
49.2	20	80			8
51.2	20	80			
51.3	76	24			5
76.0	76	24			

The procedure includes a postcleanup step to allow for multiple analyses. Reprinted from ref. 15 with permission from Elsevier

[a]The gradient curve is the line representing the change in gradient conditions. Shapes of gradient curves are defined in the GS50 Gradient Pump Operator's Manual, pp. 37–38 (Dionex Document No. 031612, Revision 3). Curve 5 is linear. Curve 8 is concave with 20% of change at about 60% of a time segment and 70% change at 90% of the same time segment

Table 2
Detection waveform for amino acid analysis by HPAEC-IPAD (reprinted from ref. 15 with permission from Elsevier)

Time (ms)	Potential (V) versus pH	Current integration
0	0.13	
40	0.13	
50	0.33	
210	0.33	Begin
220	0.60	
460	0.60	
470	0.33	
560	0.33	End
570	−1.67	
580	−1.67	
590	0.93	
600	0.13	

Inject the standard mix at least three times and calculate the average RF for each compound (see Note 7).

2. CF: correction factor.
   CF = 1.00 for all amino acids but Val (CF = 1.07), Ile (CF = 1.13), and Ser (CF = 1.32) (see Note 8).
3. $(\text{Conc of IS})_{\text{sample}}$: concentration of L-Norleucine in the sample (μM).

$$(\text{Conc of IS})_{\text{sample}} = \frac{\text{mass of IS}}{131.18\ \text{g/mol} \times 0.050\ \text{L}} \times 0.5$$

(with mass of IS = mass of L-norleucine in IS stock solution in mg).

4. Amino acid levels in sample, expressed on dry matter protein (μmol/g).

$$\text{AA level} = \frac{(\text{Area of AA/Area of IS})_{\text{sample}} \times \text{RF} \times (\text{Conc of IS})_{\text{sample}} \times \text{CF} \times 200}{\text{Samplemass} \times (\text{proteincontent}/100)}$$

(with sample mass =sample mass in mg; *protein content* in %).

4. Notes

1. Correction factors account for losses during evaporation, and compensate for errors due to incomplete cleavage or degradation of certain amino acids during acid hydrolysis. As these correction factors vary for each protein, an accurate estimation of amino acid levels in hydrolyzates can only be guaranteed when correction factors are derived based on a time-of-hydrolysis test of the protein to be analyzed.

   Also, with other proteins, other protective measures may be required. For instance, proteins containing high levels of Trp are sensitive to acid-induced degradation of the sulfur-containing amino acids (16). Underestimation of Cys and Met was observed for gluten proteins when complete flour samples were hydrolyzed. In such cases, methods to accurately determine Met and Cys include oxidation of Cys and Met prior to hydrolysis with performic acid (17), derivatization prior to hydrolysis with iodoacetic acid or iodoacetamide (18), and derivatization during hydrolysis with 3,3′-dithiodipropionic acid (19).
2. Take the proper health and safety precautions (fume hood, gloves, and safety glasses) when working with HCl and phenol.
3. Be aware of the waste disposal regulations concerning sodium azide.

4. It is important that the test tubes are closed airtight during heating to prevent amino acid degradation and to ensure adequate heating. Samples can either be heated in an oil bath or in a heating block. When a compound is analyzed for the first time, a time-of-hydrolysis test should be performed to check whether hydrolysis during 24 h yields maximum levels. Hereto, hydrolyze samples containing the compound for different times (3–96 h), and calculate the correction factor by dividing the obtained maximum level with the level obtained after 24 h hydrolysis.
5. Evaporate HCl and phenol under the fume hood. Most amino acids (Arg, Lys, Ala, Thr, Gly, Val, Pro, Ile, Leu, Met, His, Phe, Glu, Asp, cystine, Tyr, and norleucine) remain stable during this drying step, but Ser is partially degraded. Ser losses during evaporation can be accounted for by a correction factor. However, for accurate determination of Ser levels, we advise to not evaporate the sample. To that end, add 4.0 ml water to the sample after 24 h hydrolysis, filter through a filter paper (diameter 90 mm), dilute one part of the filtrate with 39 parts of water, and push the mixture through a syringe filter (0.22 μm, no cellulose) into a vial. When a compound is analyzed for the first time, check its stability during evaporation by comparing the obtained level with and without evaporation. If the obtained level is lower after evaporation, you can either avoid the evaporation step, or account for the loss using the correction factor (level obtained without evaporation divided by the level obtained with evaporation). Some considerations might be helpful in deciding whether or not to evaporate the sample. First, Arg levels cannot be determined in nonevaporated samples due to interference by HCl. Second, HCl damages the gold working electrode of the Dionex BioLC system, so nonevaporated samples should be sufficiently diluted to bring the injected HCl concentration below 0.1 M. Third, phenol is eliminated during the drying step.
6. Some of the sample possibly sticks to the bottom of the test tube after evaporation. Make sure that the entire sample is resuspended by vortexing vigorously before paper filtering the mixture.
7. Injection of the standard mix is required to calculate the response factors, which have a great impact on test results. Multiple injections (at least three), spread out over time, ensure accuracy of the response factors. Figure 1 shows the separation of some amino acids, norleucine, lysinoalanine, and lanthionine. Furthermore, when analyzing a compound for the first time, a calibration curve (0.5–100 μM) should be made to estimate the upper limit of linearity and the limits of detection and quantification. The method performance indicators for most amino

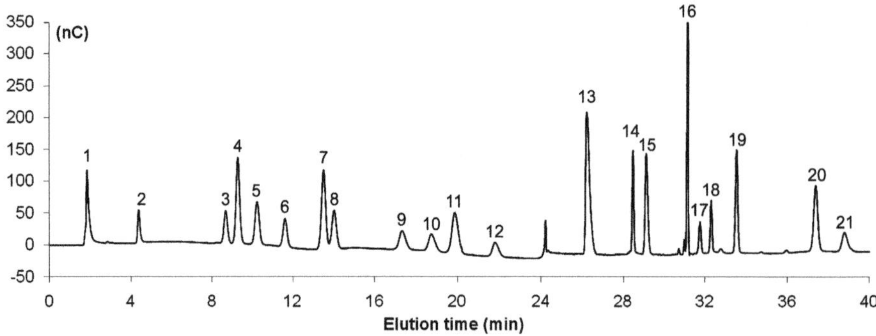

Fig. 1. Separation of an equimolar solution of different amino acids (each 20.0 μM) and norleucine (15 μM) by HPAEC-IPAD, using the postseparation cleanup gradient conditions (*1* Arg, *2* Lys, *3* Ala, *4* Thr, *5* Gly, *6* Val, *7* Ser, *8* Pro, *9* Ile, *10* Leu, *11* Met, *12* norleucine, *13* His, *14* lysinoalanine, *15* Phe, *16* lanthionine, *17* Glu, *18* Asp, *19* cystine, *20* Tyr).

Table 3
Method performance indicators for amino acid analysis by HPAEC-IPAD (25 μl sample injection)

	Limit of detection (pmol)	Limit of quantification (pmol)	Upper limit of linearity (μM)
Lanthionine	2	11	25
Lysinoalanine	40	178	25

acids and L-norleucine have been reported by Rombouts et al. (15). Table 3 shows these values for the unusual amino acids lanthionine and lysinoalanine. The limits of detection for HPAEC-IPAD are lower than for other methods (e.g., approximately 100 pmol for ninhydrin postcleanup derivatization (20)), which illustrates the sensitivity of HPAEC-IPAD.

8. For the amino acids Arg, Lys, Ala, Thr, Gly, Val, Ser, Pro, Ile, Leu, Met, His, Phe, Glu, Asp, cystine, and Tyr, it was evaluated whether hydrolysis during 24 h leads to incomplete cleavage and/or degradation, and whether evaporation leads to losses. Correction factors for Val and Ile account for incomplete cleavage of the hydrolysis-resistant peptide bonds Val–Val, Ile–Ile, and Val–Ile, while the correction factor for Ser accounts for losses during evaporation. Some amino acids (e.g., Cys, Met, Ser, Thr, and Tyr) are sensitive to degradation during hydrolysis, but only when hydrolysis times exceed 24 h. Whenever one analyzes a compound for the first time, one should perform a time-of-hydrolysis test to evaluate the impact of hydrolysis time (see Note 4), and check the stability of the compound during evaporation (see Note 5).

Acknowledgments

This work is a part of the Methusalem programme "Food for the Future" at the K.U. Leuven. B. Lagrain and I. Celus wish to acknowledge the Research Foundation – Flanders (FWO, Brussels, Belgium), and the Agency for Innovation by Science and Technology (IWT-Vlaanderen, Brussels, Belgium), respectively, for functions as postdoctoral researchers. K. Brijs wishes to acknowledge the Industrial Research Fund (K.U. Leuven, Leuven, Belgium) for a position as Industrial Research Fund fellow.

References

1. Wieser H (2007) Chemistry of gluten proteins. Food Microbiol 24:115–119
2. Veraverbeke WS, and Delcour JA (2002) Wheat protein composition and properties of wheat glutenin in relation to breadmaking functionality. Crit Rev Food Sci Nutr 42:179–208
3. Belitz HD, Grosch W, and Schieberle P (2009) Cereals and cereal products. In: Belitz HD, Grosch W, and Schieberle P (eds) Food Chemistry, 4th edn. Springer-Verlag, Berlin
4. Rombouts I et al (2010) [beta]-Elimination reactions and formation of covalent cross-links in gliadin during heating at alkaline pH. J Cereal Sci 52:362–367
5. Ruckman JE, Zscheile FP, and Qualset CO (1973) Protein, Lysine, and Grain Yields of Triticale and Wheat as Influenced by Genotype and Location. J Agric Food Chem 21:697–700
6. Sikka KC et al (1978) Comparative Nutritive-Value and Amino-Acid Content of Triticale, Wheat, and Rye. J Agric Food Chem 26:788–791
7. Del Moral LFG et al (2007) Environmentally induced changes in amino acid composition in the grain of durum wheat grown under different water and temperature regimes in a Mediterranean environment. J Agric Food Chem 55:8144–8151
8. Jiang XL, Hao Z, and Tian JC (2008) Variations in amino acid and protein contents of wheat during milling and northern-style steamed breadmaking. Cereal Chem 85:504–508
9. Shewry PR et al (2009) Wheat Grain Proteins. In: Khan K and Shewry PR (eds) Wheat: Chemistry and Technology, 4th edn. AACC International, St. Paul, MN, USA
10. Darragh AJ et al (1996) Correction for amino acid Loss during acid hydrolysis of a purified protein. Anal Biochem 236:199–207
11. Hanko VP, and Rohrer JS (2004) Determination of amino acids in cell culture and fermentation broth media using anion-exchange chromatography with integrated pulsed amperometric detection. Anal Biochem 324:29–38
12. Jandik P et al (2001) Simplified in-line sample preparation for amino acid analysis in carbohydrate containing samples. J Chromatogr B 758:189–196
13. Ding YS, Yu H, Mou SF (2002) Direct determination of free amino acids and sugars in green tea by anion-exchange chromatography with integrated pulsed amperometric detection. J Chromatogr A 982:237–244
14. Thiele C, Ganzle MG, Vogel RF (2002) Sample preparation for amino acid determination by integrated pulsed amperometric detection in foods. Anal Biochem 310: 171–178
15. Rombouts I et al (2009) Wheat gluten amino acid composition analysis by high-performance anion-exchange chromatography with integrated pulsed amperometric detection. J Chromatogr A 1216:5557–5562
16. Olcott HS, and Fraenkel-Conrat H (1947) Formation and loss of cysteine during acid hydrolysis of proteins. Role of tryptophan. J Biol Chem 171:583–594
17. Fountoulakis M, and Lahm HW (1998) Hydrolysis and amino acid composition analysis of proteins. J Chromatogr A 826:109–134
18. Aitken A, and Learmonth M (2002) Carboxymethylation of Cysteine Using Iodoacetamide/Iodoacetic Acid. In: Walker JM (ed) The Protein Protocols Handbook, 2nd edn. Humana Press, NJ
19. Barkholt V, and Jensen AL (1989) Amino acid analysis: Determination of cysteine plus half-cystine in proteins after hydrochloric acid hydrolysis with a disulfide compound as additive. Anal Biochem 177:318–322
20. Koehler P (2003) Amino Acid and Protein Sequence Analysis. In: Shewry PR and Lookhart GL (eds) Wheat Gluten Protein Analysis, AACC Press, Washington

Chapter 27

Preparative HPLC Separation of Underivatized Amino Acids for Isotopic Analysis

Jennifer A. Tripp and James S.O. McCullagh

Abstract

Single-compound analysis of stable or radio-isotopes has found application in a number of fields ranging from archaeology to forensics. Often, the most difficult part of these analyses is the development of a method for isolating the compounds of interest.

Here, we describe three complementary preparative HPLC procedures suitable for separating and isolating single amino acids from bone collagen or hair keratin with minimal isotopic contamination. Using preparative reversed-phase, ion-pair, or mixed-mode chromatography of underivatized amino acids in aqueous mobile phases, single amino acids can be isolated and further analyzed using mass spectrometric techniques.

Key words: Amino acid, High performance liquid chromatography, Ion-pair chromatography (IP-RP-HPLC), Mixed-mode chromatography (MM-HPLC), Preparative chromatography, Isotope ratio mass spectrometry (IRMS), Accelerator mass spectrometry (AMS), Liquid chromatography isotope ratio mass spectrometry (LC-IRMS), $\delta^{13}C$, ^{14}C, Compound-specific isotope analysis, Stable isotope, Radiocarbon dating, Compound-specific dating

1. Introduction

Single-compound isotope analysis has emerged as a useful technique for fields as diverse as archaeology, geology, ecology, and forensics, with applications covering a broad range including radiocarbon dating, dietary analysis, environmental reconstructions, studies of metabolism, and sourcing of agricultural products (1–6). While mass spectrometric methods for studying stable and radioisotopes are established, separation techniques for isolating the desired compounds are less mainstream; these methods often involve preparative gas or high-performance liquid chromatography (GC or HPLC),

Michail A. Alterman and Peter Hunziker (eds.), *Amino Acid Analysis: Methods and Protocols*, Methods in Molecular Biology, vol. 828, DOI 10.1007/978-1-61779-445-2_27,

followed by isolation of the compounds and subsequent measurement of isotopic ratios using mass spectrometry. Many archaeological applications such as radiocarbon dating and stable isotope analysis for study of diet utilize bone collagen, so for our studies (7, 8) we focused on the separation and isolation of bone collagen amino acids, for the purpose of measuring carbon and nitrogen isotope ratios. The method also successfully separates amino acids in hair keratin, so we have included that procedure as well (9).

Chromatographic techniques developed for isolation of single compounds for isotopic measurements must be compatible with the isotope measurement method and provide appropriate analyte resolution. Contamination of the compound with substances containing isotopes of interest must be avoided; for instance, if carbon-13 or carbon-14 is to be measured, carbon-containing components in the mobile phase must be avoided or completely removed prior to the isotope measurement, and the stationary phase must be stable with the chosen separation conditions to prevent, for example, column bleed contaminating the eluent with C18 alkyl chains. Derivatization of the compounds, especially common in analytical separations of amino acids, must be avoided, reversed, or corrected. Lastly, unless baseline separation is achieved and the entire eluting peak can be collected, isotopic fractionation of the peak during the separation must be avoided or, if possible, corrected. If the edges of the peak are enriched or depleted in heavy isotope, and the entire peak cannot be collected, the measured isotopic ratio will differ significantly from the actual ratio of the compound.

With these considerations in mind, we have developed several preparative HPLC protocols that result in sufficient amounts of isolated amino acids for analysis of their $\delta^{13}C$ and $\delta^{15}N$ by isotope ratio mass spectrometry (IRMS), or compound-specific radiocarbon dating using accelerator mass spectrometry (AMS). A two-step separation procedure was initially developed that allowed for isolation of a number of single amino acids from collagen and keratin (7). The first separation in this procedure involves reversed-phase (RP) HPLC using a C18 column and only water as the mobile phase. While the more polar amino acids elute in the void volume with no separation, the amino acids with nonpolar side chains are efficiently separated. The initial peak from the first separation is collected, concentrated, and reinjected into a RP column packed with a perfluorinated stationary phase, using an aqueous mobile phase containing a surfactant in a technique called ion-pair (IP) chromatography (10). This surfactant, or IP reagent, possesses a nonpolar tail that interacts strongly with the RP column, and an ionic head group that can exchange with the amino acids. For our separation, we chose pentadecafluorooctanoic acid (PDFOA) as the IP reagent. Perfluorinated compounds such as this have been widely used as IP reagents in preparative chromatography due to their lack of UV absorption and volatility. The combination of these two chromatographic processes allowed for the efficient separation of bone collagen amino acids, which minimized contamination from allochthonous carbon. One advantage of this separation method is that the initial RP-HPLC procedure results in quick, simple, and nearly complete separation of essential from nonessential amino acids, a useful distinction for dietary studies. Overall, the method provided accurate results for single-compound carbon and nitrogen IRMS measurements, but AMS dates on single amino acids proved problematic, likely due to small amounts of contamination from column bleed.

Recent advances in instrumentation, in particular the introduction of the IsoLink system by Thermo Electron, have provided an alternative to preparative chromatography followed by isolation of the amino acids and measurement of isotopic ratios. The IsoLink is an integrated LC-IRMS system, and isotope ratios of amino acids separated on an analytical scale can be measured by direct injection of the eluent into the IRMS. This instrument is suitable for direct measurement of stable carbon isotope ratios ($\delta^{13}C$), but the IP separation procedure was not suitable for use with this system. Thus, a mixed mode separation, utilizing a column containing both ionic and nonpolar groups and an aqueous mobile phase, was developed for use with the IsoLink. The analytical-scale work is described in detail elsewhere (2, 11–14), but a preparative version of the mixed-mode separation was recently reported for use in separating amino acids for radiocarbon dating (8). This procedure provides accurate $\delta^{13}C$ measurements and AMS dates on amino acids isolated in a single chromatographic separation. Thus, we present here three complementary preparative chromatographic procedures suitable for the isotopic analysis of single amino acids.

2. Materials

Glassware used in sample preparation for radiocarbon dating is cleaned using an acid–base wash and baked at 500°C for 12 h before use.

2.1. LC Components

1. Preparative chromatography was carried out on a Varian ProStar HPLC system consisting of two 210 isocratic pumps, a 410 autosampler, a 320 dual-path length UV detector, and a 701 fraction collector, controlled by Star Workstation PC software (see Note 1). The autosampler was fitted with a 1-ml syringe and 2-ml sample loop.
2. The entire system was kept at room temperature during separations.
3. Prepare all solutions using ultrapure water from a MilliQ system. Use HPLC-grade methanol and tetrahydrofuran for the chromatography, and anhydrous reagent-grade chloroform for the extractions.

2.2. Extraction of Collagen from Bone

1. 2:1 methanol–chloroform.
2. 0.5 M HCl.
3. HCl solution at pH 3, prepared by the careful addition of 6 M HCl to an excess of water with stirring. An accurate pH meter should be used to adjust the concentration to pH 3.

2.3. Preparation of Hair (Keratin) Samples

1. 2:1 methanol–chloroform.
2. 1:2 methanol–chloroform.

2.4. Protein Hydrolysis

1. 6 M HCl.
2. Nitrogen gas (or another inert gas).

2.5. Preparative Reversed-Phase HPLC

1. Column: Waters Symmetry C18, 19 × 150 mm, 7 μm particles.
2. Mobile phase: water.
3. Additional solvent: HPLC-grade methanol.

2.6. Preparative Ion-Pair HPLC

1. Column: Supelco Discovery HS F5, 10 × 250 mm, 5 μm particles (see Note 2).
2. Mobile phase: 0.5 mM aqueous pentadecafluorooctanoic acid (PDFOA), prepared by mixing 0.207 g of PDFOA with 1 L of water.
3. Additional solvents: HPLC-grade methanol and tetrahydrofuran.
4. 6 M HCl.

2.7. Preparative Mixed-Mode Chromatography

1. Column: Primsep A (22 × 250 mm), 5 μm particle size (SIELC Technologies, Prospect Heights, IL, USA).
2. Mobile phases: purified water and 0.3% (*v*/*v*) phosphoric acid.

2.8. Accelerator Mass Spectrometry and Isotope Ratio Mass Spectrometry

1. Chromosorb.
2. Tin capsules to hold samples.

3. Methods

3.1. Extraction of Collagen from Bone (see Note 3)

1. *Bone sample preparation*:
   (a) For archaeological samples: clean bones by sandblasting, ensuring that all ink and other surface contaminants are removed. Remove any extraneous cartilage.
   (b) For modern samples: place the bone in a beaker and submerge in 2:1 methanol–chloroform; place sample in an ultrasonic bath for 30 min, then allow to soak for 24 h. Remove the bone from the solution using tweezers (see Note 4), and repeat the procedure once more. After the second wash,

examine the bottom of the beaker to see if any lipid is present. If some lipid remains, repeat the wash until no lipids are seen in the beaker. Finally, rinse three times with purified water, ultrasonicating for 30 min each time.

2. *Removal of the mineral phase from bone*:
   (a) Crush bone to small pieces roughly 0.5–2 cm^3 and place in a test tube.
   (b) Add 10 ml cold 0.5 M HCl to cover bone chunks.
   (c) Place tubes in a test tube rack, cover them with aluminum foil and leave in a refrigerator at <10°C for several days (see Note 5). Shake once/twice daily. Change acid every 2 days.
   (d) When sample is soft or floats, the bone has been sufficiently demineralized. Decant the supernatant liquor and then rinse the sample with distilled water three times (see Note 6).
3. *Gelatinization*:
   (a) The remaining bone pellets contain mostly insoluble protein. Place these pellets in a glass test tube and submerge them in dilute HCl at pH 3. Heat to 75°C for 24–48 h. This process denatures the collagen triple helical structure making it more soluble. The protein pellets should now go into solution (see Note 7).
   (b) Filter the supernatant liquor to remove particulates. The solution now contains mostly dissolved collagen and some salts. Freeze-dry the solution overnight in preweighed tubes. The resulting material should be mostly denatured collagen. Weigh after drying to determine the collagen yield.

3.2. Hair (Keratin) Preparation for Isotopic Analysis

Hair is mainly composed of protein and the majority of this protein is keratin, hence no protein extraction procedure is required. However, because hair is entirely exposed to the external environment it is liable to contamination. Cleaning hair samples prior to analysis is vital (see Note 8).

1. Put hair in 10-ml test tube. Add 8 ml purified water and place in an ultrasonic bath for 30 min.
2. Decant water away from sample. Add 8 ml 2:1 chloroform–methanol and place in an ultrasonic bath for 30 min. Decant solution, add fresh 2:1 chloroform–methanol, and repeat the ultrasonication.
3. Decant the solution. Add 8 ml 1:2 chloroform–methanol and place in an ultrasonic bath for 30 min.
4. Decant solution. Add 8 ml water and place in an ultrasonic bath for 30 min. Decant water, add fresh water, and repeat the ultrasonication. Decant the solution. The hair is now ready for hydrolysis.

3.3. Protein Hydrolysis (see Note 9)

1. Add 6 M HCl (1 ml/mg of protein) to the sample in a glass hydrolysis tube and seal under vacuum or nitrogen atmosphere (see Note 10).
2. Heat the tube to 110°C for 24 h.
3. Cool the sample under an inert atmosphere, and then remove the HCl by vacuum evaporation.
4. Resuspend the sample in pure water to the appropriate concentration for the chromatographic procedure being performed and filter with 2 μm filters. The filtrate is now ready for preparative chromatography and isotopic analysis (see Note 11).

3.4. Preparative Reversed-Phase HPLC

1. Dissolve the isolated amino acid residue in purified water to a concentration of 10 mg amino acids per 1 ml of water to prepare the sample solution.
2. Wash the HPLC column at a flow rate of 8 ml/min, first with methanol for 30 min and then with purified water for at least 5 h.
3. Set the detector to 205 nm (see Note 12).
4. Inject 1 ml of sample solution into the column. Elute for 45 min, while collecting fractions at 30-s intervals. An annotated chromatogram of hydrolyzed bovine bone collagen is shown in Fig. 1.

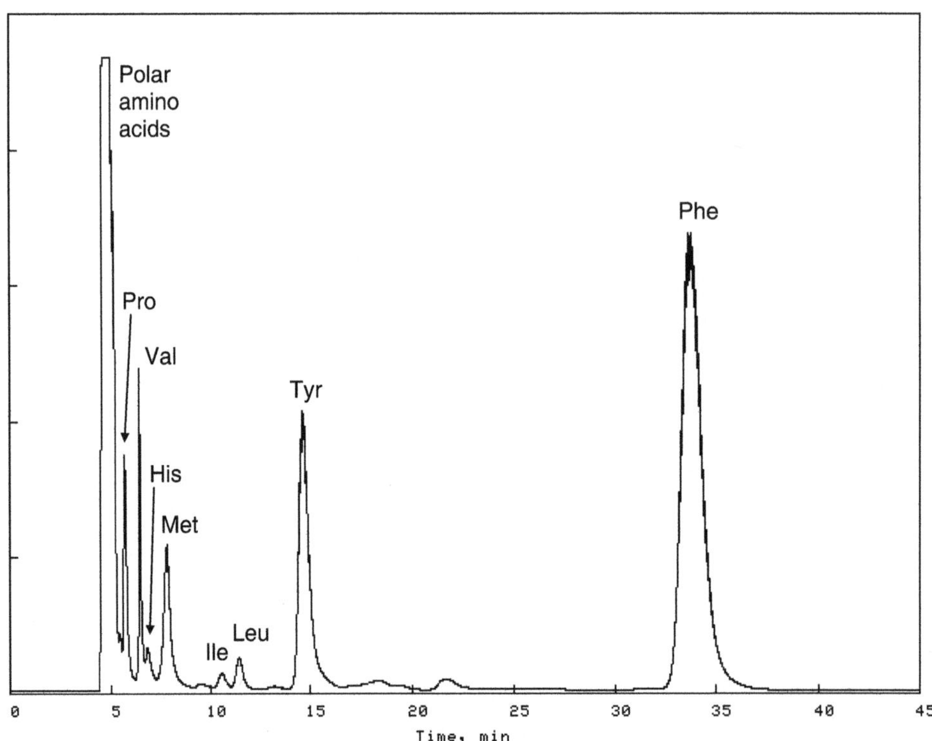

Fig. 1. Chromatogram from preparative RP-HPLC separation of modern bovine collagen. Reproduced with permission from ref. 7.

5. The fractions containing peaks eluting after 5 min contain the amino acids with nonpolar side groups. After identifying these fractions, combine the tubes containing the same amino acid, freeze the solutions, and remove the water by lyophilization.
6. The initial one or two peaks on the chromatogram contain the rest of the amino acids. Combine these fractions, lyophilize, and then dissolve the residue in purified water to a concentration of 10 mg/ml. This becomes the sample solution for the subsequent ion-pair separation.

3.5. Preparative Ion-Pair HPLC

1. Wash the column at a flow rate of 8 ml/min with methanol, tetrahydrofuran, then methanol again, for 30 min each.
2. Equilibrate the column by pumping 0.5 mM PDFOA solution through the column at a flow rate of 8 ml/min for at least 48 h (see Note 13).
3. Inject 1 ml of sample solution and allow eluting for 60 min, collecting fractions every 1 min (see Note 14). An annotated chromatogram of the ion-pair separation of hydrolyzed bovine bone collagen is shown in Fig. 2.

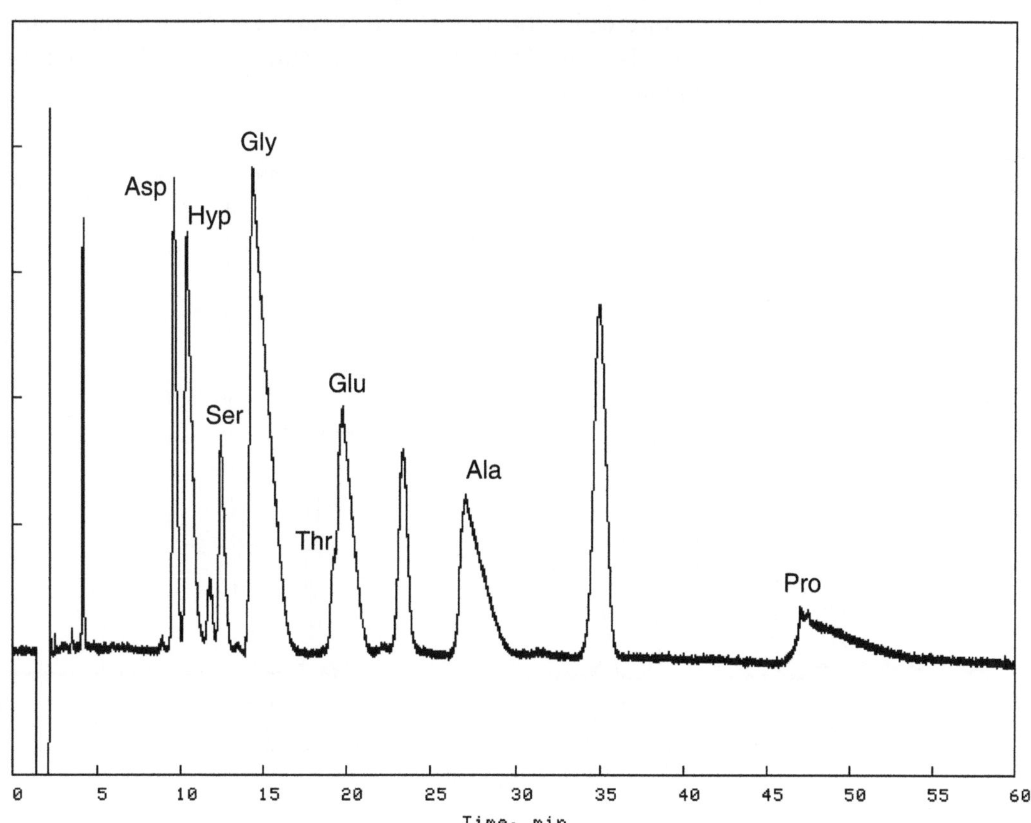

Fig. 2. Chromatogram from preparative IP-HPLC separation of modern bovine collagen. Reproduced with permission from ref. 7.

4. Combine fractions containing the same amino acid and add 1 ml of 6 M HCl per 20 ml of collected sample (see Note 15).
5. Remove the mobile phase by lyophilization.

3.6. Preparative Mixed-Mode Chromatography

1. Prepare the sample solution by adding purified water to the lyophilized amino acid mixture to make a concentration of approximately 15 mg/ml.
2. Equilibrate the column with purified water for 40 min at a flow rate of 6 ml/min.
3. Inject 1 ml of sample solution and elute with pure water for 40 min, then run a linear gradient from 40 until 70 min from 100% water to 100% phosphoric acid (0.3% by volume). A representative chromatogram showing the separation of archaeological bone collagen hydrolyzate, annotated with the amino acid labels, is shown in Fig. 3.
4. Combine fractions containing the same amino acid into a single flask and remove the water from the sample using rotary evaporation, gyro-vacuum evaporation, or lyophilization (see Note 16).
5. To remove the remaining phosphoric acid, reinject the sample into the same HPLC column using isocratic conditions (mobile phase purified water). The phosphoric acid elutes in the void volume, while the amino acid is retained. Collect the amino acid using the fraction collector, and remove the water by lyophilization.

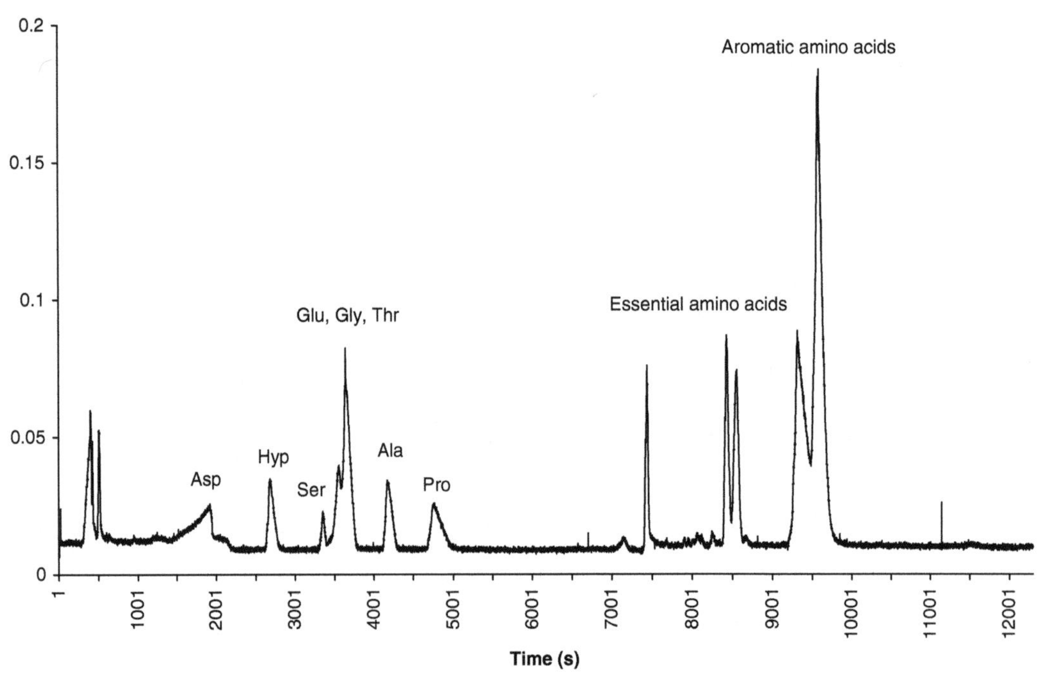

Fig. 3. Chromatogram from preparative MM-HPLC separation of archarological porcine collagen from the Mary Rose. Reproduced with permission from 2.

6. Before beginning the next separation, equilibrate column with water for 40 min.

3.7. Isotope Ratio Mass Spectrometry

Small amounts of amino acid are often difficult to weigh into the tin capsules used for IRMS measurements. To deal with this, the following procedure was used.

1. Measure a small amount (tip of a spatula) of Chromosorb into the tin capsule.
2. Dissolve the amino acid in a minimal amount of purified water. Drop the solution onto the Chromosorb within the tin capsule.
3. Roll the tin to encapsulate the sample and Chromosorb, and transfer to the IRMS instrument for measurement (see Note 17).

3.8. Accelerator Mass Spectrometry

1. Samples were prepared for AMS dating in the same manner as described for IRMS measurements, except the Chromosorb and tin capsules were baked at 500°C for 12 h prior to use.
2. ^{14}C dates were measured on an accelerator mass spectrometer system using a cesium ion source for solid graphite sample.
3. Graphitized samples were prepared by reduction of CO_2 over an iron catalyst in an excess H_2 atmosphere at 560°C prior to AMS ^{14}C measurement (17).
4. Calibration and statistical analyses were performed on OxCal software (18, 19) using the IntCal04 calibration curve data (20).

4. Notes

1. Any HPLC system consisting of at least two isocratic pumps that can achieve flow rates up to 10 ml/min and a UV detector would be suitable for these methods. The autosampler and fraction collector are not necessary but make the injection of samples and collection of multiple fractions much easier. Detection methods other than UV, such as refractive index or light scattering, were not tried but may work as well.
2. This separation was also done using the same column used in the reversed-phase separation. While offering higher loading, separation on the C18 column between critical pairs of amino acids (such as hydroxyproline-aspartic acid) was not as good. Chromatograms can be compared in ref. 7.
3. The presented method is used to extract collagen from modern and archaeological bone in preparation for isotopic analysis. Note the different sample preparation procedure for archaeological and modern bone. Modern bone contains lipids that can

interfere with isotope measurements if they are not removed, so a solvent wash is typically performed. Archaeological bones, while containing a trivial amount of lipid, normally have sediment stuck to the outer surface that is removed by sandblasting. The extraction procedure after the initial sample preparation is the same for both sample types. For conventional isotopic analysis in triplicate, about 0.2–0.6 g of bone should be sufficient. For preparative scale radiocarbon dating of individual amino acids, it is recommended that 10–20 g of bone be used to account for variability in amino acid abundance and collagen yield. The extraction procedure reported here is based on work published by Longin (15).

4. Do not decant the liquid as lipids may have sunk to bottom of the tube.
5. If the bone is powdered instead of left in chunks, the acid demineralization step will take a few hours only.
6. Use an inert filter to remove any small particles during the water wash.
7. Make sure the test tubes have a plastic cover to prevent evaporation, and ensure that the lids stay on during heating as pressure can build up.
8. Hair grows at a rate of approximately 1 cm/month. As there is no protein turnover associated with hair keratin, a chronological isotopic record of dietary history is represented along the length of the hair ending at the present in the hair root. The whole hair may be analyzed or it may be sectioned prior to hydrolysis to provide a chronological record of changes in the amino acid isotope ratios.
9. The amount of hydrolysates required will depend upon the desired measurement. IRMS requires about 10 mg of collagen or hair which should be ample for triplicate analysis. Dating individual amino acids using accelerator mass spectrometry required higher concentrations of individual amino acids and ultimately between 1 and 2 mg of carbon from the analyte of interest. Aim to isolate approximately 2–4 mg of the amino acid of interest to ensure between 1 and 2 mg of carbon is used for dating (17). A graph showing the approximate amino acid abundances in collagen and keratin can be found in Fig. 4.
10. Use clean glassware (not plastic) that can be sealed and heated to 110°C ensuring no evaporation.
11. Careful note of pH limits for filters used should be made to ensure they do not degrade and contaminate the sample.
12. 205 nm was chosen as the detection wavelength for the separation because all of the amino acids absorb at that wavelength, but water has very low absorption. Other detection methods, if available, would likely also give suitable results.

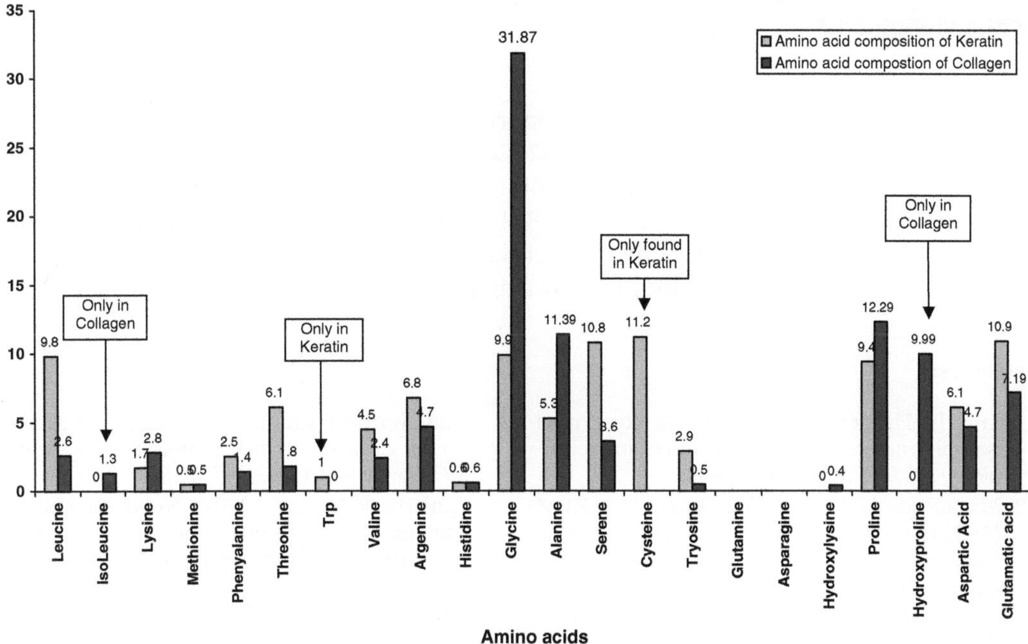

Fig. 4. Approximate relative abundance of individual amino acids in collagen and keratin (16).

13. Monitoring the baseline during the equilibration will enable you to determine when the column is fully equilibrated. The baseline will remain steady as the PDFOA becomes adsorbed onto the column. At the moment of equilibration, once the column is fully saturated with PDFOA, the baseline absorption rises suddenly as the PDFOA breaks through and elutes from the column. At this point, the column is ready to be used for the separation.
14. Separation on the C18 column, if that is used, takes 120 min.
15. Adding HCl protonates the PDFOA to ensure that it is volatile (its anion is not). Fluorine analysis was used to monitor the removal of the PDFOA from the amino acid sample, and this method resulted in fluorine levels below detection limits (7).
16. These three methods of water removal were used interchangeably and as available in the laboratory. Each was tested to show that they did not contribute exogenous carbon to the sample.
17. Stable isotope ratio ($\delta^{13}C$ and $\delta^{15}N$) measurements for our studies were made using an elemental analyzer (Carlo Erba, Milan, Italy) coupled to a 20–20 isotope ratio mass spectrometer (Sercon Ltd., Crewe, UK). The isotopic reference used was Vienna Pee Dee Belemnite (VPDB) for $\delta^{13}C$ and air for $\delta^{15}N$.

Acknowledgments

This work was supported by a grant from the National Science Foundation (INT-0202648). Assistance with isotopic measurements and analysis by personnel from the Oxford Radiocarbon Accelerator Unit, particularly Robert Hedges, Angela Bowles, Christopher Bronk Ramsey, Michael Dee, Peter Ditchfield, and Tom Higham, is gratefully acknowledged.

References

1. Tripp JA, Hedges REM (2004) Single-compound isotopic analysis of organic materials in archaeology. LC GC Eur 17:358–364
2. McCullagh JSO, Juchelka D, Hedges REM (2006) Analysis of amino acid C-13 abundance from human and faunal bone collagen using liquid chromatography/isotope ratio mass spectrometry. Rapid Comm Mass Spectrom 20:2761–2768
3. Chikaraishi Y et al (2009) Determination of aquatic food-web structure based on compound-specific nitrogen isotopic composition of amino acids. Limnol Oceanogr Meth 7:740–750
4. Naito YI et al (2010) Quantitative Evaluation of Marine Protein Contribution in Ancient Diets Based on Nitrogen Isotope Ratios of Individual Amino Acids in Bone Collagen: An Investigation at the Kitakogane Jomon Site. Amer J Phys Anthropol 143:31–40
5. Zhang L, Altabet MA (2008) Amino-group-specific natural abundance nitrogen isotope ratio analysis in amino acids. Rapid Comm Mass Spectrom 22:559–566
6. Gibbs MM (2008) Identifying source soils in contemporary estuarine sediments: A new compound-specific isotope method. Est Coasts 31:344–359
7. Tripp JA, McCullagh JSO, Hedges REM (2006) Preparative separation of underivatized amino acids for compound-specific stable isotope analysis and radiocarbon dating of hydrolyzed bone collagen. J Sep Sci 29:41–48
8. McCullagh JSO, Marom A, Hedges REM (2010) Radiocarbon dating of individual amino acids from archaeological bone collagen. Radiocarbon 52:620–634
9. McCullagh JSO, Tripp JA, Hedges REM (2005) Carbon isotope analysis of bulk keratin and single amino acids from British and North American hair. Rapid Comm Mass Spectrom 19:3227–3231
10. Petritis KN et al (1999) Ion-pair reversed-phase liquid chromatography for determination of polar underivatized amino acids using perfluorinated carboxylic acids as ion pairing agent. J Chromatogr A 833:147–155
11. Schierbeek H et al (2007) Novel method for measurement of glutathione kinetics in neonates using liquid chromatography coupled to isotope ratio mass spectrometry. Rapid Comm Mass Spectrom 21:2805–2812
12. McCullagh JSO, Gaye-Siessegger J, Focken U (2008) Determination of underivatized amino acid δ13C by liquid chromatography/isotope ratio mass spectrometry for nutritional studies: the effect of dietary non-essential amino acid profile on the isotopic signature of individual amino acids in fish. Rapid Comm Mass Spectrom 22:1817–1822
13. Smith CI et al (2009) A three phase liquid chromatographic method for δ13C analysis of amino acids from biological protein hydrolysates using LC-IRMS. Anal Biochem 390:165–172
14. McCullagh JSO (2010) Mixed mode chromatography isotope ratio mass spectrometry. Rapid Comm Mass Spectrom 24:483–494
15. Longin R (1971) New Method of Collagen Extraction for Radiocarbon Dating. Nature 203: 241–2
16. Eastoe JE (1955) The amino acid composition of mammalian collagen and gelatin. Biochem J 61:589–600
17. Dee M, Bronk Ramsey C (2000) Refinement of graphite target production at ORAU. Nucl Instr Meth Phys Res B: Beam Int Mater Atoms 172:449–453
18. Bronk Ramsey C (1995) Radiocarbon calibration and analysis of stratigraphy: The OxCal program. Radiocarbon 37:425–430
19. Bronk Ramsey C (2001) Development of the radiocarbon calibration program OxCal. Radiocarbon 43:355–363
20. Reimer PJ et al (2004) IntCal04 terrestrial radiocarbon age calibration, 0–26 cal kyr BP. Radiocarbon 46:1029–58

Chapter 28

Quantification of Amino Acids in a Single Cell by Microchip Electrophoresis with Chemiluminescence Detection

Yi-Ming Liu and Shulin Zhao

Abstract

Analyzing individual cells allows detecting a minor group of abnormal cells present in a large population of normal cells. This ability can be essential to understanding diseases, such as cancer and diabetes. Microchip electrophoresis (MCE) is the technique of choice for single-cell analysis. However, since the channels in microfluidic devices are very small, achieving the desired assay sensitivity on a microfluidic platform remains a challenge. Here, we describe an MCE method with highly sensitive chemiluminescence detection for simultaneous determination of multiple amino acids present in single cells.

Key words: Microchip electrophoresis, Chemiluminescence detection, Single-cell analysis, Amino acids quantification, Amino acid analysis

1. Introduction

Quantification of intracellular constituents is essential for a better understanding of basic cellular functions and intra- and intercellular communications (1, 2). Such demand has promoted the development of analytical techniques for single-cell analysis. Methods developed so far are based on flow cytometry (3, 4), open tubular liquid chromatography (5), electrochemical method (6), fluorescence microscopy (7), mass spectrometry (8), and capillary electrophoresis (CE) (9–14). Microchip electrophoresis (MCE), which is considered as a miniaturized version of CE, is well-suited for single-cell analysis because it offers the possibility of integrating sample injection, preconcentration, digestion/cell lysing, separation, and detection onto one single microchip (15, 16). However, since the channels in microfluidic devices are very small, a sensitive detection is required to follow an MCE separation. Up to date, in

Michail A. Alterman and Peter Hunziker (eds.), *Amino Acid Analysis: Methods and Protocols*, Methods in Molecular Biology, vol. 828, DOI 10.1007/978-1-61779-445-2_28, © Springer Science+Business Media, LLC 2012

most of the MCE-based methods developed for single-cell analysis, laser-induced fluorescence detection (LIFD) is employed (17–19). Although LIFD is highly sensitive, it requires precolumn tagging of the analytes with a fluorophore in many cases, including the analysis of amino acids. On the other hand, chemiluminescence (CL) detection, not widely used in MCE, is another most sensitive detection approach (20, 21). A major advantage of CL over LIFD is that it does not require an external light source. This is attractive, particularly when multiple functions are integrated onto a small microfluidic chip. Here, we demonstrate an MCE-CL assay of amino acids present in single cells. Individual rat hepatocytes are analyzed as the proof-of-concept samples. Amino acids, including Trp, Gly, and Ala, are detected at the fmol/cell level. Compared with MCE-LIFD methods, the present MCE-CL method is simple in instrumental setup and offers a comparable sensitivity (the limit of detection is 2.1 μM or 0.46 fmol for Trp). In addition, no precolumn derivatization of the sample is required.

2. Materials

1. Prepare all solutions using Milli-Q ultrapure water (18 MΩcm at 25°C) and analytical-grade reagents.
2. All solutions used in MCE are filtered through a 0.22-μm nylon membrane (see Note 1).
3. Amino acid stock solutions: 1.00 mM, prepared in water.
4. MCE running buffer: 20 mM Na_2HPO_4 buffer solution (pH 10.0, adjusted with a 1 M NaOH solution) containing 2.5 mM luminol and 40 mM NaBr (see Note 2).
5. CL reaction buffer: 50 mM $NaHCO_3$ solution (pH 12.5, adjusted with a 1 M NaOH solution) containing 0.8 mM $K_3Fe(CN)_6$.
6. *Rat hepatocytes*: Prepared from SD rats (~300 g, male) following the procedure described by Jin et al. (22). Hepatocytes are suspended in 20 mM PBS at pH 7.4.
7. *MCE* Microchip: The design of the glass/PDMS microchip (95 × 25 mm) is illustrated in Fig. 1. The glass layer with microchannels is fabricated by using a standard photolithography and wet chemical etching technique (23). The PDMS surface for bonding to the etched glass slide is prepared from Sylgard 184 (PDMS) silicone elastomer mixed with its curing agent at 10:1 (*w*/*w*).
8. *Setup for MCE-CL assay*: Fig. 2 illustrates the setup. The microchip is mounted on the X-Y translational stage of an inverted microscope that also serves as a platform of CL detection. Use

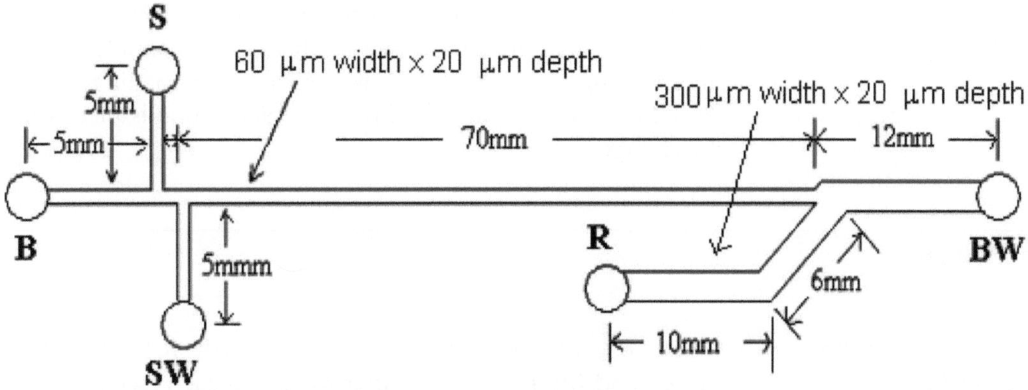

Fig. 1. Layout and dimensions of the glass/PDMS microfluidic chip used in this work. *S* sample reservoir, *B* buffer reservoir, *SW* sample waste reservoir, *BW* buffer waste reservoir, *R* oxidizer solution reservoir.

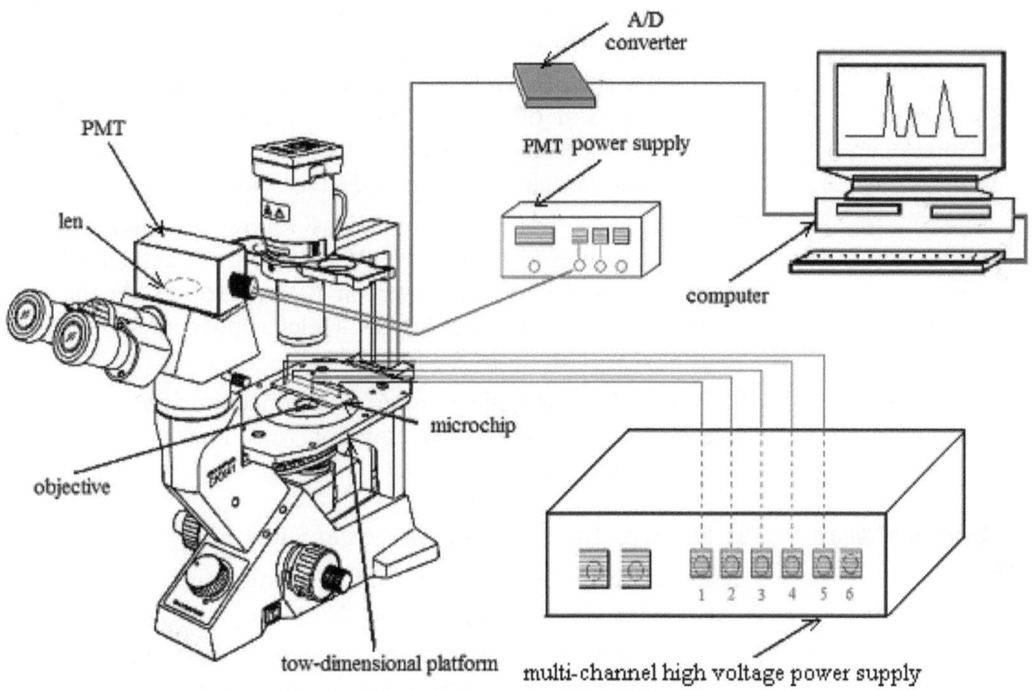

Fig. 2. Schematic of the integrated MCE-CL system for single-cell analysis.

of the X-Y translational stage allows viewing any point of the microchannel for introducing a cell. CL signal is collected by means of a microscope objective. After passing a dichroic mirror and a lens, CL emission with a wavelength maximum at 425 nm is detected by a photomultiplier (PMT, Hamamatsu R105). The PMT is mounted in an integrated detection module, including HV power supply, voltage divider, and amplifier. The output signal of PMT is recorded and processed with a computer. A multiterminal high-voltage power supply, variable in the range of 0–8,000 V, is used for cell loading, lysing, and MCE separation. The inverted microscope is placed in a black box.

3. Methods

3.1. Calibration of the MCE-CL System

1. Rinse the microchannels sequentially with 0.1 M NaOH, water, and MCE running buffer for 1 min each (see Note 3).
2. Fill reservoirs B, S, and SW with the MCE running buffer.
3. Fill reservoir R with the CL reaction buffer.
4. Apply vacuum to reservoir BW (see Note 4).
5. Replace the MCE running buffer solution in reservoir S with an amino acid mixture standard solution (see Note 5).
6. Apply a set of electrical potentials to reservoirs as following: reservoir S at 600 V, reservoir B at 150 V, reservoir BW at 200 V, reservoir SW at grounded, and reservoir R at 0 V to inject the sample. Duration: 20 s (see Note 6).
7. Change the potentials applied as following: reservoir B at 2,050 V, reservoir S at 1,250 V, reservoir SW at 1,250 V, reservoir R at 350 V, and reservoir BW at ground. At the same time, start to record the MCE-CL electropherogram (as shown in Fig. 3). Duration: 3 min.
8. Plot peak areas against amino acid concentrations to obtain a calibration curve and the equation from linear regression for each amino acid.

3.2. Analysis of Single Cells

1. Do steps 1 through 4 described in Subheading 3.1.
2. Replace the MCE running buffer solution in reservoir S with the cell suspension sample (see Note 7).
3. Apply a voltage of 100 at reservoir S with reservoir SW grounded and other reservoirs floating. Monitor cells flowing from reservoir S to SW (see Note 8).
4. Change the potentials as following: reservoir B at 160 V, reservoir S and SW at 75 V, reservoir BW at ground, and reservoir R floating as soon as a single cell moves into the cross intersection of the channels. Duration: 10 s.

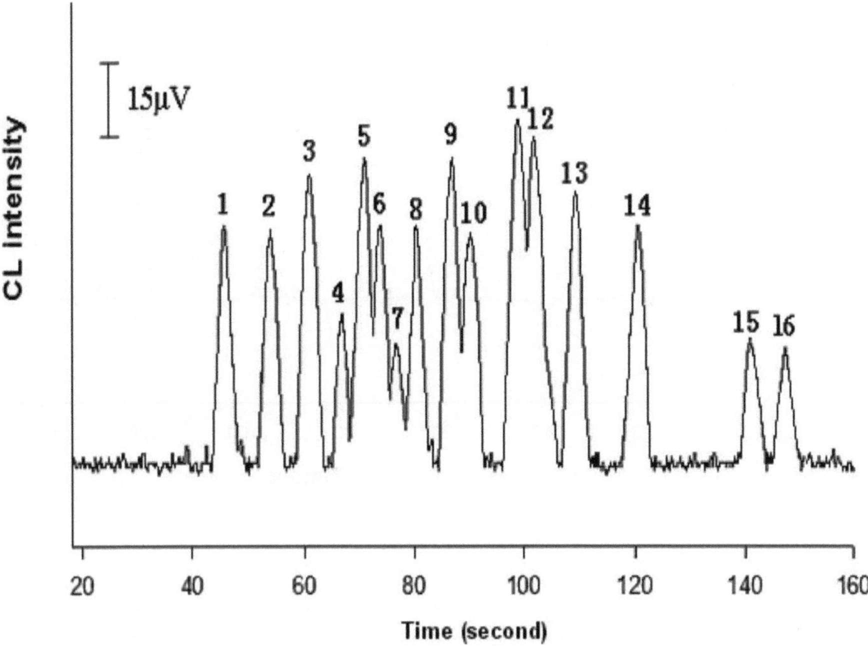

Fig. 3. Electropherogram obtained from the separation of a standard mixture containing amino acids and ascorbic acid. Each analyte is at a concentration of 2.0×10^{-5} M. Peak identification: (*1*) Arg + Lys; (*2*) Ala; (*3*) Gly; (*4*) Val + Pro; (*5*) Asn + Met; (*6*) Gln; (*7*) Ser; (*8*) Cys; (*9*) Phe + Leu; (*10*) Ile + His; (*11*) Thr; (*12*) Tyr; (*13*) Trp; (*14*) ascorbic acid; (*15*) Glu; and (*16*) Asp.

5. Set all potentials to zero to let the sampled cell settle down and adhere to the channel wall.
6. Shift the microchip from the cross intersection viewing position to the detection point (i.e., the join point of the oxidizer introduction channel with the separation channel).
7. Apply a set of electrical shocks to reservoirs as following: reservoir B at 1,800 V, reservoir BW at ground, and other reservoirs floating. Duration: 0.5 s for six times.
8. Apply potentials to reservoirs as following: reservoir B at 2,050 V, reservoir S at 1,250 V, reservoir SW at 1,250 V, reservoir R at 350 V, and reservoir BW at ground. At the same time, start to record the MCE-CL electropherogram (as shown in Fig. 4). Duration: 3 min.
9. Determine the concentration (at fmol/cell) from the peak area using the calibration equation obtained above for each amino acid.

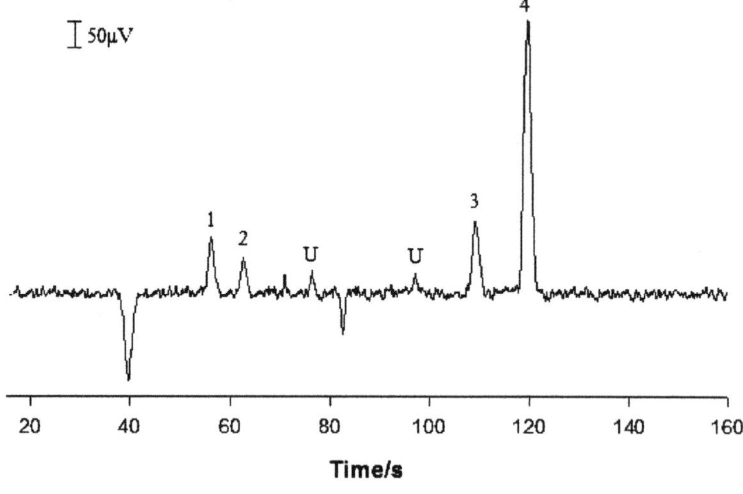

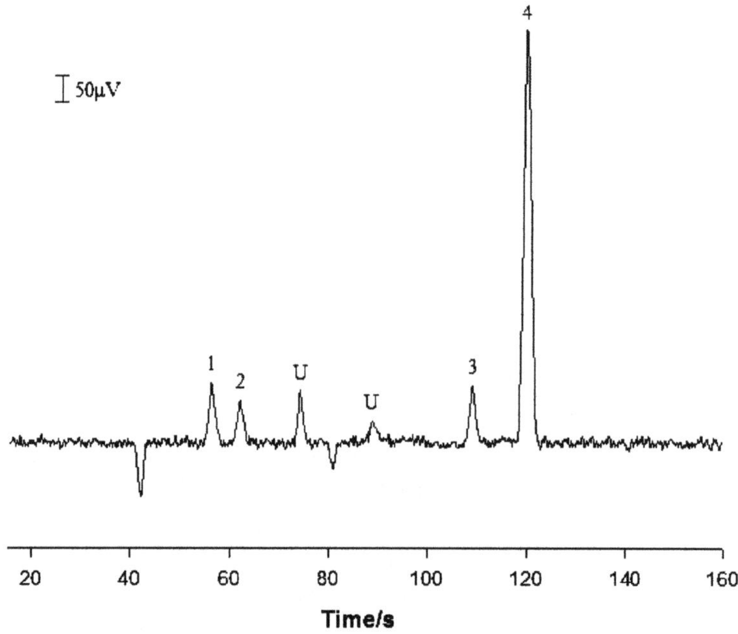

Fig. 4. Electropherograms obtained from analyzing two individual rat hepatocytes. Peak identification: (*1*) Ala; (*2*) Gly; (*3*) Trp; (*4*) ascorbic acid; and (*U*) unknown. The average intracellular contents of Trp, Gly, Ala, and ascorbic acid are found to be 5.15, 3.78, 3.84, and 38.3 fmol/cell, respectively.

4. Notes

1. Filtering all solutions before use in MCE is very important since channels in the microchip are very small in size and they are easily blocked by minute particles in solutions.
2. In the presence of NaBr, the CL detection is about 40 times more sensitive.

3. Rinse the microchannels with 1 M NaOH for 30 min for the first use of the microchip.
4. Apply a vacuum to reservoir BW to fill the channels with the electrophoresis running buffer. Check to make sure that there are no air bubbles in the channels.
5. Several amino acids in a mixture solution can be simultaneously quantified in one MCE-CL run.
6. Under the potential conditions applied, the amino acid standard solution is transported from reservoir S to SW. That is, the sampling channel that involves a small section of the separation channel is filled with the amino acid standard solution. The sample volume injected for MCE separation is determined by the volume of the cross intersection of the sampling channel and the separation channel. It is calculated to be $\sim 2.6 \times 10^{-9}$ ml with the microchip used in this work.
7. The cell suspension must be diluted appropriately to facilitate single-cell loading and docking.
8. The traveling speed of cells is dependent upon the electrical field strengths applied. At 100 V as applied, the speed is about 0.1 mm/s.

Acknowledgments

Financial support from National Institutes of Health (SC1 GM089557) is gratefully acknowledged.

References

1. McClain MA et al (2003) Microfluidic devices for the high-throughput chemical analysis of cells. Anal Chem 75:5646–5655
2. Dufva M (2009) Microchips for cell-based assays. Methods Mol Biol 509:135–144
3. Kruth HS (1982) Flow cytometry: rapid biochemical analysis of single cells. Anal Biochem 125:225–242
4. Darzynkiewicz Z, Halicka HD, Zhao H (2010) Analysis of cellular DNA content by flow and laser scanning cytometry. Adv Exp Med Biol 676:137–47
5. Oates MD, Cooper BR, Jorgenson JW (1990) Quantitative amino acid analysis of individual snail neurons by open tubular liquid chromatography. Anal Chem 62:1573–1577
6. Troyer KP, Wightman RM (2002) Dopamine transport into a single cell in a picoliter vial. Anal Chem 74:5370–5375
7. Ishijima A, Yanagida T (2001) Single molecule nanobioscience. Trends Biochem Sci 26:438–444
8. Amantonico A et al (2010) Single-cell MALDI-MS as an analytical tool for studying intrapopulation metabolic heterogeneity of unicellular organisms. Anal Chem 82:7394–7400
9. Hogan BL, Yeung ES (1992) Determination of intracellular species at the level of a single erythrocyte via capillary electrophoresis with direct and indirect fluorescence detection. Anal Chem 64:2841–2845
10. Han FT, Lillard SJ (2002) Monitoring differential synthesis of RNA in individual cells by capillary electrophoresis. Anal Biochem 302:136–143
11. Hu S et al (2001) Capillary sodium dodecyl sulfate-DALT electrophoresis of proteins in a single human cancer cell. Electrophoresis 22:3677–3682

12. Zhang H, Jin WR (2004) Analysis of amino acids in individual human erythrocytes by capillary electrophoresis with electroporation for intracellular derivatization and laser-induced fluorescence detection. Electrophoresis 25:480–486
13. Fuller KM, Arriaga E A (2003) Analysis of individual acidic organelles by capillary electrophoresis with laser-induced fluorescence detection facilitated by the endocytosis of fluorescently labeled microspheres. Anal Chem 75:2123–2130
14. Zhi Q et al (2007) Coupling chemiluminescence with capillary electrophoresis to analyze single human red blood cells. Anal Chim Acta 583:217–222
15. Sims CE, Allbritton NL (2007) Analysis of single mammalian cells on-chip. Lab Chip 7: 423–440
16. Huang WH et al (2008) Recent advances in single-cell analysis using capillary electrophoresis and microfluidic devices. J Chromatogr B 866:104–122
17. Wu H, Wheeler A, Zare RN (2004) Chemical cytometry on a picoliter-scale integrated microfluidic chip. Proc Natl Acad Sci USA 101: 12809–12813
18. Ling YY, Yin XF, Fang ZL (2005) Simultaneous determination of glutathione and reactive oxygen species in individual cells by microchip electrophoresis. Electrophoresis 26:4759–4766
19. Klepárník K, Horký M (2003) Detection of DNA fragmentation in a single apoptotic cardiomyocyte by electrophoresis on a microfluidic device. Electrophoresis 24:3778–3783
20. Zhao S, Li X, Liu YM (2009) Integrated microfluidic system with chemiluminescence detection for single cell analysis after intracellular labeling. Anal Chem 81:3873–3878
21. Zhao S et al (2010) Determination of intracellular sulphydryl compounds by microchip electrophoresis with selective chemiluminescence detection. J Chromatogr A 1217:5732–5736
22. Jin W, Li X, Gao N (2003) Simultaneous determination of tryptophan and glutathione in individual rat hepatocytes by capillary zone electrophoresis with electrochemical detection at a carbon fiber bundle–Au/Hg dual electrode. Anal Chem 75:3859–3864
23. Kim M-S et al (2005) Fabrication of microchip electrophoresis devices and effects of channel surface properties on separation efficiency. Sensors and Actuators B 107:818–824.

INDEX

Michail A. Alterman and Peter Hunziker (eds.), *Amino Acid Analysis: Methods and Protocols*, Methods in Molecular Biology, vol. 828, DOI 10.1007/978-1-61779-445-2, © Springer Science+Business Media, LLC 2012

E

F

G

H

Z

MIX
Papier aus verantwortungsvollen Quellen
Paper from responsible sources
FSC® C105338

If you have any concerns about our products,
you can contact us on
ProductSafety@springernature.com

In case Publisher is established outside the EU,
the EU authorized representative is:
Springer Nature Customer Service Center GmbH
Europaplatz 3, 69115 Heidelberg, Germany

Printed by Libri Plureos GmbH
in Hamburg, Germany